CONCERNING THE TWO CHIEF WORLD SYSTEMS

GALILEO GALILEI LINCEO FILOSOFO E MATEMATICO DEL SER.mo GRAN DVCA DI TOSC. GALILEO

GALILEO GALILEI

Dialogue Concerning the Two Chief World Systems—Ptolemaic & Copernican *translated by Stillman Drake, foreword by Albert Einstein*

SECOND EDITION

UNIVERSITY OF CALIFORNIA PRESS

BERKELEY AND LOS ANGELES 1970

SECOND REVISED EDITION, 1967
SECOND PRINTING, 1970
STANDARD BOOK NUMBER 520-00450-7
LIBRARY OF CONGRESS CATALOG CARD NUMBER 53–11238

UNIVERSITY OF CALIFORNIA PRESS, BERKELEY AND LOS ANGELES, CALIFORNIA
UNIVERSITY OF CALIFORNIA PRESS, LTD., LONDON, ENGLAND
COPYRIGHT, 1953, 1962, AND 1967, BY THE REGENTS OF THE UNIVERSITY OF CALIFORNIA
PRINTED IN THE UNITED STATES OF AMERICA   DESIGNED BY JOHN B. GOETZ

Take note, theologians, that in your desire to make matters of faith out of propositions relating to the fixity of sun and earth you run the risk of eventually having to condemn as heretics those who would declare the earth to stand still and the sun to change position — eventually, I say, at such a time as it might be physically or logically proved that the earth moves and the sun stands still.

—Note added by Galileo in the preliminary leaves of his own copy of the Dialogue.

VORWORT

GALILEOS *Dialog über die beiden hauptsächlichen Welt-*
systeme ist eine Fundgrube für jeden, der sich für die Geistes-
geschichte des Westens und für deren Rückwirkung auf die
ökonomische und politische Entwicklung interessiert.

Da offenbart sich ein Mann, der den leidenschaftlichen Willen,
die Intelligenz und den Mut hat, sich als Vertreter des ver-
nünftigen Denkens der Schar derjenigen entgegenzustellen, die
auf die Unwissenheit des Volkes und die Indolenz der Lehren-
den in Priester- und Professoren-Gewande sich stützend, ihre
Machtpositionen einnehmen und verteidigen. Seine ungewöhn-
liche schriftstellerische Begabung erlaubt es ihm, zu den Gebilde-
ten seiner Zeit so klar und eindrucksvoll zu sprechen, dass er
das anthropozentrische und mythische Denken der Zeitgenossen
überwand und sie zu einer objektiven, kausalen Einstellung zum
Kosmos zurückführte, die mit der Blüte der griechischen Kultur
der Menschheit verloren gegangen war.

Wenn ich dies so ausspreche, sehe ich zugleich, dass ich der
weitverbreiteten Schwäche aller derer zum Opfer falle, die trun-
ken von einer übermässigen Verliebtheit die Statur ihrer Heroen
übertrieben darstellen. Es mag sein, dass die Lähmung der Geis-
ter durch starre autoritäre Tradition des dunklen Zeitalters im
siebzehnten Jahrhundert bereits so weit gemildert war, dass die
Fesseln einer überlebten intellektuellen Tradition nicht mehr
für die Dauer standhalten konnten — mit oder ohne Galileo.

Nun, dieser Zweifel betrifft ja nur einen Sonderfall der Frage,
inwieweit der Verlauf der menschlichen Geschichte durch ein-

FOREWORD

GALILEO'S *Dialogue Concerning the Two Chief World
Systems* is a mine of information for anyone interested in the
cultural history of the Western world and its influence upon
economic and political development.

A man is here revealed who possesses the passionate will, the
intelligence, and the courage to stand up as the representative
of rational thinking against the host of those who, relying on the
ignorance of the people and the indolence of teachers in priest's
and scholar's garb, maintain and defend their positions of authority. His unusual literary gift enables him to address the
educated men of his age in such clear and impressive language
as to overcome the anthropocentric and mythical thinking of his
contemporaries and to lead them back to an objective and causal
attitude toward the cosmos, an attitude which had become lost
to humanity with the decline of Greek culture.

In speaking this way I notice that I, too, am falling in with
the general weakness of those who, intoxicated with devotion,
exaggerate the stature of their heroes. It may well be that during
the seventeenth century the paralysis of mind brought about by
the rigid authoritarian tradition of the Dark Ages had already
so far abated that the fetters of an obsolete intellectual tradition
could not have held much longer — with or without Galileo.

Yet these doubts concern only a particular case of the general
problem concerning the extent to which the course of history
can be decisively influenced by single individuals whose qualities
impress us as accidental and unique. As is understandable, our

zelne Individuen und deren als zufällig und einmalig empfundene Qualitäten entscheidend beeinflusst werden kann. Unsere Zeit steht solchen Auffassungen skeptischer gegenüber als das achtzehnte Jahrhundert und die erste Hälfte des neunzehnten Jahrhunderts — begreiflicherweise. Denn die weitgehende Spezialisierung der Berufe und des Wissens lässt den Einzelnen gewissermassen als „auswechselbar" erscheinen wie den Einzelteil einer durch Massenfabrikation hergestellten Maschine.

Der Wert des *Dialogs* als Dokument ist glücklicherweise von der Stellung zu solch prekären Fragen unabhängig. Vor allem gibt der Dialog eine überaus lebendige und überzeugende Darstellung der herrschenden Ansichten über den Bau des Kosmos im Grossen. Die im früheren Mittelalter herrschende kindliche Auffassung der Erde als einer flachen Scheibe, verknüpft mit ganz unklaren Ideen über den von den Sternen erfüllten Raum und die Bewegung der Gestirne, waren längst durch das Weltbild der Griechen, speziell durch Ideen des Aristoteles und durch die ptolemäische konsequente räumliche Auffassung der Gestirne und deren Bewegung verbessert. Das Weltbild, welches zur Zeit Galileos noch vorherrschte, war etwa folgendes:

Es gibt einen Raum, der einen bevorzugten Punkt, den Weltmittelpunkt besitzt. Die Materie — wenigstens der dichtere Teil derselben — sucht sich diesem Punkt möglichst zu nähern. Sie hat demzufolge ungefähr Kugelgestalt angenommen (Erde). Vermöge dieser Entstehung der Erde fällt der Mittelpunkt dieser Erdkugel praktisch mit den Weltmittelpunkt zusammen. Sonne, Mond und Sterne sind, damit sie nicht nach dem Weltmittelpunkt fallen, auf (durchsichtigen) starren Kugelschalen befestigt, deren Mittelpunkt mit den Weltmittelpunkt (oder Raummittelpunkt) zusammenfällt. Diese Kugelschalen drehen sich um den ruhenden Erdball (bezw. um den Weltmittelpunkt) mit etwas verschiedenen Winkelgeschwindigkeiten. Die Mondschale hat den kleinsten Radius; sie umschliesst alles „Irdische". Die äusseren Schalen mit ihren Gestirnen repräsentieren die „himmlische Sphäre", deren Objekte als ewig, unzerstörbar und unveränderlich gedacht sind, im Gegensatz zur „unteren, irdischen Sphäre", die durch die Mondschale umschlossen wird und alles enthält, was vergänglich, hinfällig und „sündhaft" ist.

Natürlich ist diese kindliche Konstruktion nicht den griechischen Astronomen zur Last zu legen, die sich bei ihrer Dar-

age takes a more sceptical view of the role of the individual than did the eighteenth and the first half of the nineteenth century. For the extensive specialization of the professions and of knowledge lets the individual appear "replaceable," as it were, like a part of a mass-produced machine.

Fortunately, our appreciation of the *Dialogue* as a historical document does not depend upon our attitude toward such precarious questions. To begin with, the *Dialogue* gives an extremely lively and persuasive exposition of the then prevailing views on the structure of the cosmos in the large. The naïve picture of the earth as a flat disc, combined with obscure ideas about star-filled space and the motions of the celestial bodies, prevalent in the early Middle Ages, represented a deterioration of the much earlier conceptions of the Greeks, and in particular of Aristotle's ideas and of Ptolemy's consistent spatial concept of the celestial bodies and their motions. The conception of the world still prevailing at Galileo's time may be described as follows:

There is space, and within it there is a preferred point, the center of the universe. Matter — at least its denser portion — tends to approach this point as closely as possible. Consequently, matter has assumed approximately spherical shape (earth). Owing to this formation of the earth the center of the terrestrial sphere practically coincides with that of the universe. Sun, moon, and stars are prevented from falling toward the center of the universe by being fastened onto rigid (transparent) spherical shells whose centers are identical with that of the universe (or space). These spherical shells revolve around the immovable globe (or center of the universe) with slightly differing angular velocities. The lunar shell has the smallest radius; it encloses everything "terrestrial." The outer shells with their heavenly bodies represent the "celestial sphere" whose objects are envisaged as eternal, indestructible, and inalterable, in contrast to the "lower, terrestrial sphere" which is enclosed by the lunar shell and contains everything that is transitory, perishable, and "corruptible."

Naturally, this naïve picture cannot be blamed on the Greek astronomers who, in representing the motions of the celestial bodies, used abstract geometrical constructions which grew more and more complicated with the increasing precision of astronomical observations. Lacking a theory of mechanics they tried to

stellung der Sternbewegungen abstrakter geometrischer Konstruktionen bedienten, die mit wachsender Genauigkeit der Gestirn-Beobachtungen immer komplizierter wurden. In Ermangelung einer Mechanik suchte man alle die komplizierten (scheinbaren) Bewegungen auf die für die denkbar einfachst gehaltene zurückzuführen, nämlich auf die gleichförmige Kreisbewegung und die Superposition solcher Bewegungen. (Die Anhänglichkeit an die Idee der Kreisbewegung als der wahrhaft natürlichen spürt man noch sehr wohl bei Galileo; sie hat es wohl verhindert, dass er das Trägheitsprinzip und dessen zentrale Bedeutung völlig erkannte.)

Die obige Skizze stellt eine der barbarischen, primitiven Denkweise der damaligen Europäer angepasste Vergröberung der spät-griechischen Ideen dar, welch letztere zwar unkausal, aber doch objektiv und frei von animistischen Auffassungen waren — ein Vorzug, den man der aristotelischen Kosmologie allerdings nur bedingt zubilligen kann.

Wenn Galileo für die Lehre des Kopernikus eintrat und kämpfte, so war es ihm nicht etwa nur darum zu tun, eine Vereinfachung der Darstellung der Sternbewegungen zu erzielen. Sein Ziel war es, eine erstarrte und unfruchtbar gewordene Ideenwelt zu ersetzen durch das vorurteilslose, mühevolle Ringen um eine tiefere und konsequentere Erfassung der physikalischen und astronomischen Tatsachen.

Die Dialogform des Werkes mag zum Teil auf Platos leuchtendes Vorbild zurückzuführen sein; sie erlaubte Galileos ungewöhnlicher literarischer Begabung eine scharfe und lebendige Gegenüberstellung der Meinungen. Freilich mag auch das Bedürfnis mitgewirkt haben, es auf diese Weise zu vermeiden, in eigener Person eine Entscheidung in den strittigen Fragen treffen zu müssen, die ihn der Vernichtung durch die Inquisition ausgeliefert hätte. Es war Galileo ja sogar direkt verboten worden, für die Lehre des Kopernikus einzutreten. Der *Dialog* stellt, abgesehen von seinem bahnbrechenden sachlichen Gehalt, einen geradezu schalkhaften Versuch dar, dies Gebot scheinbar zu befolgen, sich *de facto* jedoch darüber hinwegzusetzen. Es zeigte sich aber leider, dass die heilige Inquisition für solch feinen Humor nicht das adäquate Verständnis aufzubringen vermochte.

Die Theorie der ruhenden Erde stützte sich auf die Hypothese von der Existenz eines abstrakten Weltmittelpunktes. Dieser

reduce all complicated (apparent) motions to the simplest motions they could conceive, namely, uniform circular motions and
superpositions thereof. Attachment to the idea of circular motion as the truly natural one is still clearly discernible in Galileo;
probably it is responsible for the fact that he did not *fully* recognize the law of inertia and its fundamental significance.

Thus, briefly, had the ideas of later Greece been crudely
adapted to the barbarian, primitive mentality of the Europeans
of that time. Though not causal, those Hellenistic ideas had
nevertheless been objective and free from animistic views — a
merit which, however, can be only conditionally conceded to
Aristotelian cosmology.

In advocating and fighting for the Copernican theory Galileo
was not only motivated by a striving to simplify the representation of the celestial motions. His aim was to substitute for a petri-
fied and barren system of ideas the unbiased and strenuous quest
for a deeper and more consistent comprehension of the physical
and astronomical facts.

The form of dialogue used in his work may be partly due to
Plato's shining example; it enabled Galileo to apply his extraordinary literary talent to the sharp and vivid confrontation of
opinions. To be sure, he wanted to avoid an open commitment in
these controversial questions that would have delivered him to
destruction by the Inquisition. Galileo had, in fact, been expressly forbidden to advocate the Copernican theory. Apart from
its revolutionary factual content the *Dialogue* represents a down-
right roguish attempt to comply with this order in appearance
and yet in fact to disregard it. Unfortunately, it turned out that
the Holy Inquisition was unable to appreciate adequately such
subtle humor.

The theory of the immovable earth was based on the hypothesis that an abstract center of the universe exists. Supposedly, this
center causes the fall of heavy bodies at the earth's surface, since
material bodies have the tendency to approach the center of the
universe as far as the earth's impenetrability permits. This leads
to the approximately spherical shape of the earth.

Galileo opposes the introduction of this "nothing" (center of
the universe) that is yet supposed to act on material bodies; he
considers this quite unsatisfactory.

But he also draws attention to the fact that this unsatisfactory

sollte den Fall der schweren Körper an der Erdoberfläche bewirken, indem die Körper das Streben haben sollen, sich diesem Weltmittelpunkte soweit zu nähern, als es die Undurchdringlichkeit zulässt. Dies Streben führt dann zu der annähernden Kugelgestalt der Erde.

Galileo wendet sich gegen die Einführung dieses „Nichts" (Weltmittelpunkt), das doch auf die materiellen Dinge wirken soll; dies findet er ganz unbefriedigend.

Ferner aber macht er darauf aufmerksam, dass diese unbefriedigende Hypothese auch zu wenig leistet. Sie erklärt nämlich zwar die Kugelgestalt der Erde, aber nicht die Kugelgestalt der übrigen Himmelskörper. Die Mondphasen und die von ihm durch das neuentdeckte Fernrohr entdeckten Phasen der Venus bewiesen aber die Kugelgestalt dieser beiden Himmelskörper, die genauere Beobachtung der Sonnenflecken die Kugelgestalt der Sonne. Ueberhaupt war damals wohl ein Zweifel an der Kugelgestalt der Planeten und der Sterne überhaupt kaum mehr möglich.

Die Hypothese des Weltmittelpunktes war daher durch eine solche zu ersetzen, welche die Kugelgestalt der Sterne überhaupt und nicht nur der Erde verstehen lässt. Galileo sagt klar, dass dies eine Art Wechselwirkung (Bestreben gegenseitiger Näherung) der den Stern konstituierenden Materie sein muss. Diese selbe Ursache musste nun (nach Aufgeben des Weltmittelpunktes) auch den freien Fall der Körper an der Erdoberfläche bewirken.

Ich möchte hier — in Form einer Einschaltung — darauf aufmerksam machen, dass eine weitgehende Analogie besteht zwischen Galileos Ablehnung der Setzung eines Weltmittelpunktes zur Erklärung des Fallens der Körper und der Ablehnung der Setzung des Inertialsystems zur Erklärung des Trägheitsverhaltens der Körper (welche Ablehnung der allgemeinen Relativitätstheorie zugrunde liegt). Beiden Setzungen gemeinsam ist nämlich die Einführung eines begrifflichen Dinges mit folgenden Eigenschaften:

(1). Es ist nicht als etwas Reales gedacht, von der Art der ponderablen Materie (bezw. des „Feldes").

(2). Es ist massgebend für das Verhalten der realen Dinge, ist aber umgekehrt keiner Einwirkung durch die realen Dinge unterworfen.

hypothesis accomplishes too little. Although it accounts for the spherical shape of the earth it does not explain the spherical shape of the other heavenly bodies. However, the lunar phases and the phases of Venus, which latter he had discovered with the newly invented telescope, proved the spherical shape of these two celestial bodies; and the detailed observation of the sunspots proved the same for the sun. Actually, at Galileo's time there was hardly any doubt left as to the spherical shape of the planets and stars.

Therefore, the hypothesis of the "center of the universe" had to be replaced by one which would explain the spherical shape of the stars, and not only that of the earth. Galileo says quite clearly that there must exist some kind of interaction (tendency to mutual approach) of the matter constituting a star. The same cause has to be responsible (after relinquishing the "center of the universe") for the free fall of heavy bodies at the earth's surface.

Let me interpolate here that a close analogy exists between Galileo's rejection of the hypothesis of a center of the universe for the explanation of the fall of heavy bodies, and the rejection of the hypothesis of an inertial system for the explanation of the inertial behavior of matter. (The latter is the basis of the theory of general relativity.) Common to both hypotheses is the introduction of a conceptual object with the following properties:

(1). It is not assumed to be real, like ponderable matter (or a "field").

(2). It determines the behavior of real objects, but it is in no way affected by them.

The introduction of such conceptual elements, though not exactly inadmissible from a purely logical point of view, is repugnant to the scientific instinct.

Galileo also recognized that the effect of gravity on freely falling bodies manifests itself in a vertical acceleration of constant value; likewise that an unaccelerated horizontal motion can be superposed on this vertical accelerated motion.

These discoveries contain essentially — at least qualitatively — the basis of the theory later formulated by Newton. But first of all the general formulation of the principle of inertia is lacking, although this would have been easy to obtain from Galileo's law of falling bodies by a limiting process. (Transition to vanishing vertical acceleration.) Lacking also is the idea that

Die Einführung derartiger begrifflichen Elemente ist zwar vom rein logischen Gesichtspunkte nicht schlechthin unzulässig, widerstrebt aber dem wissenschaftlichen Instinkt.

Galileo erkannte auch, dass die Wirkung der Schwere auf frei fallende Körper in dem Auftreten einer vertikalen Beschleunigung von festem Werte sich manifestiere, und dass dieser vertikalen Fallbewegung sich eine unbeschleunigte Horizontalbewegung superponieren lasse.

In diesen Erkenntnissen ist wenigstens qualitativ die Basis der später von Newton formulierten Theorie im Wesentlichen bereits enthalten. Es fehlt aber bei Galileo erstens die allgemeine Formulierung des Trägheitsprinzipes, obwohl dieses durch Grenzübergang aus den von ihm gefundenen Gesetzen des freien Falles ganz leicht zu gewinnen war. (Uebergang zu verschwindender Vertikalbeschleunigung.) Es fehlte insbesondere noch die Idee, dass dieselbe Materie eines Himmelskörpers, welche an dessen Oberfläche eine Fallbeschleunigung erzeugt, auch imstande wäre einem anderen Himmelskörper eine Beschleunigung zu erteilen, und dass solche Beschleunigungen in Verbindung mit der Trägheit Umlaufsbewegungen erzeugen können. Was aber gewonnen war, war die Erkenntnis, dass die Anwesenheit von Massen (Erde) eine Beschleunigung freier Körper (an der Erdoberfläche) bewirke.

Man kann sich heute nicht mehr vorstellen, was für eine grosse Phantasieleistung in der klaren Bildung des Begriffes der Beschleunigung und in der Erkenntnis der physikalischen Bedeutung dieses Begriffes lag.

Mit der wohlbegründeten Ablehnung der Idee von der Existenz eines Weltmittelpunktes war auch der Idee der ruhenden Erde und überhaupt die Idee einer Sonderstellung der Erde die innere Berechtigung genommen. Die Frage, was man bei der Darstellung der Bewegung der Himmelskörper als „ruhend" zu betrachten habe, wurde dadurch zu einer Zweckmässigkeitsfrage. In Anlehnung an Aristarch-Kopernikus werden die Vorteile dargelegt, die man dadurch erzielt, dass man die Sonne als ruhend annimmt (nach Galileo nicht etwa eine blosse Konvention, sondern eine Hypothese, die „wahr" oder „falsch" ist). Da wird natürlich angeführt, dass die Annahme der Drehung der Erde um ihre Achse einfacher ist als eine gemeinsame Drehbewegung aller Fixsterne um die Erde. Ferner wird natürlich darauf hinge-

the same matter which causes a vertical acceleration at the surface of a heavenly body can also accelerate another heavenly
body; and that such accelerations together with inertia can
produce revolving motions. There was achieved, however, the
knowledge that the presence of matter (earth) causes an acceleration of free bodies (at the surface of the earth).

It is difficult for us today to appreciate the imaginative power
made manifest in the precise formulation of the concept of acceleration and in the recognition of its physical significance.

Once the conception of the center of the universe had, with
good reason, been rejected, the idea of the immovable earth, and,
generally, of an exceptional role of the earth, was deprived of
its justification. The question of what, in describing the motion
of heavenly bodies, should be considered "at rest" became thus a
question of convenience. Following Aristarchus and Copernicus,
the advantages of assuming the sun to be at rest are set forth
(according to Galileo not a pure convention but a hypothesis
which is either "true" or "false"). Naturally, it is argued that
it is simpler to assume a rotation of the earth around its axis
than a common revolution of all fixed stars around the earth.
Furthermore, the assumption of a revolution of the earth around
the sun makes the motions of the inner and outer planets appear
similar and does away with the troublesome retrograde motions
of the outer planets, or rather explains them by the motion of the
earth around the sun.

Convincing as these arguments may be — in particular coupled with the circumstance, detected by Galileo, that Jupiter
with its moons represents so to speak a Copernican system in
miniature — they still are only of a qualitative nature. For since
we human beings are tied to the earth, our observations will
never directly reveal to us the "true" planetary motions, but
only the intersections of the lines of sight (earth–planet) with
the "fixed-star sphere." A support of the Copernican system over
and above qualitative arguments was possible only by determining the "true orbits" of the planets — a problem of almost in-
surmountable difficulty, which, however, was solved by Kepler
(during Galileo's lifetime) in a truly ingenious fashion. But this
decisive progress did not leave any traces in Galileo's life work —
a grotesque illustration of the fact that creative individuals are
often not receptive.

wiesen, dass bei Annahme der Erdbewegung um die Sonne die Bewegungen der inneren und äusseren Planeten als gleichartig erscheinen und dass die so störenden rückläufigen Bewegungen der äusseren Planeten in Wegfall kommen, bezw. durch die Erdbewegung um die Sonne erklärt werden.

So stark diese Argumente sind, besonders in Verbindung mit dem von Galileo entdeckten Umstand, dass Jupiter mit seinen Monden gewissermassen ein kopernikanisches System in Miniatur uns vor Augen stellt, so sind doch alle diese Argumente nur qualitativer Art. Denn da wir Menschen auf der Erde festsitzen, so geben uns unsere Beobachtungen keineswegs die „wirklichen" Bewegungen der Planeten, sondern nur die Schnittpunkte der Blickrichtungen Erde—Planet mit der „Fixstern-Sphäre". Eine Stützung des kopernikanischen Systems, die über das Qualitative hinausging, war nur möglich, wenn die „wahren Bahnen" der Planeten ermittelt wurden—ein fast unlösbar scheinendes Problem, das aber von Kepler zu Galileos Zeiten in wahrhaft genialer Weise gelöst wurde. Dass in Galileos Lebenswerk dieser entscheidende Fortschritt keine Spuren hinterlassen hat, ist ein groteskes Beispiel dafür, dass schöpferische Menschen oft nicht rezeptiv orientiert sind.

Grosse Anstrengung wird von Galileo darauf verwendet, zu zeigen, dass die Hypothese von der Dreh- und Umlauf-Bewegung der Erde nicht dadurch widerlegt wird, dass wir keine m e c h a n i s c h e n Wirkungen dieser Bewegung wahrnehmen. Es war dies ein Vorhaben, das, genau betrachtet, mangels einer vollständigen Mechanik unlösbar war. Ich finde, dass gerade in dem Ringen mit diesem Problem Galileos Originalität sich besonders imponierend zeigt. Es ist Galileo natürlich auch wichtig zu zeigen, dass die Fixsterne so weit weg sind, dass die durch die jährliche Bewegung der Erde erzeugten Parallaxen für die damalige Messgenauigkeit unmessbar klein sein müssen. Auch diese Untersuchung ist genial bei aller Primitivität.

Zu seiner unrichtigen Theorie von Ebbe und Flut wurde Galileo verführt durch seine Sehnsucht nach einem mechanischen Beweis für die Erdbewegung. Die faszinierende Ueberlegung, welche hierüber im letzten Gespräch gegeben wird, würde wohl von Galileo selbst als nicht beweisend erkannt worden sein, wenn sein Temperament nicht mit ihm durchgegangen wäre. Ich widerstehe nur mühsam der Versuchung, darauf näher einzugehen.

Galileo takes great pains to demonstrate that the hypothesis of the rotation and revolution of the earth is not refuted by the fact that we do not observe any mechanical effects of these motions. Strictly speaking, such a demonstration was impossible because a complete theory of mechanics was lacking. I think it is just in the struggle with this problem that Galileo's originality is demonstrated with particular force. Galileo is, of course, also concerned to show that the fixed stars are too remote for parallaxes produced by the yearly motion of the earth to be detectable with the measuring instruments of his time. This investigation also is ingenious, notwithstanding its primitiveness.

It was Galileo's longing for a mechanical proof of the motion of the earth which misled him into formulating a wrong theory of the tides. The fascinating arguments in the last conversation would hardly have been accepted as proofs by Galileo, had his temperament not got the better of him. It is hard for me to resist the temptation to deal with this subject more fully.

The *leitmotif* which I recognize in Galileo's work is the passionate fight against any kind of dogma based on authority. Only experience and careful reflection are accepted by him as criteria of truth. Nowadays it is hard for us to grasp how sinister and revolutionary such an attitude appeared at Galileo's time, when merely to doubt the truth of opinions which had no basis but authority was considered a capital crime and punished accordingly. Actually we are by no means so far removed from such a situation even today as many of us would like to flatter ourselves; but in theory, at least, the principle of unbiased thought has won out, and most people are willing to pay lip service to this principle.

It has often been maintained that Galileo became the father of modern science by replacing the speculative, deductive method with the empirical, experimental method. I believe, however, that this interpretation would not stand close scrutiny. There is no empirical method without speculative concepts and systems; and there is no speculative thinking whose concepts do not reveal, on closer investigation, the empirical material from which they stem. To put into sharp contrast the empirical and the deductive attitude is misleading, and was entirely foreign to Galileo. Actually it was not until the nineteenth century that logical (mathematical) systems whose structures were com-

Das Leitmotiv von Galileos Schaffen sehe ich in dem leidenschaftlichen Kampf gegen jeglichen auf Autorität sich stützenden Glauben. Erfahrung und sorgfältige Ueberlegung allein lässt
er als Kriterien der Wahrheit gelten. Wir können uns heute
schwer vorstellen, wie unheimlich und revolutionär eine solche
Einstellung zu Galileos Zeit erschien, in welcher der blosse Zweifel an der Wahrheit von auf blosse Autorität sich stützenden
Meinungen als todeswürdiges Verbrechen betrachtet und be-
straft wurde. Wir sind zwar auch heute keineswegs so weit von
einer solchen Situation entfernt, als sich viele von uns schmei-
cheln mögen; aber der Grundsatz, dass das Denken vorurteils-
frei sein soll, hat sich inzwischen wenigstens in der Theorie
durchgesetzt, und die meisten sind bereit, diesem Grundsatz
Lippendienste zu leisten.

Es ist oft behauptet worden, dass Galileo insofern der Vater
der modernen Naturwissenschaft sei, als er die empiristische,
experimentelle Methode gegenüber der spekulativen, deduktiven
Methode durchgesetzt habe. Ich denke jedoch, dass diese Auffassung genauerer Ueberlegung nicht standhält. Es gibt keine empirische Methode ohne spekulative Begriffs- und System-Kon-
struktion; und es gibt kein spekulatives Denken, dessen Begriffe
bei genauerem Hinsehen nicht das empirische Material verraten,
dem sie ihren Ursprung verdanken. Solche scharfe Gegenüberstellung des empirischen und deduktiven Standpunktes ist ir-
releitend, und sie lag Galileo ganz ferne. Dies hängt schon
damit zusammen, dass logische (mathematische) Systeme, deren
Struktur völlig getrennt ist von jeglichem empirischen Gehalt,
erst im neunzehnten Jahrhundert reinlich herausdestilliert wurden. Ausserdem waren die Galileo zur Verfügung stehenden
experimentellen Methoden so unvollkommen, dass es nur gewag-
ter Spekulation möglich war, die Lücken in den empirischen
Daten zu überbrücken. (So gab es z.B. kein Mittel um Zeiten
unter einer Sekunde zu messen.) Die Antithese Empirismus–
Rationalismus erscheint bei Galileo nicht als Streitpunkt. Galileo
tritt bei Aristoteles und seinen Schülern deduktiven Schlusswei-
sen nur dann entgegen, wenn deren Prämissen ihm willkürlich
oder unhaltbar erscheinen, aber er tadelt seine Gegner nicht weil
sie sich überhaupt deduktiver Methoden bedienen. Er betont in
mehreren Stellen im ersten Dialog, dass auch gemäss Aristoteles
jede — auch die plausibelste — Ueberlegung fallen gelassen

pletely independent of any empirical content had been cleanly extracted. Moreover, the experimental methods at Galileo's disposal were so imperfect that only the boldest speculation could possibly bridge the gaps between the empirical data. (For example, there existed no means to measure times shorter than a second.) The antithesis Empiricism *vs.* Rationalism does not appear as a controversial point in Galileo's work. Galileo opposes the deductive methods of Aristotle and his adherents only when he considers their premises arbitrary or untenable, and he does not rebuke his opponents for the mere fact of using deductive methods. In the first dialogue, he emphasizes in several passages that according to Aristotle, too, even the most plausible deduction must be put aside if it is incompatible with empirical findings. And on the other hand, Galileo himself makes considerable use of logical deduction. His endeavors are not so much directed at "factual knowledge" as at "comprehension." But to comprehend is essentially to draw conclusions from an already accepted logical system.

ALBERT EINSTEIN

Authorized translation by Sonja Bargmann.

werden müsse, wenn sie mit empirischen Befunden unvereinbar ist. Anderseits spielt auch bei Galileo die logische Deduktion eine wichtige Rolle; seine Bemühungen sind weniger auf das „Wissen" als auf das „Begreifen" gerichtet. Begreifen aber ist nichts anderes als aus einem bereits akzeptierten logischen Systeme zu folgern.

ALBERT EINSTEIN

Princeton, Juli 1952

THE TRANSLATOR'S PREFACE

GALILEO'S *Dialogue* ranks high among the classics of science, and is deservedly even more famous as a chapter in the struggle for freedom of thought. It was not Galileo's greatest contribution to the body of scientific knowledge, and yet in a sense it was his most significant service to science itself, for it effectively made clear to scientists and nonscientists alike the claims of experiment and observation as against those of authority and tradition. As Professor Einstein has remarked, this would have been done anyway, even if Galileo had not accomplished it, and might perhaps have been not much longer delayed if he had never lived. Yet the fact remains that this is the book which historically did the most toward breaking down the religious and academic barriers against free scientific thought. Moreover, unlike most scientific classics, it is a book which was capable of interesting the layman and which still is today. Despite all this, the *Dialogue* has remained practically unavailable to the English reader for nearly three centuries. It is now some two decades since I first noticed this extraordinary breach in our literature of the history and philosophy of science, and more than a decade since I commenced the task of repairing it.

The story of Galileo and of this book has been told frequently and well. Born at Pisa in 1564, of noble but impoverished parents, Galileo received his childhood instruction from a talented father who, besides being well versed in mathematics, was a very accomplished musician and the author of a *Dialogue on Ancient and Modern Music*. From him Galileo learned to play the organ and other instruments, among which the lute remained his favorite and gave him solace in his final years of blindness.

At the age of seventeen, Galileo was sent to study medicine at the University of Pisa. Instruction in mathematics which he contrived to get from a tutor at the Tuscan court soon caused him to lose interest in his medical course, and by 1586 he had composed his first scientific work, an essay on the hydrostatic balance (not published until 1644). He had already noticed the isochronism of the pendulum and suggested its use as a timekeeper. By 1589 he had achieved the professorship of mathematics at Pisa. Tradition, unsupported by contemporary records, says that about this time he publicly demonstrated the unreliability of Aristotle's physical views by simultaneously dropping two balls of very unequal weight from the Leaning Tower. In 1592 he obtained the mathematics professorship at Padua, and there he remained until 1610. His invention of a primitive thermometer and his application of the telescope to the heavens belong to the Padua period, which was one of unceasing activity.

From 1611 until his imprisonment by the Inquisition, Galileo served as mathematician and philosopher to the Grand Duke at Florence. The five years following his new appointment were very fruitful; Galileo composed papers on the roughness of the moon's surface, on floating bodies, on the sunspots, and on the tides. Most interesting for us was his *Letter to the Grand Duchess Christina* on the reconciliation of the Scriptures with his new astronomical discoveries and with the Copernican system. Many of his friends urged him to stay out of controversies and to refrain from publishing: "Why court martyrdom for the sake of winning fools from their folly?" But Galileo persisted, and late in 1615 he went to Rome in an effort to win favor for the Copernican view from high officials of the Church. Although received in a friendly manner by several cardinals and by the pope, he was not only unsuccessful in his purpose, but worse; the Congregation of the Index decreed instead to ban Copernicus's book until certain "corrections" were made in it, and Galileo was cautioned not to hold or defend its doctrines any longer. He may also have been ordered not to teach them, though the only evidence for this is an unsigned memorandum (at one time believed to have been a forgery) which was produced at his ultimate trial and condemnation in 1633.

By 1620 Galileo had made enemies among the Jesuits and the Dominicans, chiefly as a result of numerous controversies in print over the nature of sunspots (Galileo being in the right) and of comets (Galileo being in the wrong). In 1623 he published *Il Saggiatore,* his masterpiece in the philosophy of science, designed as an answer to a Jesuit's book on the three comets of 1618 and containing a detailed exposition of the underlying principles of experimental science and the empiricist philosophy. Shortly before its publication Pope Gregory XV died and Cardinal Maffeo Barberini was elected, taking the name Urban VIII; Barberini had a real love for art and science, had been opposed to the decree of 1616, and was personally favorable to Galileo. Galileo promptly dedicated his book to the new pope, putting his opponents on the defensive. He now threw himself into the task of composing the *Dialogue,* believing that it would be assured of license for publication. Yet when it was completed, late in 1629, a whole new series of obstacles appeared, and more than a year was spent in either meeting the various conditions imposed upon him, or evading them wherever circumstances made this possible.

The *Dialogue* was an immense success upon publication, and by the time Galileo's enemies had succeeded in banning it some five months later, there was scarcely a copy to be found with the booksellers. Galileo was ordered to Rome despite his plea of age (nearly seventy years) and infirmity. His treatment there was humane, but the Inquisition was unyielding in its demand that Galileo abjure his error in holding and teaching the Copernican view. In their zeal, the inquisitors themselves committed an error of considerable moment by declaring the view that the sun is immovable to be formally heretical — a status which it never had and never has achieved. This error was one which Galileo himself had earnestly hoped the Church would never make, in view of the serious consequences to religion of branding as heretical an opinion which might eventually be physically proven to be true. A note to this effect, found among the fragments written in his personal copy of the *Dialogue,* serves as the motto for this edition. Arguments in a similar vein make up a considerable part of the *Letter to the Grand Duchess.* An amusing and ironical touch to the entire proceeding is that although the anger of Pope Urban VIII toward Galileo at the time of the trial had its origin in Galileo's having meddled in "high matters" (i.e., having tried to argue the theological merits as well as the scientific aspects of the case), the Church not only eventually conceded the correctness of Galileo's science, but has recently adopted views very similar to his theological arguments also. Thus in the papal encyclical *Humani Generis* of August 12, 1950, we read of the "naif, symbolical way of talking [in the first eleven books of the Bible] well suited to the understanding of a primitive people," and in Galileo's *Letter* of 1615 the following passage occurs: "Since it is very obvious that it was necessary to attribute motion to the sun and rest to the earth, in order not to confound the shallow understanding of the common people and make them obstinate and perverse about believing in the principal articles of the faith, it is no wonder that this was very wisely done in the divine Scriptures."

Some critics have portrayed Galileo as a coward in his abjuration, comparing him unfavorably with Giordano Bruno, who had been burned at the stake in 1600 rather than recant his belief in the Copernican system and the plurality of worlds. Others have represented Galileo as a martyr of science. Both views appear to me to be curiously unrealistic. Neither a coward nor a martyr, Galileo acted as would any shrewd and practical man. Having secured the publication of his *opus majus,* having seen it widely distributed and received with acclaim, knowing that the facts and the ideas in it would work for themselves regardless of his subsequent actions, and not wishing to submit to torture and execution for no purpose, he consented to sign a statement which had been prepared for him and in which the most significant passage was essentially true:

"But whereas — after an injunction had been judicially intimated to me by this Holy Office, to the effect that I must altogether abandon the false opinion that the sun is the center of the world and immovable, and that the earth is not the center of the world, and moves, and that I must not hold, defend, or teach in any way whatsoever, verbally or in writing, the said doctrine, and after it had been notified to me that the said doctrine was contrary to Holy Scripture — I wrote and printed a book in which I discuss this doctrine already condemned, and adduced arguments of great cogency in its favor, without presenting any solution of these; and for this cause I have been pronounced by the Holy Office to be vehemently suspected of heresy — that is to say, of having held and believed that the sun is the center of the world and immovable, and that the earth is not the center, and moves:

"Therefore, desiring to remove from the minds of your Eminences, and of all faithful Christians, this strong suspicion reasonably conceived against me, with sincere heart and unfeigned faith I abjure, curse, and detest the aforesaid errors and heresies, and generally every other error and sect whatsoever contrary to the said Holy Church. . . ."

Although technically imprisoned for the balance of his life, Galileo was in fact treated humanely and considerately, was housed in comfortable surroundings and was permitted to pursue his researches in the company of his favorite pupils. During his remaining years he wrote the *Discourses and Demonstrations Concerning Two New Sciences,* his supreme contribution to physics, published at Leyden in 1638. By January of that year he was totally blind. John Milton, who visited him a few months later, wrote: "There it was that I found and visited the famous Galileo, grown old, a prisoner to the Inquisition for thinking in Astronomy otherwise than the Franciscan and Dominican licensers of thought." Galileo's death occurred on January 8, 1642.

The *Dialogue* was written in colloquial Italian rather than in Latin (into which it was shortly translated) in order to reach the widest possible audience. Within thirty years it had been put into English by Thomas Salusbury *(Mathematical Collections and Translations,* London, 1661), but only a few copies of this work survive, probably because of the great fire of London five years later. It was never reprinted,* and would present difficulties to the modern English reader despite the careful and conscientious work of Salusbury, who attempted to preserve in English the very long and involved sentences of the Italian original. In my opinion the spirit and the historical role of the work demand reasonably easy reading in preference to strict literalness, even at the price of taking certain liberties with the text. This I have not hesitated to do, being heartened by Galileo's known abhorrence of pedantry; in the margin of his copy of Antonio Rocco's book attacking the *Dialogue* he wrote:". . . if I

---

*Salusbury's translation was revised and edited by Giorgio de Santillana and published by the University of Chicago Press in 1953. A facsimile reprint of Salusbury's *Collections* is now in preparation in London.

had been writing for pedants, I should have spoken like a pedant, as you do; but writing for those who are accustomed to reading serious authors, I have spoken as the latter speak." My wish was to make Salviati speak so to modern ears.

The present translation has been made entirely anew, using the definitive National Edition prepared under the direction of Antonio Favaro and published at Florence in 1897. The material specifically added to the text by Galileo himself after publication of the first edition (1632) has been included, and indicated by enclosure in square brackets. Galileo's postils (running notes in the margin) have been placed as nearly as possible beside their textual references. The portrait of Galileo used as a frontispiece is reproduced from *Il Saggiatore*.

There is an excellent German translation by Emil Strauss (Leipzig, 1891) which has been my guide in conjunction with Salusbury's book during the course of this translation. In writing my notes I have drawn heavily upon the erudition of Strauss and of Professor Pietro Pagnini, who wrote the notes for an excellent modern Italian edition published by the Casa Editrice Adriano Salani (Florence, 1935). In order not to disturb the reader's eye unnecessarily, all notes (with the exception of Galileo's postils) have been placed at the end of the text, where they are arranged in order of the pages to which they refer. Biographical and bibliographical notes, as well as translations of foreign phrases and identifications of quotations are supplied without special indication in the text, while other notes are indicated in the text by a dagger (†).

I am much indebted to Sig. Vittorio di Suvero of San Francisco, who has been kind enough to check the translation for me and has made a number of essential corrections and valuable suggestions. The English version has been read at my request by Mr. Daniel Belmont of San Francisco and by Dr. Mark Eudey and Professor Ralph Hultgren of Berkeley; to each I am obliged for various corrections and improvements which have been incorporated into the present text and notes. Mr. Stephen Heller of Ross has prepared many of the illustrations. Professor William Hardy Alexander of Berkeley has given me valuable aid on both the Latin translations and the English text. Mr. Maxwell E. Knight and Mr. John Jennings have assisted me greatly in giving the work its final form. The famous engraved title page is

reproduced from the first edition in the collection of John Howell.
I wish also to acknowledge with thanks the kind permission of
Sig. Mario Salani to borrow from Professor Pagnini's notes.
Grateful acknowledgment is also made to the Clarendon Press
for the privilege of quoting directly from *The Oxford Transla-
tion of Aristotle,* and in particular from the translation of *De
Caelo* by J. L. Stocks and of *Physica* by R. P. Hardie and
R. K. Gaye.

PREFACE TO THE SECOND EDITION

The foregoing preface is virtually unchanged from that of the
1953 edition. I no longer believe that Galileo was entirely in
the wrong concerning comets; my present views are set forth in
the introduction to *The Controversy on the Comets of 1618*
(Philadelphia, 1960). My reconstruction of the events of 1615–
1616 and 1633 which resulted in Galileo's abjuration and sen-
tencing has been published as an appendix to Ludovico Gey-
monat's *Galileo Galilei* (New York, 1965).

In the present edition, many corrections or revisions of text
and notes have been made. Some additional notes which are not
indicated in the text will be found in their proper order at the
end, keyed as before by page number and catchword.

San Francisco
November 25, 1966

DIALOGO
DI
GALILEO GALILEI LINCEO
MATEMATICO SOPRAORDINARIO
DELLO STVDIO DI PISA.

E Filosofo, *e Matematico primario del*

SERENISSIMO

GR. DVCA DI TOSCANA.

Doue ne i congreſſi di quattro giornate ſi diſcorre
ſopra i due

MASSIMI SISTEMI DEL MONDO
TOLEMAICO, E COPERNICANO;

*Proponendo indeterminatamente le ragioni Filoſofiche, e Naturali
tanto per l'vna, quanto per l'altra parte.*

CON PRI   VILEGI.

IN FIORENZA, Per Gio:Batiſta Landini MDCXXXII.

CON LICENZA DE' SVPERIORI.

Imprimatur ſi videbitur Reuerendiſſ. P. Magiſtro Sacri
Palatij Apoſtolici.
A. Epiſcopus Bellicaſtenſis Vicesgerens.

Imprimatur
Fr. Nicolaus Riccardius
Sacri Palatij Apoſtolici Magiſter.

Imprimatur Florentiæ ordinibus conſuetis ſeruatis.
11. Septembris 1630.
Petrus Nicolinus Vic. Gener. Florentiæ.

Imprimatur die 11. Septembris 1630.
Fr. Clemens Egidius Inqu. Gener. Florentiæ.

Stampiſi adi 12. di Settembre 1630.
Niccolò dell'Altella.

THE AUTHOR'S DEDICATION TO
THE GRAND DUKE OF TUSCANY

M OST SERENE GRAND DUKE:
Though the difference between man and the other animals
is enormous, yet one might say reasonably that it is little less
than the difference among men themselves. What is the ratio
of one to a thousand? Yet it is proverbial that one man is
worth a thousand where a thousand are of less value than a
single one. Such differences depend upon diverse mental
abilities, and I reduce them to the difference between being
or not being a philosopher; for philosophy, as the proper
nutriment of those who can feed upon it, does in fact distinguish that single man from the common herd in a greater
or less degree of merit according as his diet varies.

He who looks the higher is the more highly distinguished,
and turning over the great book of nature (which is the
proper object of philosophy) is the way to elevate one's
gaze. And though whatever we read in that book is the
creation of the omnipotent Craftsman, and is accordingly
excellently proportioned, nevertheless that part is most suitable and most worthy which makes His work and His crafts-
manship most evident to our view. The constitution of the
universe I believe may be set in first place among all natural things that can be known, for coming before all others
in grandeur by reason of its universal content, it must also

stand above them all in nobility as their rule and standard. Therefore if any men might claim extreme distinction in intellect above all mankind, Ptolemy and Copernicus were such men, whose gaze was thus raised on high and who philosophized about the constitution of the world. These dialogues of mine revolving principally around their works, it seemed to me that I should not dedicate them to anyone except Your Highness. For they set forth the teaching of these two men whom I consider the greatest minds ever to have left us such contemplations in their works; and, in order to avoid any loss of greatness, must be placed under the protection of the greatest support I know from which they can receive fame and patronage. And if those two men have shed so much light upon my understanding that this work of mine can in large part be called theirs, it may properly be said also to belong to Your Highness, whose liberal munificence has not only given me leisure and peace for writing, but whose effective assistance, never tired of favoring me, is the means by which it finally reaches publication.

Therefore may Your Highness accept it with your customary beneficence; and if anything is to be found in it from which lovers of truth can draw the fruit of greater knowledge and utility, let them acknowledge it as coming from you who are so accustomed to being of assistance that in your happy dominions no man feels the widespread distress existing in the world or suffers anything that disturbs him. Wishing you prosperity and continual increase in your pious and magnanimous practices, I most humbly offer you reverence.

> Your Most Serene Highness's most humble
> and most devoted servant and subject,
> GALILEO GALILEI

TO THE DISCERNING READER

SEVERAL YEARS *ago there was published in Rome a salutary edict which, in order to obviate the dangerous tendencies of our present age, imposed a seasonable silence upon the Pythagorean opinion that the earth moves. There were those who impudently asserted that this decree had its origin not in judicious inquiry, but in passion none too well informed. Complaints were to be heard that advisers who were totally unskilled at astronomical observations ought not to clip the wings of reflective intellects by means of rash prohibitions.*

Upon hearing such carping insolence, my zeal could not be contained. Being thoroughly informed about that prudent determination, I decided to appear openly in the theater of the world as a witness of the sober truth. I was at that time in Rome; I was not only received by the most eminent prelates of that Court, but had their applause; indeed, this decree was not published without some previous notice of it having been given to me. Therefore I propose in the present work to show to foreign nations that as much is understood of this matter in Italy, and particularly in Rome, as transalpine diligence can ever have imagined. Collecting all the reflections that properly concern the Copernican system, I shall make it known that everything was brought before the attention of the Roman censorship, and that there proceed from this clime not only dogmas for the welfare of the soul, but ingenious discoveries for the delight of the mind as well.

To this end I have taken the Copernican side in the discourse, proceeding as with a pure mathematical hypothesis and striving by every artifice to represent it as superior to supposing the earth motionless—not, indeed, absolutely, but as against the arguments of some professed Peripatetics.† These men indeed deserve not even that name, for they do not walk about; they are content to adore the shadows, philosophizing not with due circumspection but merely from having memorized a few ill-understood principles.

Three principal headings are treated. First, I shall try to show that all experiments practicable upon the earth are insufficient measures for proving its mobility, since they are indifferently adaptable to an earth in motion or at rest. I hope in so doing to reveal many observations unknown to the ancients. Secondly, the celestial phenomena will be examined, strengthening the Copernican hypothesis until it might seem that this must triumph absolutely. Here new reflections are adjoined which might be used in order to simplify astronomy, though not because of any necessity imposed by nature. In the third place, I shall propose an ingenious speculation. It happens that long ago I said that the unsolved problem of the ocean tides might receive some light from assuming the motion of the earth. This assertion of mine, passing by word of mouth, found loving fathers who adopted it as a child of their own ingenuity. Now, so that no stranger may ever appear who, arming himself with our weapons, shall charge us with want of attention to such an important matter, I have thought it good to reveal those probabilities which might render this plausible, given that the earth moves.

I hope that from these considerations the world will come to know that if other nations have navigated more, we have not theorized less. It is not from failing to take count of what others have thought that we have yielded to asserting that the earth is motionless, and holding the contrary to be a mere mathematical caprice, but (if for nothing else) for those reasons that are supplied by piety, religion, the knowledge of Divine Omnipotence, and a consciousness of the limitations of the human mind.

I have thought it most appropriate to explain these concepts in the form of dialogues, which, not being restricted to the rigorous observance of mathematical laws, make room also for digressions which are sometimes no less interesting than the principal argument.

Many years ago I was often to be found in the marvelous

city of Venice, in discussions with Signore Giovanni Francesco Sagredo, a man of noble extraction and trenchant wit. From Florence came Signore Filippo Salviati, the least of whose glories were the eminence of his blood and the magnificence of his fortune. His was a sublime intellect which fed no more hungrily upon any pleasure than it did upon fine meditations. I often talked with these two of such matters in the presence of a certain Peripatetic philosopher† whose greatest obstacle in apprehending the truth seemed to be the reputation he had acquired by his interpretations of Aristotle.

Now, since bitter death has deprived Venice and Florence of those two great luminaries in the very meridian of their years, I have resolved to make their fame live on in these pages, so far as my poor abilities will permit, by introducing them as interlocutors in the present argument. (Nor shall the good Peripatetic lack a place; because of his excessive affection toward the Commentaries *of Simplicius, I have thought fit to leave him under the name of the author he so much revered, without mentioning his own.) May it please those two great souls, ever venerable to my heart, to accept this public monument of my undying love. And may the memory of their eloquence assist me in delivering to posterity the promised reflections.*

It happened that several discussions had taken place casually at various times among these gentlemen, and had rather whetted than satisfied their thirst for learning. Hence very wisely they resolved to meet together on certain days during which, setting aside all other business, they might apply themselves more methodically to the contemplation of the wonders of God in the heavens and upon the earth. They met in the palace of the illustrious Sagredo; and, after the customary but brief exchange of compliments, Salviati commenced as follows.

THE FIRST DAY

INTERLOCUTORS

SALVIATI, SAGREDO, AND SIMPLICIO

SALVIATI. Yesterday we resolved to meet today and discuss as clearly and in as much detail as possible the character and the efficacy of those laws of nature which up to the present have been put forth by the partisans of the Aristotelian and Ptolemaic position on the one hand, and by the followers of the Copernican system on the other. Since Copernicus places the earth among the movable heavenly bodies, making it a globe like a planet, we may well begin our discussion by examining the Peripatetic steps in arguing the impossibility of that hypothesis; what they are, and how great is their force and effect. For this it is necessary to introduce into nature two substances which differ essentially. These are the celestial and the elemental, the former being invariant† and eternal; the latter, temporary and destructible. This argument Aristotle treats in his book *De Caelo,* introducing it with some discourses dependent upon certain general assumptions, and afterwards confirming it by experiments and specific demonstrations. Following the same method, I shall first propound, and then freely speak my opinion, submitting myself to your criticisms—particularly those of Simplicio, that stout champion and defender of Aristotelian doctrines.

> Copernicus deems the earth a globe similar to a planet.

> Inalterable celestial substances and alterable elemental substances are necessities in nature, in Aristotle's view.

The first step in the Peripatetic arguments is Aristotle's proof of the completeness and perfection of the world. For, he tells us, it is not a mere line, nor a bare surface, but a body having length, breadth, and depth. Since there are only these three dimensions, the world, having these, has them all, and, having the Whole, is perfect. To be sure, I much wish that Aristotle had proved to me

> Aristotle holds the world to be perfect because it has threefold dimensionality.

by rigorous deductions that simple length constitutes the dimension which we call a line, which by the addition of breadth becomes a surface; that by further adding altitude or depth to this there results a body, and that after these three dimensions there is no passing farther—so that by these three alone, completeness, or, so to speak, wholeness is concluded. Especially since he might have done so very plainly and speedily.

SIMP. What about the elegant demonstrations† in the second, third, and fourth texts, after the definition of "continuous"? Is it not there first proved that there are no more than three dimensions, since Three is everything, and everywhere? And is this not confirmed by the doctrine and authority of the Pythagoreans, who say that all things are determined by three—beginning, middle, and end—which is the number of the Whole? Also, why leave out another of his reasons; namely, that this number is used, as if by a law of nature, in sacrifices to the gods? Furthermore, is it not dictated by nature that we attribute the title of "all" to those things that are three, and not less? For two are called "both," and one does not say "all" unless there are three.

Aristotle's demonstrations to prove there are three dimensions and no more.

The ternary number celebrated among Pythagoreans.

You have all this doctrine in the second text. Afterwards, in the third we read, *ad pleniorem scientiam,* that All, and Whole, and Perfect† are formally one and the same; and that therefore among figures only the solid is complete. For it alone is determined by three, which is All; and, being divisible in three ways, it is divisible in every possible way. Of the other figures, one is divisible in one way, and the other in two, because they have their divisibility and their continuity according to the number of dimensions allotted to them. Thus one figure is continuous in one way, the other in two; but the third, namely the solid, is so in every way.

Moreover, in the fourth text,† after some other doctrines, does he not clinch the matter with another proof? To wit: a transition is made only according to some defect; thus there is a transition in passing from the line to the surface, because the line is lacking in breadth. But it is impossible for the perfect to lack anything, being complete in every way; therefore there is no transition beyond the solid or body to any other figure.

Do you not think that in all these places he has sufficiently proved that there is no passing beyond the three dimensions, length, breadth, and thickness; and that therefore the body, or solid, which has them all, is perfect?

SALV. To tell you the truth, I do not feel impelled by all these reasons to grant any more than this: that whatever has a beginning, middle, and end may and ought to be called perfect. I feel no compulsion to grant that the number three is a perfect number, nor that it has a faculty of conferring perfection upon its possessors. I do not even understand, let alone believe, that with respect to legs, for example, the number three is more perfect than four or two; neither do I conceive the number four to be any imperfection in the elements, nor that they would be more perfect if they were three. Therefore it would have been better for him to leave these subtleties to the rhetoricians, and to prove his point by rigorous demonstrations such as are suitable to make in the demonstrative sciences.

SIMP. It seems that you ridicule these reasons, and yet all of them are doctrines to the Pythagoreans, who attribute so much to numbers. You, who are a mathematician, and who believe many Pythagorean philosophical opinions, now seem to scorn their mysteries.

SALV. That the Pythagoreans held the science of numbers in high esteem, and that Plato himself admired the human understanding and believed it to partake of divinity simply because it understood the nature of numbers, I know very well; nor am I far from being of the same opinion. But that these mysteries which caused Pythagoras and his sect to have such veneration for the science of numbers are the follies that abound in the sayings and writings of the vulgar, I do not believe at all. Rather I know that, in order to prevent the things they admired from being exposed to the slander and scorn of the common people, the Pythagoreans condemned as sacrilegious the publication of the most hidden properties of numbers or of the incommensurable and irrational quantities which they investigated. They taught that anyone who had revealed them was tormented in the other world. Therefore I believe that some one of them, just to satisfy the common sort and free himself from their inquisitiveness, gave it out that the mysteries of numbers were those trifles which later spread among the vulgar. Such astuteness and prudence remind one of the wise young man who, in order to stop the importunity of his mother or his inquisitive wife—I forget which—who pressed him to impart the secrets of the Senate, made up some story which afterwards caused her and many other women to be the laughing-stock of that same Senate.†

Plato's opinion that the human mind partakes of divinity because it comprehends numbers.

Legendary character of Pythagorean number mysteries.

SIMP. I do not want to join the number of those who are too curious about the Pythagorean mysteries. But as to the point in hand, I reply that the reasons produced by Aristotle to prove that there are not and cannot be more than three dimensions seem to me conclusive; and I believe that if a more cogent demonstration had existed, Aristotle would not have omitted it.

SAGR. You might at least add, "if he had known it or if it had occurred to him." Salviati, you would be doing me a great favor by giving me some effective arguments, if there are any clear enough to be comprehended by me.

SALV. Not only by you, but by Simplicio too; and not merely comprehended, but already known—though perhaps without your realizing it.† And to make them easier to understand, let us take this paper and pen which I see already prepared for such occasions, and draw a few figures.

Geometrical
demonstration
of threefold
dimensionality.

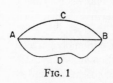

FIG. 1

First we shall mark these two points, A and B, and draw from one to the other the curved lines ACB and ADB, and the straight line AB. I ask which of them is to your mind the one that determines the distance between the ends A and B, and why?

SAGR. I should say the straight line, and not the curves, because the straight one is shorter and because it is unique, distinct, and determinate; the infinite others are indefinite, unequal, and longer. It seems to me that the choice ought to depend upon that which is unique and definite.

SALV. We have the straight line, then, as determining the distance between the two points. We now add another straight line parallel to AB—let it be CD—so that between them there lies a surface of which I want you to show the breadth.

Therefore starting from point A, tell me how and which way you will go, stopping on the line CD, so as to show me the breadth included between those lines. Would you determine it according to the measure of the curve AE, or the straight line AF, or . . . ?

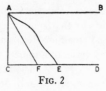

FIG. 2

SIMP. According to the straight line AF, and not according to the curve, such being already excluded for such a use.

SAGR. But I should take neither of them, seeing that the straight line AF runs obliquely. I should draw a line perpendicular to CD, for this would seem to me to be the shortest, as well as being unique among the infinite number of longer and unequal ones which may be drawn from the point A to every other point of the opposite line CD.

SALV. Your choice and the reason you adduce for it seem to me most excellent. So now we have it that the first dimension is determined by a straight line; the second (namely, breadth) by another straight line, and not only straight, but at right angles to that which determines the length. Thus we have defined the two dimensions of a surface; that is, length and breadth.

But suppose you had to determine a height—for example, how high this platform is from the pavement down below there. Seeing that from any point in the platform we may draw infinite lines, curved or straight, and all of different lengths, to the infinite points of the pavement below, which of all these lines would you make use of?

SAGR. I would fasten a string to the platform and, by hanging a plummet from it, would let it freely stretch till it reached very near to the pavement; the length of such a string being the straightest and shortest of all the lines that could possibly be drawn from the same point to the pavement, I should say that it was the true height in this case.

SALV. Very good. And if, from the point on the pavement indicated by this hanging string (taking the pavement to be level and not inclined), you should produce two other straight lines, one for the length and the other for the breadth of the surface of the pavement, what angles would they make with the thread?

SAGR. They would surely meet at right angles, since the string falls perpendicularly and the pavement is quite flat and level.

SALV. Therefore if you assign any point for the point of origin of your measurements, and from that produce a straight line as the determinant of the first measurement (that is, of the length) it will necessarily follow that the one which is to define the breadth leaves the first at a right angle. That which is to denote the altitude, which is the third dimension, going out from the same point, also forms right angles and not oblique angles with the other two. And thus by three perpendiculars you will have determined the three dimensions AB length, AC breadth, and AD height, by three unique, definite, and shortest lines. And since clearly no more lines can meet in the said point to make right angles with them, and the dimensions must be determined by the only straight lines which make right angles with each other, then the dimensions are no more than three; and whatever has the three has all of them, and that which has all of them is divisible in every way, and that which is so, is perfect, etc.†

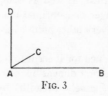

FIG. 3

SIMP. Who says that I cannot draw other lines? Why may I not bring another line from beneath to the point A, which will be perpendicular to the rest?

SALV. Surely you cannot make more than three straight lines meet in the same point and form right angles with each other!

SAGR. Yes, because it seems to me that what Simplicio means would be the same DA prolonged downward. In that way there might also be drawn two others; but they would be the same as the first three, differing only in that whereas now they merely touch, they would then intersect. But this would not produce any new dimensions.

SIMP. I shall not say that this argument of yours cannot be conclusive. But I still say, with Aristotle, that in physical (*naturali*) matters one need not always require a mathematical demonstration.

Geometrical exactitude should not be sought in physical proofs.

SAGR. Granted, where none is to be had; but when there is one at hand, why do you not wish to use it? But it would be good to spend no more words on this point, for I think that Salviati will have conceded both to Aristotle and to you, without further demonstration, that the world is a body, and perfect; yea, most perfect, being the chief work of God.

SALV. Exactly so. Therefore leaving the general contemplation of the whole, let us get to the consideration of the parts. Aristotle in his first division separates the whole into two differing and, in a way, contrary parts; namely, the celestial and the elemental, the former being ingenerable, incorruptible, inalterable, impenetrable, etc.; the latter being exposed to continual alteration, mutation, etc. He takes this difference from the diversity of local motions as his original principle. With this step he proceeds.

Aristotle's two divisions of the universe, celestial and elemental, mutually exclusive.

Leaving, so to speak, the sensible world and retiring into the ideal world, he begins architectonically to consider that, nature being the principle of motion, it is appropriate that natural bodies should be endowed with local motion. He then declares local motions to be of three kinds; namely, circular, straight, and mixed straight-and-circular. The first two he calls simple, because of all lines only the circular and the straight are simple. Hereupon, restricting himself somewhat, he newly defines among the simple motions one, the circular, to be that which is made around the center; and the other, the straight, to be upward and downward—upward, that which goes from the center; and downward, whatever goes toward the center. And from this he infers it to be necessary and proper that all simple motions are confined to these three kinds; namely, toward the center, away from the center, and around the center. This answers, he says, with a certain beautiful harmony to what has been said previously about the body; it is perfect in three things, and its motion is likewise.

Local motions of three kinds— straight, circular, and mixed.

Straight and circular motions simple, being along simple lines.

These motions being established, he goes on to say that some natural bodies being simple, and others composites of those (and he calls those bodies simple which have a natural principle of motion, such as fire and earth), it is proper that simple motions should be those of simple bodies, and that mixed motions should belong to compound bodies; in such a way, moreover, that compounds take the motion of that part which predominates in their composition.

SAGR. Wait awhile, Salviati, for in this argument I find so many doubts assailing me on all sides that I shall either have to tell them to you if I want to pay attention to what you are going to say, or withhold my attention in order to remember my doubts.

SALV. I shall willingly pause, for I run the same risk too, and am on the verge of getting shipwrecked. At present I sail between rocks and boisterous waves that are making me lose my bearings, as they say. Therefore, before I multiply your difficulties, propound them.

SAGR. With Aristotle, you began by removing me somewhat from the sensible world, to show me the architecture with which it must have been built. I thought it proper that you began by telling me that a natural body is naturally movable, since nature, as defined elsewhere,† is the principle of motion. Here I felt some doubt; why does Aristotle not say that among natural bodies some are naturally movable and others immovable, inasmuch as

The definition of
nature either de-
fective or in-
opportunely
brought up by
Aristotle.

The cylindrical
helix might be
called a simple
line.

nature is said in his definition to be the principle of motion and
of rest? For if all natural bodies have the principle of mobility,
either it was not necessary to include rest in the definition of
nature, or to introduce such a definition in this place.

Next, as to the explanation of what Aristotle means by simple
motions, and how he determines them from properties of space,
calling those simple which are made along simple lines, these
being the straight and the circular only, I accept this willingly;
nor do I care to quibble about the case of the cylindrical helix, of
which all parts are similar and which therefore seems to belong
among the simple lines. But I resent rather strongly finding myself restricted to calling the latter "motion about the center"
(while it seems that he wants to repeat the same definition in
other words) and the former *sursum* and *deorsum*—that is, "upward" and "downward." For such terms are applicable only to
the actual world, and imply it to be not only constructed, but
already inhabited by us. Now if straight motion is simple with
the simplicity of the straight line, and if simple motion is natural,
then it remains so when made in any direction whatever; to wit,
upward, downward, backward, forward, to the right, to the left;
and if any other way can be imagined, provided only that it is
straight, it will be suitable for any simple natural body. Or, if
not, then Aristotle's supposition is defective.

Moreover, it appears that Aristotle implies that only one circular motion exists in the world, and consequently only one center
to which the motions of upward and downward exclusively refer.
All of which seems to indicate that he was pulling cards out of his

Aristotle shapes
the rules of
architecture to
the construction
of the universe,
not the con-
struction to
the rules.

sleeve, and trying to accommodate the architecture to the building instead of modeling the building after the precepts of architecture. For if I should say that in the real universe there are
thousands of circular motions, and consequently thousands of
centers, there would also be thousands of motions upward and
downward. Again, he supposes (as was said) simple motions and
mixed motions, calling the circular and the straight "simple" and
motions composed of both "mixed." Now among natural bodies
he calls some simple (namely, those that have a natural principle
of simple motion) and others compound; and the simple motions
he attributes to simple bodies, and the mixed to the compound.
But by "mixed motion" he no longer means that motion in which
straight and circular are mixed, and which may exist in the world.

The mixed motion he introduces is as impossible as it is impossible to mix opposite motions in the same straight line, so as to produce a motion partly upward and partly downward. In order to moderate such an absurdity and impossibility, he is reduced to saying that such compound bodies move according to the predominant simple part. This eventually forces people to say that even motion made along the same straight line is sometimes simple, and sometimes mixed. Thus the simplicity of the motion no longer corresponds to the simplicity of the line alone.

SIMP. Isn't there enough difference between them for you when the simple and absolute movement is more swift than that which comes from predominance? Think how much faster a chunk of pure earth drops than does a stick of wood!

SAGR. All well and good, Simplicio; but if the simplicity is changed by this, then in addition to requiring a great many mixed motions, you would not be able to show me how to distinguish the simple ones. Furthermore, if a greater or lesser velocity can alter the simplicity of motion, no simple body would ever move with a simple motion, since in all natural straight motions the velocity is ever increasing and consequently always changing in simplicity—which, as simplicity, ought properly to be immutable. But what is more important, you burden Aristotle with a new flaw, since in the definition of mixed motions he made no mention of slowness and speed, whereas you now include this as a necessary and essential point. Moreover, you are no better off for having such a rule, for among compound bodies there will be some (and not a few of them) which will move more swiftly than the simple, while others move more slowly; as, for example, lead and wood in comparison with earth. Among such motions, which do you call simple and which mixed?

SIMP. I should call that motion "simple" which is made by a simple body, and that "mixed" which is made by a compound body.

SAGR. Very good indeed. But what are you saying now, Simplicio? A little while ago you would have it that simple and mixed motions would reveal to me which bodies were compound and which were simple. Now you want to use simple and compound bodies to find out which motion is simple, and which is mixed—an excellent rule for never understanding either motions or bodies. Besides which, you have just declared that greater velocity is not even sufficient, and have sought a third condition to define simple motion, while Aristotle contents himself with but one—namely, the simplicity of space. Now, according to you, simple motion will be that which is made along a simple line with a certain determinate velocity by a simple movable body.

Well, have it your own way, and let us return to Aristotle, who defines mixed motion to be that which is composed of straight plus circular — though he failed to show me any body whatever which moves naturally with any such motion.

SALV. I return then to Aristotle. Having very well and methodically begun his discourse, at this point—being more intent upon arriving at a goal previously established in his mind than upon going wherever his steps directly lead him—he cuts right across the path of his discourse and assumes it as a known and manifest thing that the motions directly upward and downward correspond to fire and earth. Therefore it is necessary that beyond these bodies, which are close to us, there must be some other body in nature to which circular motion must be suitable. This must, in turn, be as much more excellent as circular motion is more perfect than straight. Just how much more perfect the former is than the latter, he determines from the perfection of the circular line over the straight line. He calls the former perfect and the latter imperfect; imperfect, because if it is infinite, it lacks an end and termination, while if finite, there is something outside of it in which it might be prolonged. This is the cornerstone, basis, and foundation of the entire structure of the Aristotelian universe, upon which are superimposed all other celestial properties— freedom from gravity and levity, ingenerability, incorruptibility, exemption from all mutations except local ones, etc. All these properties he attributes to a simple body with circular motion. The contrary qualities of gravity or levity, corruptibility, etc., he assigns to bodies naturally movable in a straight line.

Why, to Aristotle, a circular line is perfect and a straight line imperfect.

Now whenever defects are seen in the foundations, it is reasonable to doubt everything else that is built upon them. I do not deny that what Aristotle has introduced up to this point, with a general discourse upon universal first principles, is reinforced with specific reasons and experiments later on in his argument, all of which must be separately considered and weighed. But what has already been said does present many and grave difficulties, whereas basic principles and fundamentals must be secure, firm, and well established, so that one may build confidently upon them. Hence before we multiply doubts, it would not be amiss to see whether (as I believe) we may, by taking another path, discover a more direct and certain road, and establish our basic principles with sounder architectural precepts. Therefore, suspending for the moment the Aristotelian course, which we shall resume again at the proper time and examine in detail, I say that in his conclusions up to this point I agree with him, and I admit that the world is a body endowed with all the dimensions, and therefore most perfect. And I add that as such it is of necessity most orderly, having its parts disposed in the highest and most perfect order among themselves. Which assumption I do not believe to be denied either by you or by anyone else.

The author assumes the universe to be perfectly ordered.

SIMP. Who do you think would deny it? The first point is Aristotle's own, and its very name appears to be derived from no other thing than the order which the world perfectly contains.†

SALV. This principle being established then, it may be immediately concluded that if all integral bodies in the world are by nature movable, it is impossible that their motions should be straight, or anything else but circular; and the reason is very plain and obvious. For whatever moves straight changes place and, continuing to move, goes ever farther from its starting point and from every place through which it successively passes. If that were the motion which naturally suited it, then at the beginning it was not in its proper place. So then the parts of the world were not disposed in perfect order. But we are assuming them to be perfectly in order; and in that case, it is impossible that it should be their nature to change place, and consequently to move in a straight line.

Straight motion cannot exist in a well-ordered universe.

Besides, straight motion being by nature infinite (because a straight line is infinite and indeterminate), it is impossible that anything should have by nature the principle of moving in a straight line; or, in other words, toward a place where it is impossible to arrive, there being no finite end. For nature, as Aristotle well says himself, never undertakes to do that which cannot be done, nor endeavors to move whither it is impossible to arrive.

Straight motion infinite by nature.

Straight motion impossible for nature.

Nature does not attempt the impossible.

But someone might say nevertheless that although a straight line (and consequently the motion along it) can be extended *in infinitum* (that is to say, is unending), still nature has, so to

Straight motion
possible in pri-
mordial chaos.

Straight motion
suitable for ar-
ranging badly
disposed bodies.

According to
Plato, worlds
were moved first
in straight and
afterward in
circular motions.

A body set at
rest will not
move unless it
tends toward
some partic-
ular place.

The body accel-
erates its motion
going in the di-
rection of its
tendency.

Leaving rest, the
body passes
through every
degree of
slowness.

speak, arbitrarily assigned to it some terminus, and has given
her natural bodies natural instincts to move toward that. And I
shall reply that this might perhaps be fabled to have occurred in
primordial chaos, where vague substances wandered confusedly
in disorder, to regulate which nature would very properly have
used straight motions. By means of these, just as well-arranged
bodies would become disordered in moving, so those which were
previously badly disposed might be arranged in order. But after
their optimum distribution and arrangement it is impossible that
there should remain in them natural inclinations to move any
more in straight motions, from which nothing would now follow
but their removal from their proper and natural places; which is
to say, their disordering.†

We may therefore say that straight motion serves to transport
materials for the construction of a work; but this, once constructed, is to rest immovable—or, if movable, is to move only
circularly. Unless we wish to say with Plato that these world
bodies, after their creation and the establishment of the whole,
were for a certain time set in straight motion by their Maker.
Then later, reaching certain definite places, they were set in rotation one by one, passing from straight to circular motion, and
have ever since been preserved and maintained in this. A sublime
concept, and worthy indeed of Plato, which I remember having
heard discussed by our friend, the Lincean Academician.† If I
remember correctly, his remarks were as follows:

Every body constituted in a state of rest but naturally capable
of motion will move when set at liberty only if it has a natural
tendency toward some particular place; for if it were indifferent
to all places it would remain at rest, having no more cause to
move one way than another. Having such a tendency, it naturally
follows that in its motion it will be continually accelerating. Beginning with the slowest motion, it will never acquire any degree
of speed (*velocità*)† without first having passed through all the
gradations of lesser speed—or should I say of greater slowness?
For, leaving a state of rest, which is the infinite degree of slowness, there is no way whatever for it to enter into a definite degree
of speed before having entered into a lesser, and another still less
before that. It seems much more reasonable for it to pass first
through those degrees nearest to that from which it set out, and
from this to those farther on. But the degree from which the

movable body began to move was that of most extreme slowness;
that is to say, from rest. Now this acceleration of motion occurs
only when the body in motion keeps going, and is attained only
by its approaching its goal. So wherever its natural inclination
draws it, it is conducted there by the shortest line; namely, the
straight. We may therefore reasonably say that nature, in conferring a definite speed upon a body constituted at first in rest,
gives it a straight motion through a certain time and space.

This assumed, let us suppose God to have created the planet
Jupiter, for example, upon which He had determined to confer
such-and-such a velocity, to be kept perpetually uniform forever
after. We may say with Plato that at the beginning He gave it a
straight and accelerated motion; and later, when it had arrived at that degree of velocity, converted its straight motion
into circular motion whose speed thereafter was naturally uniform.

SAGR. I take great delight in hearing this discourse, and believe
it will be greater after you have removed one difficulty for me. I
do not see how it necessarily follows that since a moving body,
departing from rest and entering into the motion for which it
has a natural inclination, passes through all the antecedent gradations of slowness that exist between a state of rest and any
assigned degree of velocity, these gradations being infinite, then
nature was not able to confer circular motion upon the newly
created body of Jupiter with such-and-such a velocity.

SALV. I did not say, nor dare I, that it was impossible for nature
or for God to confer immediately that velocity which you speak
of. I do indeed say that *de facto* nature does not do so—that the
doing of this would be something outside the course of nature,
and therefore miraculous. [Let any massive body move with any
given velocity and encounter any body at rest, even the weakest
and least resisting. The former body, meeting the latter, can
never confer upon it immediately its own velocity. An obvious
sign of this is hearing the sound of the percussion, which would
not be heard—or better, would not exist—if the body at rest were
to receive an equal velocity with that of the moving body upon
the arrival of the latter.]†

SAGR. Then you believe that a stone, leaving its state of rest and
entering into its natural motion toward the center of the earth,
passes through every degree of slowness less than any degree of
speed?

Rest is an in-
finite degree of
slowness.

The body is ac-
celerated only by
approaching
its goal.

Nature makes
the body move in
a straight line to
induce in it any
given speed.

Uniform speed
agrees with cir-
cular motion.

Between rest
and any as-
signed speed
lie infinite
degrees of
lesser speed.

Nature does not
immediately con-
fer a determinate
velocity, though
she could.

SALV. I believe it; indeed I am confident of it; confident in the certainty that I can make you also equally convinced of it.

SAGR. If in all this day's discourse I should gain no more than that knowledge, I should regard the time as well spent.

SALV. I seem to gather from your remarks that a great part of your difficulty consists in accepting this very rapid passage of the movable body through the infinite gradations of slowness antecedent to the velocity acquired during the given time. Therefore, before we go any further, I shall try to remove that difficulty. This ought to be an easy task when I tell you that the movable body does pass through the said gradations, but without pausing in any of them. So that even if the passage requires but a single instant of time, still, since every small time contains infinite instants, we shall not lack a sufficiency of them to assign to each its own part of the infinite degrees of slowness, though the time be as short as you please.

Leaving from rest, the body passes through all degrees of speed without pausing in any of them.

SAGR. So far I follow you; nevertheless it seems remarkable to me that a cannon ball (for such I imagine the falling body to be) which is seen to fall so speedily that in less than ten pulse beats it will pass two hundred yards† of altitude, should be found at so small a degree of velocity during its motion that, continuing to move at such a rate without further acceleration, it would not have passed the same distance in a whole day.

No clocks

SALV. Say rather, not in a whole year, nor in ten, no, nor in a thousand. Of this I shall try to persuade you, and with your permission I shall ask you some very simple questions. Accordingly, tell me if you have any trouble granting that the ball, in descending, is always gaining greater impetus† and velocity.

SAGR. I am quite confident of that.

SALV. And if I should say that the impetus acquired at any point in its motion is enough to carry it back to the height from which it started, would you concede that to me?

SAGR. I should concede it without objection, provided that its entire impetus could be applied without impediment to the single operation of restoring it (or an equivalent ball) to the very same height. Thus, if the earth were tunneled through the center, and the ball were let fall a hundred or a thousand yards toward the center, I verily believe that it would pass beyond the center and ascend as much as it had descended. This is shown plainly in the experiment of a plummet hanging from a cord, which, removed

The heavy falling body acquires sufficient impetus to carry it back to an equal height.

from the perpendicular (its state of rest) and then set free, falls toward the perpendicular and goes the same distance beyond it— or only so much less as the cord, the resistance of the air, and other accidents impede it. The same thing is shown by water which, descending through a siphon, climbs back up as much as it went down.

SALV. You argue very well. And I know that you will make no question of granting that the acquisition of impetus is measured by the departure of the movable body from the point of origin and its approach toward the center to which its motion tends. So will you not put an end to your difficulty by conceding that two equal movable bodies, descending by different lines and without any impediment, will have acquired equal impetus whenever their approaches to the center are equal?

SAGR. I do not quite understand the question.

SALV. I shall express it better by drawing a little sketch. Thus I take the line AB as parallel to the horizon, and at the point B I erect the perpendicular BC, and then I add this slanted line CA.

Now the line CA is meant to be an inclined plane, exquisitely polished and hard, upon which descends a perfectly round ball of some very hard substance. Suppose another ball, quite similar, to fall freely along the perpendicular CB. I ask you to concede

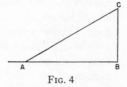

Fig. 4

that the impetus of that which descends by the plane CA, upon arriving at the point A, would be equal to the impetus acquired by the other at point B after falling along the perpendicular CB.

SAGR. I surely believe it would. In fact, they have both advanced equally toward the center; and by what I have already granted, the impetus of each should be equally sufficient to carry it back to the same height.

SALV. Now tell me what you believe the same ball would do if placed upon the horizontal plane AB?

SAGR. It would lie still, the said plane having no inclination.

SALV. But on the inclined plane CA it would descend, though with a slower motion than by the perpendicular CB?

SAGR. I was about to answer confidently in the affirmative, it seeming to me necessary that the motion by the perpendicular CB should be faster than by the inclined plane CA. Yet if that is so, how can the ball on the incline, arriving at the point A, have as much impetus (that is, the same degree of velocity) as the ball dropped along the perpendicular will have at the point B? These two propositions seem contradictory.

Speed along the inclined plane equal to that along the perpendicular, and motion along the perpendicular faster than on the incline.

SALV. Then it would seem to you still more false if I should say categorically that the speeds of the bodies falling by the perpendicular and by the incline are equal. Yet this proposition is quite true, just as it is also true that the body moves more swiftly along the perpendicular than along the incline.

SAGR. To my ears these sound like contradictory propositions. How about you, Simplicio?

SIMP. Likewise to me.

SALV. I think you are joking, pretending not to understand what you know just as well as I do. Tell me, Simplicio, when you think of one body as being faster than another, what concept do you form in your mind?

SIMP. I imagine one to pass over a greater space than the other in the same time, or to travel an equal space in less time.

SALV. Very good. Now as to bodies of equal speed, what is your idea of them?

SIMP. I conceive them to pass equal spaces in equal times.†

SALV. And nothing more than that?

SIMP. This seems to me to be the proper definition of equal motions.

Velocities called equal when spaces passed over are proportional to the times.

SAGR. Let us add another, however, and call the velocities equal when the spaces passed over are in the same proportion as the times in which they are passed. That will be a more general definition.

SALV. So it is, because it includes equal spaces passed in equal times, and also those which are unequal but are passed in times proportionate to them. Now refer again to the same figure, and applying the concept that you have formed of faster motion, tell me why you think the velocity of the body falling along CB should be greater than the velocity of that descending by CA.

SIMP. It seems so to me because in the time it takes one body to pass all CB, the other body will pass a part less than CB in CA.

SALV. Exactly so; and thus it is proved that the body along the perpendicular moves more swiftly than that along the incline. Now consider whether in this same figure one may verify the other concept, and find how the bodies have equal velocities along both the lines CA and CB.

SIMP. I see no such thing. On the contrary, this seems to me to contradict what was just said.

SALV. And what do you say, Sagredo? I do not want to teach you what you already know and have just now defined for me.

SAGR. The definition that I gave you was that movable bodies may be called equally fast when the spaces passed over have the same ratio to each other as the times of travel. Hence for the definition to apply in the present case, it would be required that the time of descent along CA to that along CB should have the same proportion as that of the lines CA and CB. But I do not understand how that can be, when the motion along CB is swifter than that along CA.

SALV. And yet there is no escaping it. Tell me, are not these motions continually accelerated?

SAGR. They are, but more so along the perpendicular than along the incline.

SALV. Well, is this acceleration along the perpendicular such, in comparison with that along the incline, that if we take two equal parts anywhere in these two lines, the motion in the section of the perpendicular will always be faster than in the section of the incline?

SAGR. No indeed. I could choose a place on the incline in which the velocity would be far greater than in an equal space taken along the perpendicular. Such would be the case if the space on the perpendicular should be taken close to the point C, and that on the incline far from it.

SALV. You see, then, that the proposition "Motion along the perpendicular is faster than that along the incline" is true not universally, but only for motions which begin from the initial point; that is, the point of rest. Without this restriction, the proposition would be so defective that its very contrary might be true; namely, that motion is faster along the incline than along the perpendicular. For it is certain that we can choose a space in the incline passed over by the movable body in less time than an equal space was passed in the perpendicular. Now, since the motion along the incline is in some places faster and in some slower than motion along the perpendicular, it follows that along some parts of the incline the time consumed by the moving body will bear a greater proportion to the time consumed by a body moving along some parts of the perpendicular than the space passed by the one bears to the space passed by the other. In other places, the ratio of the times will be less than those of the spaces. For example, consider two bodies departing from rest— that is, from the point C—one by the perpendicular CB and the other by the incline CA.

In the time that the body moving along the perpendicular will have passed all CB, the other will have passed CT, which is less. Therefore the ratio of the times along CT and CB, which

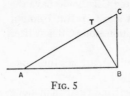

FIG. 5

times are equal, will be greater than that of the line CT to CB, since a given thing stands in greater proportion to a lesser than to a greater. On the other hand, if somewhere along CA (extended as much as may be required) a distance is taken which is equal to CB but is passed over in a shorter time, then the time along the incline will have a smaller ratio to the time along the perpendicular than one distance will to the other. Now since we can conceive distances and velocities along the incline and along the perpendicular such that the proportion between the distances will be now greater and now less than the proportion of the times, we may very reasonably admit that there are also spaces along which the times of the motions bear the same proportion as the distances.

SAGR. I am already freed from my main doubt, and perceive that something which appeared to me a contradiction is not only possible but necessary. But I still do not see that one of these possible or necessary cases is what we need at present to make it true that the time of descent by CA and the time of fall by CB shall have the same ratio as the lines CA and CB, so that we may say without contradiction that the velocity along the incline CA and that along the perpendicular CB are equal.

SALV. Be content for now that I have removed your incredulity. As for the exact knowledge, wait until some other time, when you shall see what our Academic friend has demonstrated concerning local motions. There you shall find it proved that in the time that the one body falls all the distance CB, the other descends along CA only to the point T, upon which falls the perpendicular drawn from the point B. To find where the body falling along the perpendicular would be when the other arrives at the point A, draw from A the line perpendicular to CA, producing both it and CB to their intersection, which will be the point sought. From this you see how it is true that the motion along CB is swifter than that along the incline CA, taking C for the beginning of the motions compared. For the line CB is longer than CT, and the line CB produced to its intersection with the perpendicular to CA drawn from A is greater than CA; therefore the motion along it is swifter than that along CA. But when we compare the motion along CA not with the entire motion made in the same time along the perpendicular as extended, but only with that made in a part of the time and along the portion CB alone, then it is not absurd that the body moving along CA, continuing to descend beyond T, may arrive at A in such a time that the ratio of the lines CA and CB will be the ratio of the times.

Now let us return to our original purpose, which was to show that a heavy body departing from rest passes, in descent, through all the gradations of slowness antecedent to whatever degree of velocity it acquires.

Referring once more to the same figure, let us remember that we agreed that bodies descending along the perpendicular CB and the incline CA were found to have acquired equal degrees of velocity at the points B and A. Now, proceeding from there, I believe you will have no difficulty in granting that upon another plane less steep than AC—for example, AD— the motion of the descending body would be still slower than along the plane AC. Hence one cannot doubt the possibility of planes so little elevated above the horizontal AB that the ball may take any amount of time to reach the point A. If it moved along the plane BA, an infinite time would not suffice, and the motion is retarded according as the slope is diminished. You must therefore admit that

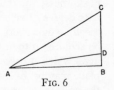

Fig. 6

a point may be taken above the point B and so near to it that if we were to draw a plane from it to the point A, the ball would not pass over it even in a whole year.

Next you must know that the impetus (that is, the degree of velocity) the ball is found to have acquired when it arrives at the point A is such that if the ball continued to move uniformly with this same speed, without accelerating or being retarded, it

would travel twice the length of the inclined plane in a time equal
to that of its descent along the incline. For example, if the ball
has passed over the plane DA in an hour, and continues to move
uniformly with the speed it is found to have upon its arrival at
the point A, it will pass in the next hour a distance double the
length DA. And, as we have said, equal degrees of velocity are
always acquired at the points B and A by bodies departing from
any point taken in the perpendicular CB and descending, the one
by the inclined plane and the other by the perpendicular. Now
the body falling along the perpendicular may leave from a point
so near to B that the degree of velocity acquired at B would not
be enough (if kept always constant) to conduct the body through
a distance double the length of the inclined plane in a year, nor
in ten, nor in a hundred.

We may therefore suppose it to be true that in the ordinary
course of nature a body with all external and accidental impediments removed travels along an inclined plane with greater and
greater slowness according as the inclination is less, until finally
the slowness comes to be infinite when the inclination ends by
coincidence with the horizontal plane. We may likewise suppose
that the degree of velocity acquired at a given point of the inclined plane is equal to the velocity of the body falling along
the perpendicular to its point of intersection with a parallel to
the horizon through the given point of the inclined plane. And if
these two propositions be true, it follows necessarily that a falling
body starting from rest passes through all the infinite gradations
of slowness; and that consequently in order to acquire a determinate degree of velocity it must first move in a straight line, descending by a short or long distance according as the velocity to
be acquired is to be lesser or greater, and according as the plane
upon which it descends is slightly or greatly inclined. Hence a
plane may be given so small an inclination that to acquire in it
the assigned degree of velocity, a body must first move a very
great distance and take a very long time. In the horizontal plane

Circular motion
can never be
acquired nat-
urally without
straight motion
preceding it.

Circular motion
perpetually
uniform.

no velocity whatever would ever be naturally acquired, since the
body in this position will never move. But motion in a horizontal
line which is tilted neither up nor down is circular motion about
the center; therefore circular motion is never acquired naturally
without straight motion to precede it; but, being once acquired,
it will continue perpetually with uniform velocity.

This same truth I could explain to you, and even demonstrate for you, by means of other arguments, but I do not want to interrupt our principal discourse with so great a digression. Rather, I shall return to it upon some other occasion—especially since we have explored this point not to use it as a necessary demonstration, but to illustrate a Platonic concept. And here I wish to add one particular observation of our Academic friend which is quite remarkable. Let us suppose that among the decrees of the divine Architect was the thought of creating in the universe those globes which we behold continually revolving, and of establishing a center of their rotations in which the sun was located immovably. Next, suppose all the said globes to have been created in the same place, and there assigned tendencies of motion, descending toward the center until they had acquired those degrees of velocity which originally seemed good to the Divine mind. These velocities being acquired, we lastly suppose that the globes were set in rotation, each retaining in its orbit (*cerchio*) its predetermined velocity. Now, at what altitude and distance from the sun would have been the place where the said globes were first created, and could they all have been created in the same place?

To make this investigation, we must take from the most skillful astronomers the sizes of the orbits in which the planets revolve, and likewise the times of their revolutions. From these data we deduce how much faster Jupiter (for example) moves than Saturn; and it being found (as in fact it is) that Jupiter does move more swiftly, it is necessary that Jupiter, departing from the same height, descended more than Saturn—as we know is actually the case, its orbit being inferior to that of Saturn. And going still further one may determine, from the proportions of the two velocities of Jupiter and Saturn and from the distance between their orbits, and from the natural ratio of acceleration of natural motion, at what altitude and distance from the center of their revolutions must have been the place from which they originally departed. This place determined and agreed upon, it is asked whether Mars, descending from there to its orbit, is found to agree in size of orbit and velocity of motion with what is found by calculation; and the same is done for the earth, Venus, and Mercury, the size of whose orbits and the velocities of whose motions agree so closely with those given by the computations that the matter is truly wonderful.†

Sizes of orbits and speeds of planetary motions accord in ratio with descent from the same place.

SAGR. I have heard this idea with extreme delight, and if I did not believe that making these calculations accurately would be a long and painful task, and perhaps one too difficult for me to understand, I should ask to see them.

SALV. The procedure is indeed long and difficult, and besides I am not sure I could reconstruct it offhand. Therefore we shall keep it for another time.

[SIMP. Please put it down to my lack of practice in the mathematical sciences if I say freely that your arguments, based upon "greater ratios" and "lesser proportions" and other terms which I do not sufficiently well understand, have not removed my doubt—or rather my incredulity—that a heavy ball of lead, weighing some 100 pounds, leaving its state of rest and being let fall from on high, must pass through every gradation of slowness, when one may see it to have moved more than 100 yards in four pulse beats.† The latter fact makes me entirely incredulous that there is any moment at which it is to be found in such a state of slowness that, continuing therein, it would not move over the half of an inch in a thousand years. Now if this is indeed so, I should like to be convinced of it.

SAGR. Salviati, being profoundly learned, often assumes that these technical expressions which are very well known and familiar to him are equally so to others, and hence at times he forgets that, in speaking with us, someone is needed to assist our incapacity with less abstruse arguments. Therefore I, not being in so lofty a position, will try (with Salviati's permission) to relieve Simplicio of his incredulity, at least in part, by using tangible evidence. Now confining ourselves to the matter of the cannon ball, please tell me, Simplicio: Do you not grant that passage from one state to another is more naturally and readily made to a closer than to a more remote one?

SIMP. I understand that, and I admit it. No doubt a heated iron, for instance, upon cooling down, passes from the tenth degree of heat to the ninth rather than going from the tenth to the sixth.

SAGR. Very good. Then tell me: Does not this cannon ball, sent perpendicularly upward by the force of the charge, continually decelerate in its motion until finally it reaches its ultimate height, where it comes to rest? And in diminishing its velocity—or I mean in increasing its slowness—is it not reasonable that it makes the change from 10 degrees to 11 sooner than from 10 to 12? And from 1,000 to 1,001 sooner than 1,002? And, in short, from any degree to a closer one rather than to one more remote?

SIMP. That is reasonable.

SAGR. But what degree of slowness is there that is so distant from any degree of motion that the state of rest (which is infinite slowness) is not still farther from it? Whence no doubt can remain that the ball, before reaching the point of rest, passes through all the greater and greater gradations of slowness, and consequently through that one at which it would not traverse the distance of one inch in a thousand years. Such being the case, as it certainly is, it should not seem improbable to you, Simplicio, that the same ball in returning downward, leaving rest, recovers the velocity of its motion by returning through those same degrees of slowness through which it passed going up; nor should it, on leaving the larger degrees of slowness which are closer to the state of rest, pass by a jump to those farther away.

SIMP. This argument convinces me much more than the previous mathematical subtleties. Therefore Salviati may resume once more, and continue his argument.]

SALV. Let us return then to our original purpose, taking up where we digressed. If I remember correctly, we were proving that motion in a straight line could be of no use to the well-ordered parts of the world. We went on to say that the same did not hold of circular motions, of which the one made by the moving body upon itself keeps it ever in the same place, and the one which carries the moving body along the circumference of a circle about a fixed center puts neither it nor those about it in disorder, for such motion is essentially finite and terminate. Not only that, but there is no point in the circumference which is not a first as well as a last point in the rotation, so it stays in the circumference assigned to it, leaving everything else inside and outside of that free for the use of others without ever impeding or disordering them. This being the motion that makes the moving body continually leave and continually arrive at the end, it alone can be essentially uniform.[†] For acceleration occurs in a moving body when it is approaching the point toward which it has a tendency, and retardation occurs because of its reluctance to leave and go away from that point; and since in circular motion the moving body is continually going away from and approaching its natural terminus, the repulsion and the inclination are always of equal

Finite and bounded circular motions do not disorder the parts of the universe.

In circular motion every point of the circumference is a beginning and an end.

Only circular motion is uniform.

Circular motion
can be perpetu-
ally maintained

Straight motion
cannot be
naturally
perpetual.

strengths (*forze*) in it. This equality gives rise to a speed which is neither retarded nor accelerated; that is, a uniformity of motion. From this uniformity, and from the motion being finite, there follows its perpetual continuation by a successive repetition of rotations, which cannot exist naturally along an unbounded line or in a motion continually accelerated or retarded. I say "naturally," because straight motion which is retarded is forced (*violento*) motion, which cannot be perpetual, while accelerated motion arrives necessarily at an end if there is one. And if none exists, there cannot be motion toward it, for nature does not move whither it is impossible to arrive.

Straight motion
assigned to nat-
ural bodies to
restore them to
perfect order
when they are
disordered.

I therefore conclude that only circular motion can naturally suit bodies which are integral parts of the universe as constituted in the best arrangement, and that the most which can be said for straight motion is that it is assigned by nature to its bodies (and their parts) whenever these are to be found outside their proper places, arranged badly, and are therefore in need of being restored to their natural state by the shortest path. From which it seems to me one may reasonably conclude that for the maintenance of perfect order among the parts of the universe, it is necessary to say that movable bodies are movable only circularly; if there are any that do not move circularly, these are

Only rest and
circular motion
suitable to main-
tain order.

necessarily immovable, nothing but rest and circular motion being suitable to the preservation of order. And I wonder not a little that Aristotle, who held that the terrestrial globe was located in the center of the universe and remained there immovably, did not say that of natural bodies some were naturally movable and others immovable, especially since he had previously defined nature to be the principle of motion and of rest.

Sensible experi-
ence must be
preferred to
human reason.

Who denies the
senses deserves
to lose them.

Our senses show
that heavy
bodies move
toward the
center, and light
ones toward the
moon's orbit.

SIMP. Aristotle would not give assurance from his reasoning of more than was proper, despite his great genius. He held in his philosophizing that sensible experiments were to be preferred above any argument built by human ingenuity, and he said that those who would contradict the evidence of any sense deserved to be punished by the loss of that sense. Now who is there so blind as not to see that earthy and watery parts, as heavy things, move naturally downward — that is to say toward the center of the universe, assigned by nature itself as the end and terminus of straight motion *deorsum?* Who does not likewise see fire and air move directly upward toward the arc of the moon's orbit,[†] as

the natural end of motion *sursum?* This being so obviously seen,
and it being certain that *eadem est ratio totius et partium,* why
should he not call it a true and evident proposition that the natural motion of earth is straight motion *ad medium,* and that of
fire, straight *a medio?*

SALV. The most that ought to be conceded to you by virtue of
this argument of yours is that just as parts of the earth, removed
from the whole (that is, from the place where they naturally rest)
and, in short, reduced to a bad and disordered arrangement, return to their places naturally and spontaneously in a straight
motion, so it may be inferred (granted that *eadem sit ratio totius
et partium*) that if the terrestrial globe were forcibly removed
from the place assigned to it by nature, it would return by a
straight line. This, as I said, is the most that can be granted to
you, even after giving you every sort of consideration. Anyone
who wants to review these matters rigorously will deny at the
outset that the parts of the earth, when returning to its whole,
do move in a straight line and not in a circular or mixed one.
You would surely have plenty of trouble demonstrating the contrary, as you will clearly see from the answers to the reasons and
the particular experiments adopted by Ptolemy and Aristotle.

There is some
doubt whether
falling heavy
bodies move in
straight lines.

Secondly, if it should be said that the parts of the earth do not
move so as to go toward the center of the universe, but so as to
unite with the whole earth (and that consequently they have a
natural tendency toward the center of the terrestrial globe, by
which tendency they coöperate to form and preserve it), then
what other "whole" and what other "center" would you find for
the universe, to which the entire terrestrial globe would seek to
return if removed therefrom, so that the rationale of the whole
might still be like that of its parts?

The earth spher-
ical because its
parts tend to-
ward the center.

I might add that neither Aristotle nor you can ever prove that
the earth is *de facto* the center of the universe; if any center may
be assigned to the universe, we shall rather find the sun to be
placed there, as you will understand in due course.

The sun more
probably at the
center of the
universe than
the earth.

Now just as all the parts of the earth mutually coöperate to
form its whole, from which it follows that they have equal tendencies to come together in order to unite in the best possible way
and adapt themselves by taking a spherical shape, why may we
not believe that the sun, moon, and other world bodies are also
round in shape merely by a concordant instinct and natural tend-

Natural tendency
of the parts of
all world globes
is toward their
centers.

ency of all their component parts? If at any time one of these parts were forcibly separated from the whole, is it not reasonable to believe that it would return spontaneously and by natural tendency? And in this manner we should conclude that straight motion is equally suitable to all world bodies.

SIMP. There is no doubt whatever that since you wish to deny not only the principles of the sciences, but palpable experience and the very senses themselves, you can never be convinced, nor relieved from any preconceived opinion. Therefore I shall hold my peace because *contra negantes principia non est disputandum,* and not because I am persuaded by your reasoning.

Concerning the things you have just said, questioning even whether the motion of heavy bodies is straight or not, how can you ever reasonably deny that parts of the earth — that is to say, heavy bodies — descend straight toward the center? For if you let a rock fall from a very high tower whose walls are straight and plumb, it will go down grazing the tower to the earth, and strike in the same place where a plummet would come to rest if hung on a cord fastened above, exactly where the rock was let drop. Isn't this only too obvious an argument that such motion is straight and toward the center?

In the second place you question whether parts of the earth move so as to go toward the center of the universe, as Aristotle affirms. As if he had not proved this conclusively by the doctrine of contrary motions, arguing as follows. The motion of heavy bodies is contrary to that of light ones. But the motion of light ones is seen to be directly upward; that is, toward the circumference of the universe. Therefore the motion of heavy bodies is directly toward the center of the universe, and it happens *per accidens* that this is toward the center of the earth, because the latter coincides with the former and is united to it.

Next it is vain to inquire, as you do, what a part of the globe of the sun or moon would do if separated from its whole, because what you inquire into would be the consequence of an impossibility. For, as Aristotle also demonstrates, celestial bodies are invariant, impenetrable, and unbreakable; hence such a case could never arise. And even if it should, and the separated part did return to its whole, it would not return thus because of being heavy or light, since Aristotle also proves that celestial bodies are neither heavy nor light.

Straight motion of heavy bodies is known through the senses.

Aristotle's argument to prove that heavy bodies move so as to go toward the center of the universe.

Heavy bodies move toward the center of the earth only by coincidence.

It is folly to inquire into the consequences of an impossibility.

According to Aristotle, heavenly bodies are neither heavy nor light.

SALV. As I said before, you will learn how reasonable it is for me to doubt whether heavy bodies move by a straight and perpendicular line when I examine that particular argument. As to the second point, I am surprised that you should need to have Aristotle's fallacy revealed, it being so obvious, and I wonder at your failure to perceive that Aristotle assumes what is in question. For observe that . . .

SIMP. Please, Salviati, speak more respectfully of Aristotle. He having been the first, only, and admirable expounder of the syllogistic forms, of proofs, of disproofs, of the manner of discovering sophisms and fallacies — in short, of all logic — how can you ever convince anyone that he would subsequently equivocate so seriously as to take for granted that which is in question? Gentlemen, it would be better first to understand him perfectly, and then see whether you want to refute him.

Aristotle cannot
equivocate, being
the inventor
of logic.

SALV. Simplicio, we are engaging in friendly discussion among ourselves in order to investigate certain truths. I shall never take it ill that you expose my errors; when I have not followed the thought of Aristotle, rebuke me freely, and I shall take it in good part. Only let me expound my doubts and reply somewhat to your last remarks. Logic, as it is generally understood, is the organ with which we philosophize. But just as it may be possible for a craftsman to excel in making organs and yet not know how to play them, so one might be a great logician and still be inexpert in making use of logic. Thus we have many people who theoretically understand the whole art of poetry and yet are inept at composing mere quatrains; others enjoy all the precepts of da Vinci and yet do not know how to paint a stool. Playing the organ is taught not by those who make organs, but by those who know how to play them; poetry is learned by continual reading of the poets; painting is acquired by continual painting and designing; the art of proof, by the reading of books filled with demonstrations—and these are exclusively mathematical works, not logical ones.

Now, returning to our purpose, I say all that Aristotle sees of the motion of light bodies is that fire leaves any part of the surface of the terrestrial globe and goes directly away from it, rising upward; this indeed is to move toward a circumference greater than that of the earth. Aristotle has it move to the arc of the moon's path. But he cannot affirm that this is the circumference

Aristotle's fal-
lacy in proving
the earth to be
at the center of
the universe.

of the universe, or is concentric with that, so that to move toward it is to move toward the circumference of the universe. To do so he must suppose that the center of the earth, from which we see these ascending light bodies depart, is the same as the center of the universe; which is as much as to say that the terrestrial globe is located in the center of the universe. Now that is just what we were questioning, and what Aristotle intended to prove. You say that this is not an obvious fallacy?

SAGR. This argument of Aristotle's appeared to me defective and inconclusive also in another respect, even if one concedes to him that the circumference toward which fire moves in straight lines is that which encloses the universe. For leaving not only from the center, but from any other point in a circle, every body moving in a straight line toward any point whatever will doubtless go toward the circumference and, continuing its motion, will arrive there. Thus we may say truly that this moves toward the circumference, but it will not always be true that anything moving by the same line in the opposite direction would go toward the center. This will be true only if the point taken is itself the center, or if the motion is made along that single line which, produced from the given point, passes through the center. So that to say "Fire moving in a straight line goes toward the circumference of the universe; therefore particles of earth, which move with a contrary motion along the same lines, go toward the center of the universe" is valid only when it is assumed that such lines of fire, produced, pass through the center of the universe. And though we know for certain that they pass through the center of the terrestrial globe (being perpendicular to its surface, not inclined), to draw any conclusion we must suppose that the center of the earth is the same as the center of the universe, or else that particles of fire and earth ascend and descend only by one particular line, which passes through the center of the universe. Now this is false and repugnant to experience, which shows us that particles of fire ascend always by lines perpendicular to the surface of the terrestrial globe, and not by any one line alone, but by infinitely various lines extending from the center of the earth toward every part of the universe.

Disclosure of
Aristotle's
quibble in
another way.

SALV. You most ingeniously lead Aristotle into the same difficulty, Sagredo, showing the obvious mistake, and adding to it yet another inconsistency. We observe the earth to be spherical, and therefore we are certain that it has a center, toward which we see that all its parts move. We are compelled to speak in this way, since their motions are all perpendicular to the surface of the earth, and we understand that as they move toward the center of the earth, they move to their whole, their universal mother. Now let us have the grace to abandon the argument that their natural instinct is to go not toward the center of the earth, but toward the center of the universe; for we do not know where that may be, or whether it exists at all. Even if it exists, it is but an imaginary point; a nothing, without any quality.

Proof it is more reasonable to say that heavy bodies tend to the earth's center than to that of the universe.

Now as to Simplicio's final remark. He says that it is folly to debate whether parts of the sun, moon, or any other celestial body, if separated from their whole, would naturally return to it; because (he says) the case is impossible, it being clear from Aristotle's proofs that celestial bodies are invariant, impenetrable, indivisible, etc. I answer that none of the conditions by which Aristotle distinguishes celestial from elemental bodies has any other foundation than what he deduces from the difference in natural motion between the former and the latter. In that case, if it is denied that circular motion is peculiar to celestial bodies, and affirmed to belong to all naturally movable bodies, then one must choose one of two necessary consequences. Either the attributes of generable-ingenerable, alterable-inalterable, divisible-indivisible, etc. suit equally and commonly all world bodies — as much the celestial as the elemental — or Aristotle has wrongly and erroneously deduced, from circular motion, those attributes which he has assigned to celestial bodies.

Ways in which celestial bodies differ from elemental depend on the motions assigned to them by Aristotle.

SIMP. This way of philosophizing tends to subvert all natural philosophy, and to disorder and set in confusion heaven and earth and the whole universe. However, I believe the fundamental principles of the Peripatetics to be such that there is no danger of new sciences being erected upon their ruins.

SALV. Do not worry yourself about heaven and earth, nor fear either their subversion or the ruin of philosophy. As to heaven, it is in vain that you fear for that which you yourself hold to be inalterable and invariant. As for the earth, we seek rather to ennoble and perfect it when we strive to make it like the celestial bodies, and, as it were, place it in heaven, from which your philosophers have banished it. Philosophy itself cannot but benefit from our disputes, for if our conceptions prove true, new achieve-

Philosophy may gain from the disputes and disagreements of philosophers.

ments will be made; if false, their rebuttal will further confirm the original doctrines. No, save your concern for certain philosophers; come to their aid and defend them. As to science itself, it can only improve.

Now, getting back to the point, please set forth freely whatever presents itself to you in confirmation of that great difference which Aristotle establishes between celestial and elemental bodies by making the former ingenerable, incorruptible, inalterable, etc., and the latter corruptible, alterable, etc.

SIMP. So far I do not see that Aristotle needs any help, standing stoutly and strongly on his feet and not being even attacked yet, much less defeated by you. Nay, what will be your defense against this initial assault?

Aristotle writes:† "That which is generated becomes so from a contrary existing in some subject, and likewise is corrupted in some subject by a contrary into a contrary." Observe thus that corruption and generation occur only if there are contraries. "But the movements of contraries are contrary. If therefore to a celestial body no contrary can be assigned (and circular motion has no other motion for its contrary) then nature did very well to keep contraries out of that which was to be ingenerable and incorruptible." This principle established, it follows immediately as a consequence that such a body is inaugmentable, inalterable, invariant, and finally eternal, and a suitable habitation for the immortal gods — which conforms to the opinion also of all men who have any concept of the gods. He then confirms the same conclusion by the senses; for in all times past, according to memory and tradition, nothing is seen to be altered, either in the remotest heavens, or in any integral part of heaven.

As to there being no motion contrary to the circular, Aristotle proves this in many ways. Without repeating them all, it is very clearly proved because the simple motions are but three — to the middle, from the middle, and around the middle — and of these, the two straight motions, *sursum* and *deorsum,* are obviously contrary. Since one thing can have only one contrary, there is no other motion left which may be contrary to the circular. Oh, contemplate this most subtle and conclusive argument of Aristotle by which he proves the incorruptibility of heaven!

SALV. Well, this is nothing more than that very step of Aristotle's which I just hinted at, and your inference from it remains futile if I deny that the motion which you attribute to celestial bodies is not also suited to the earth. But I tell you that the circular motion which you assign to celestial bodies is also suited to the earth; from which, supposing the rest of your discourse to be conclusive, will follow one of three things, as I just finished telling you, and shall now repeat. Either the earth itself is also ingenerable and incorruptible, as are celestial bodies; or celestial bodies are, like the elemental, generable and alterable; or this difference of motion has nothing to do with generation and corruption. Aristotle's argument and yours contain many propositions that are not to be lightly taken for granted, and in order to examine them better it will be good to make them as clear and distinct as possible. — Excuse me, Sagredo, if perhaps I bore you by repeating myself; pretend that you are hearing me take up the argument in a public debate.

You say: "Generation and corruption occur only where there are contraries; contraries exist only among simple natural bodies, movable in contrary motions; contrary motions include only those made in straight lines between opposite ends; of these there are but two, namely, from the middle and toward the middle; and such motions belong to no natural bodies except earth, fire, and the two other elements; therefore generation and corruption exist only among the elements. And because the third simple motion, namely, the circular, about the middle, has no contrary (because the other two are contraries, and one thing has but one contrary) therefore that natural body to which such motion belongs lacks a contrary and, having no contrary, is ingenerable and incorruptible, etc., because where there are no contraries there is no generation, corruption, etc. But such motion belongs to celestial bodies alone; therefore only these are ingenerable, incorruptible, etc."

Now in the first place, I should think it would be an easier thing to determine whether the earth, a most vast body and very convenient because of its nearness to us, moves so rapidly as to revolve about its axis every twenty-four hours than it would be to understand and determine whether generation and corruption arise from contraries, or indeed whether generation, corruption, and contraries have any existence in nature. And if you, Simplicio, know how to teach me nature's method of operation in quickly begetting a hundred thousand flies from a small quantity

Easier to discover whether the earth moves than whether corruption is wrought by contraries.

of musty wine fumes, showing me what the contraries are in that case, and what thing corrupts, and how, I should esteem you even more than I do; for I comprehend these matters not at all. In addition I should very much like to understand how and why these corrupting contraries are so favorable to daws and so cruel to doves, so indulgent to stags and so impatient with horses, as to allow the former many more years of life (that is to say, of incorruptibility) than they give weeks to the latter. Peach and olive trees are planted in the same soil, exposed to the same cold and heat, to the same rains and wind, and in brief to the same contraries; yet the former decay in a short time, and the latter live many hundreds of years. Besides, I never was thoroughly convinced of any transmutation of substance (always confining ourselves to strictly natural phenomena) according to which matter becomes transformed in such a way that it is utterly destroyed, so that nothing remains of its original being, and another quite different body is produced in its place. If I fancy to myself a body under one aspect, and then under another quite different, I do not think it impossible for transformation to occur by a simple transposition of parts, without any corruption or the generation of anything new, for we see similar metamorphoses daily.

A simple re-
arrangement of
parts can show
bodies in various
aspects.

So, to repeat once more, I answer you that inasmuch as you wish to persuade me that the earth cannot move circularly because of its corruptibility and generability, you will have a much greater task than I, who will prove the opposite to you with arguments that are indeed more difficult but no less conclusive.

SAGR. Pardon me if I interrupt your discourse, Salviati, but much as it delights me (since I, too, find myself entangled in the same difficulties) I doubt that your remarks can be brought to a conclusion without our altogether laying aside the main subject. Therefore if it is possible to go ahead with our first argument, I think it would be better to leave this question of generation and corruption to another separate and exclusive session. Also, if it suits you and Simplicio, we may do the same with other special questions which come up in the course of discussion. These I shall try to keep in mind separately, so as to propose them some other day for careful examination.

Now as to the present question: You say that if one denies Aristotle's statement that circular motion does not belong to the earth as it does to celestial bodies, then it follows that whatever is true of the earth as to its being generable, alterable, and so forth, is true also of the heavens. Let us then inquire no further whether or not such things as generation and corruption exist in nature, but turn to investigating what the terrestrial globe actually does.

SIMP. I cannot accustom my ears to hearing it questioned whether generation and corruption exist in nature, this being a thing which is continually before our eyes, and one about which Aristotle has written two whole books. But once you have denied the principles of the sciences and have cast doubt upon the most evident things, everybody knows that you may prove whatever you will, and maintain any paradox. If you do not daily see herbs, plants, and animals generate and decay, what on earth do you see? Do you not continually behold contraries contending together, the earth changing into water, water into air, air into fire, and air again condensing into clouds, rains, hail, and tempests?

By denying scientific principles one may maintain any paradox.

SAGR. Of course we see those things, and we are willing to grant you Aristotle's argument as to this aspect of the generation and corruption produced by contraries. But what if I should prove to you, on the basis of the very propositions conceded to Aristotle, that celestial bodies themselves are no less generable and corruptible than elemental? What would you say to that?

SIMP. I should say that you would have accomplished the impossible.

SAGR. Tell me, Simplicio, aren't these qualities contrary to one another?

SIMP. Which?

SAGR. Why, these: alterable, inalterable; variable, invariant; generable, ingenerable; corruptible, incorruptible.

SIMP. They are quite contrary.

SAGR. As this is so, and it is also true that celestial bodies are ingenerable and incorruptible, I shall prove to you that celestial bodies must necessarily be generable and corruptible.

SIMP. This cannot be anything but a sophism.

SAGR. First listen to my argument; then criticize it and resolve it.

Celestial bodies, since they are ingenerable and incorruptible, have their contraries in nature, these being such bodies as are generable and corruptible. But where there is contrariety, there are also generation and corruption. Therefore celestial bodies are generable and corruptible.

Celestial bodies are generable and corruptible because they are ingenerable and incorruptible.

Forked argu-
ment, otherwise
called sorites.

SIMP. Did I not tell you it could be nothing but a sophism? This is one of those forked arguments called "sorites,"† like that of the Cretan who said that all Cretans were liars. Therefore, being a Cretan, he had told a lie in saying that Cretans were liars. It follows therefore that the Cretans were not liars, and conse-quently that he, being a Cretan, had spoken truth. And since in saying that Cretans were liars he had spoken truly, including himself as a Cretan, he must consequently be a liar. And thus, in such sophisms, a man may go round and round forever and never come to any conclusion.

SAGR. So far you have given it a name; it now remains for you to unravel it and reveal the fallacy.

Between celestial
bodies no con-
trariety exists.

SIMP. As to its solution and the showing of the fallacy, do you not in the first place see an obvious contradiction in it? Celestial bodies are ingenerable and incorruptible; therefore, celestial bodies are generable and corruptible! Besides, contrariety does not exist between celestial bodies, but between the elements, which have the contrariety of motion *sursum* and *deorsum,* and of levity and gravity. But the heavens, which move circularly (to which motion no other motion is contrary) lack contrariety and are therefore incorruptible, etc.

Contraries caus-
ing corruption do
not reside in the
same body which
becomes
corrupted.

SAGR. Go easy now, Simplicio. Does this contrariety, which makes you call some simple bodies corruptible, reside in the very body that is corrupted, or merely in its relation to some other? I mean, for example, does the moisture by which a piece of earth is corrupted reside in that same earth, or in some other body, which would be either air or water? I believe you will say that just as motion upward and downward, or gravity and levity, which you make out to be the original contraries, cannot both exist in the same subject, neither can moist and dry, nor hot and cold. You must therefore say that when a body becomes cor-rupted, this is occasioned by the quality contrary to its own re-siding in another body. Therefore, to make a celestial body corruptible, it is sufficient that there are in nature bodies having a contrariety to the celestial bodies. And such are the elements, if it is true that corruptibility is contrary to incorruptibility.

Heavenly bodies
touch, but are
not touched by,
elemental ones.

SIMP. No, this is not sufficient, my dear sir. The elements become altered and corrupted because they contact and mix with one another, and thus can exercise their contrariety. But celestial bodies are separated from the elemental, by which they are not even touched — though they, indeed, do influence the elements.

If you want to establish generation and corruption in celestial bodies, you must show that contrariety exists among them. SAGR. Here is how I shall find it among them. The original source from which you derive the contrarieties of the elements is the contrariety of their motions upward and downward. Therefore it must be that whatever principles those motions depend upon are likewise contrary to each other. Now since whatever moves upward does so because of lightness, and whatever downward does so because of heaviness,† lightness and heaviness must be contrary to each other. No less ought we to consider as contraries any other principles that are causes of one thing being heavy and another light. According to you yourself, levity and gravity occur in consequence of rarity and density; therefore density and rarity will be contraries. Now these qualities are to be found so abundantly in celestial bodies that you deem the stars to be merely denser parts of their heaven. If that is so, it follows that the density of the stars exceeds that of the rest of heaven almost infinitely. (This is obvious from the heavens being extremely transparent and the stars extremely opaque, and from there being no qualities except greater or less density and rarity which can be causes of greater or less transparency.) There being, then, such contrariety between celestial bodies, they must necessarily be generable and corruptible in the same way that elemental bodies are, or else contrariety is not the cause of corruptibility, etc.

43

Heaviness and lightness, rarity and density, are contrary qualities.

Stars infinitely exceed in density the material of the rest of the heavens.

SIMP. Neither alternative is necessary, because in celestial bodies density and rarity are not contraries to each other as they are in elemental bodies. For there they do not depend upon the primary qualities, cold and heat, which are contrary, but upon greater or less matter in proportion to size. Now "much" and "little" have only a relative opposition, which is the most trifling there is and has nothing to do with generation and corruption.

Rarity and density in heavenly bodies different from those of the elements. —*Cremonino.*

SAGR. So that to have dense and rare† be the cause of heaviness and lightness in the elements, which in turn are able to cause the contrary motions *sursum* and *deorsum*, upon which next depend the contrarieties for generation and corruption, it is not sufficient for these elements to be "dense" and "rare" from enclosing much or little matter within the same size, or bulk; they must be "dense" and "rare" thanks to the primary qualities of heat and cold. Otherwise they accomplish nothing.

But if that is so, Aristotle has deceived us. He should have

Aristotle
diminished in
stature by assign-
ment of causes
for the elements
being generable
and corruptible.

said this in the first place, and should have written that those
simple bodies are generable and corruptible which are movable
with simple motions upward and downward, depending upon
levity and gravity, caused by rarity and density, made by much
or little matter, resulting from heat and cold. He ought not to
have stopped at simple motion *sursum* and *deorsum*, for I assure
you that as to bodies being heavy or light so that they come to be
moved with contrary motions, any kind of density and rarity
will do, whether it comes from heat and cold or anything else
you please. Heat and cold have nothing to do with this matter.
You will find upon experiment that a glowing iron, which can
surely be called hot, weighs the same and moves in the same
manner as when it is cold. But all this aside, how do you know
that celestial rarity and density do not depend upon heat and
cold?

SIMP. I know it because those qualities do not exist among celestial bodies, which are neither hot nor cold.

SALV. I see we are once more going to engulf ourselves in a boundless sea from which there is no getting out, ever. This is navigating without compass, stars, oars, or rudder, in which we must
needs either pass from bank to bank or run aground, or sail forever lost. If, as you suggested, we are to get on with our main
subject, it is necessary for the present to put aside the general
question whether straight motion is necessary in nature and is
proper to some bodies, and proceed to demonstrations, observations, and particular experiments. First we must propound all
those that have been put forward to prove the earth's stability by
Aristotle, Ptolemy, and others, trying next to resolve them.
Finally we must produce those by which a person may become
persuaded that the earth, no less than the moon or any other
planet, is to be numbered among the natural bodies that move
circularly.

SAGR. I submit to the latter more willingly, as I am better satisfied with your architectonic and general discourse than with that
of Aristotle. For yours satisfies me without the least misgiving,
while the other blocks me in some way at every turn. Nor do I
know why Simplicio should not be quickly satisfied with the
argument you put forward to prove that motion in a straight line
can have no place in nature, so long as we suppose the parts of
the universe to be disposed in the best arrangement and perfectly
ordered.

SALV. Stop there, Sagredo, for now a way occurs to me in which Simplicio may be given satisfaction, provided only that he does not wish to stay so closely tied to every phrase of Aristotle's as to hold it sacrilege to depart from a single one of them.

There is no doubt that to maintain the optimum placement and perfect order of the parts of the universe as to local situation, nothing will do but circular motion or rest. As to motion by a straight line, I do not see how it can be of use for anything except to restore to their natural location such integral bodies as have been accidentally removed and separated from their whole, as we have just said.

Let us now consider the whole terrestrial globe, and let us see what can happen to make it and the other world bodies keep themselves in the natural and best disposition. One must either say that it is at rest and remains perpetually immovable in its place, or else that it stays always in its place but revolves itself, or finally that it goes about a center, moving along the circumference of a circle. Of these events, Aristotle and Ptolemy and all their followers say that it is the first which has always been observed and which will be forever maintained; that is, perpetual rest in the same place. Now why, then, should they not have said from the start that its natural property is to remain motionless, rather than making its natural motion downward, a motion with which it never did and never will move? And as to motion by a straight line, let it be granted to us that nature makes use of this to restore particles of earth, water, air, fire, and every other integral mundane body to their whole, when any of them find themselves separated and transported into some improper place — unless this restoration can also be made by finding some more appropriate circular motion. It seems to me that this original position fits all the consequences much better, even by Aristotle's own method, than to attribute straight motion as an intrinsic and natural principle of the elements. This is obvious; for let me ask the Peripatetic if, being of the opinion that celestial bodies are incorruptible and eternal, he believes that the terrestrial globe is not so, but corruptible and mortal, so that there will come a time when, the sun and moon and other stars continuing their existence and their operations, the earth will not be found in the universe but will be annihilated along with the rest of the elements, and I am certain that he would answer,

Aristotle and Ptolemy suppose the terrestrial globe immovable.

Rest rather than downward motion must be considered natural to the terrestrial globe.

Uer
of
argument

No. Therefore generation and corruption belong to the parts
and not to the whole; indeed, to very small and superficial parts
which are insensible in comparison to the whole mass. Now since
Aristotle argues generation and corruption from the contrariety
of straight motions, let us grant such motions to the parts, which
alone change and decay. But to the whole globe and sphere of
the elements will be ascribed either circular motion or perpetual
continuance in its proper place — the only tendencies fitted for
the perpetuation and maintenance of perfect order.

What is thus said of earth may be said as reasonably of fire
and of the greater part of the air, to which elements the Peripatetics are forced to assign as an intrinsic and natural motion one
with which they were never moved and never will be, and to
abolish from nature that motion with which they move, have
moved, and are to be moved perpetually. I say this because they
assign an upward motion to air and fire, which is a motion that
never belongs to the said elements, but only to some of their
particles — and even then only to restore them to perfect arrangement when they are out of their natural places. On the
other hand, they call circular motion (with which they are incessantly moved) preternatural to them, forgetting what Aristotle
has said many times, that nothing violent can last very long.

SIMP. To all these things we have the most suitable answers,
which I omit for the present in order that we may come to the
particular reasons and sensible experiments which ought to be
finally preferred, as Aristotle well says, above anything that can
be supplied by human argument.

SAGR. Then what has been said up to now will serve to place
under consideration which of two general arguments has the
more probability. First there is that of Aristotle, who would persuade us that sublunar bodies are by nature generable and corruptible, etc., and are therefore very different in essence from
celestial bodies, these being invariant, ingenerable, incorruptible,
etc. This argument is deduced from differences of simple motions. Second is that of Salviati, who assumes the integral parts
of the world to be disposed in the best order, and as a necessary
consequence excludes straight motions for simple natural bodies
as being of no use in nature; he takes the earth to be another of
the celestial bodies, endowed with all the prerogatives that belong
to them. The latter reasoning suits me better up to this point

than the other. Therefore let Simplicio be good enough to pro- duce all the specific arguments, experiments, and observations, both physical and astronomical, by which one may be fully persuaded that the earth differs from the celestial bodies, is immovable, and is located in the center of the universe, or anything else that would exclude the earth from being movable like a planet such as Jupiter, or the moon, etc. And you, Salviati, have the kindness to reply step by step.

SIMP. For a beginning, then, here are two powerful demonstrations proving the earth to be very different from celestial bodies. First, bodies that are generable, corruptible, alterable, etc., are quite different from those that are ingenerable, incorruptible, inalterable, etc. The earth is generable, corruptible, alterable, etc., while celestial bodies are ingenerable, incorruptible, inalterable, etc. Therefore the earth is very different from the celestial bodies.

SAGR. With your first argument, you bring back to the table what has been standing there all day and has just now been carried away.

SIMP. Softly, sir; hear the rest, and you will see how different it is from that. Formerly the minor premise was proved *a priori*, and now I wish to prove it *a posteriori*. See for yourself whether this is the same thing. I shall prove the minor, because the major is obvious.

Sensible experience shows that on earth there are continual generations, corruptions, alterations, etc., the like of which neither our senses nor the traditions or memories of our ancestors have ever detected in heaven; hence heaven is inalterable, etc., and the earth alterable, etc., and therefore different from the heavens.

The heavens immutable because no change has ever been seen in them.

The second argument I take from a principal and essential property, which is this: whatever body is naturally dark and devoid of light is different from luminous and resplendent bodies; the earth is dark and without light, and celestial bodies are splendid and full of light; therefore, etc. Answer these, so that too great a pile does not accumulate, and then I will add others.

Bodies naturally giving light are different from dark ones.

SALV. As to the first, for whose force you appeal to experience, I wish you would tell me precisely what these alterations are that you see on the earth and not in the heavens, and on account of which you call the earth alterable and the heavens not.

SIMP. On earth I continually see herbs, plants, animals generating and decaying; winds, rains, tempests, storms arising; in a word, the appearance of the earth undergoing perpetual change. None of these changes are to be discerned in celestial bodies, whose positions and configurations correspond exactly with everything men remember, without the generation of anything new there or the corruption of anything old.

SALV. But if you have to content yourself with these visible, or rather these seen experiences, you must consider China and America celestial bodies, since you surely have never seen in them these alterations which you see in Italy. Therefore, in your sense, they must be inalterable.

SIMP. Even if I have never seen such alterations in those places with my own senses, there are reliable accounts of them; besides which, *cum eadem sit ratio totius et partium,* those countries being a part of the earth like ours, they must be alterable like this.

SALV. But why have you not observed this, instead of reducing yourself to having to believe the tales of others? Why not see it with your own eyes?

SIMP. Because those countries are far from being exposed to view; they are so distant that our sight could not discover such alterations in them.

SALV. Now see for yourself how you have inadvertently revealed the fallacy of your argument. You say that alterations which may be seen near at hand on earth cannot be seen in America because of the great distance. Well, so much the less could they be seen in the moon, which is many hundreds of times more distant. And if you believe in alterations in Mexico on the basis of news from there, what reports do you have from the moon to convince you that there are no alterations there? From your not seeing alterations in heaven (where if any occurred you would not be able to see them by reason of the distance, and from whence no news is to be had), you cannot deduce that there are none, in the same way as from seeing and recognizing them on earth you correctly deduce that they do exist here.

SIMP. Among the changes that have taken place on earth I can find some so great that if they had occurred on the moon they could very well have been observed here below. From the oldest records we have it that formerly, at the Straits of Gibraltar, Abila and Calpe were joined together with some lesser mountains

Mediterranean
formed by the
division between
Abila and Calpe.

which held the ocean in check; but these mountains being sep- arated by some cause, the opening admitted the sea, which flooded in so as to form the Mediterranean. When we consider the immensity of this, and the difference in appearance which must have been made in the water and land seen from afar, there is no doubt that such a change could easily have been seen by anyone then on the moon. Just so would the inhabitants of earth have discovered any such alteration in the moon; yet there is no history of such a thing being seen. Hence there remains no basis for saying that anything in the heavenly bodies is alterable, etc.

SALV. I do not make bold to say that such great changes have taken place in the moon, but neither am I sure that they could not have happened. Such a mutation could be represented to us only by some variation between the lighter and the darker parts of the moon, and I doubt whether we have had observant selenographers on earth who have for any considerable number of years provided us with such exact selenography as would make us reasonably conclude that no such change has come about in the face of the moon. Of the moon's appearance, I find no more exact description than that some say it represents a human face; others, that it is like the muzzle of a lion; still others, that it is Cain with a bundle of thorns on his back. So to say "Heaven is inalterable, because neither in the moon nor in other celestial bodies are such alterations seen as are discovered upon the earth" has no power to prove anything.

SAGR. This first argument of Simplicio's leaves me with another haunting doubt which I should like to have removed. Accordingly I ask him whether the earth was generable and corruptible before the Mediterranean inundation, or whether it began to be so then?

SIMP. It was without doubt generable and corruptible before, as well; but that was so vast a mutation that it might have been observed as far as the moon.

SAGR. Well, now; if the earth was generable and corruptible before that flood, why may not the moon be equally so without any such change? Why is something necessary in the moon which means nothing on the earth?

SALV. A very penetrating remark. But I am afraid that Simplicio is altering the meaning a bit in this text of Aristotle and the other

Peripatetics. They say that they hold the heavens to be inalterable because not one star there has ever been seen to be generated or corrupted, such being probably a lesser part of heaven than a city is of the earth; yet innumerable of the latter have been destroyed so that not a trace of them remains.

SAGR. Really, I thought otherwise, believing that Simplicio distorted this exposition of the text so that he might not burden the Master and his disciples with a notion even more fantastic than the other. What folly it is to say, "The heavens are inalterable because stars are not generated or corrupted in them." Is there perhaps someone who has seen one terrestrial globe decay and another regenerated in its place? Is it not accepted by all philosophers that very few stars in the heavens are smaller than the earth, while a great many are much bigger? So the decay of a star in heaven would be no less momentous than for the whole terrestrial globe to be destroyed! Now if, in order to be able to introduce generation and corruption into the universe with certainty, it is necessary that as vast a body as a star must be corrupted and regenerated, then you had better give up the whole matter; for I assure you that you will never see the terrestrial globe or any other integral body in the universe so corrupted that, after having been seen for many ages past, it dissolves without leaving a trace behind.

It is no less impossible for a star to be corrupted than the whole terrestrial globe.

SALV. But to give Simplicio more than satisfaction, and to reclaim him if possible from his error, I declare that we do have in our age new events and observations such that if Aristotle were now alive, I have no doubt he would change his opinion. This is easily inferred from his own manner of philosophizing, for when he writes of considering the heavens inalterable, etc., because no new thing is seen to be generated there or any old one dissolved, he seems implicitly to let us understand that if he had seen any such event he would have reversed his opinion, and properly preferred the sensible experience to natural reason. Unless he had taken the senses into account, he would not have argued immutability from sensible mutations not being seen.

Aristotle would alter his opinion upon seeing the new things of our century.

SIMP. Aristotle first laid the basis of his argument *a priori,* showing the necessity of the inalterability of heaven by means of natural, evident, and clear principles. He afterward supported the same *a posteriori,* by the senses and by the traditions of the ancients.

SALV. What you refer to is the method he uses in writing his doctrine, but I do not believe it to be that with which he investigated it. Rather, I think it certain that he first obtained it by means of the senses, experiments, and observations, to assure himself as much as possible of his conclusions. Afterward he sought means to make them demonstrable. That is what is done for the most part in the demonstrative sciences; this comes about because when the conclusion is true, one may by making use of analytical methods hit upon some proposition which is already demonstrated, or arrive at some axiomatic principle; but if the conclusion is false, one can go on forever without ever finding any known truth — if indeed one does not encounter some impossibility or manifest absurdity. And you may be sure that Pythagoras, long before he discovered the proof for which he sacrificed a hecatomb, was sure that the square on the side opposite the right angle in a right triangle was equal to the squares on the other two sides. The certainty of a conclusion assists not a little in the discovery of its proof — meaning always in the demonstrative sciences. But however Aristotle may have proceeded, whether the reason *a priori* came before the sense perception *a posteriori* or the other way round, it is enough that Aristotle, as he said many times, preferred sensible experience to any argument. Besides, the strength of the arguments *a priori* has already been examined.

Now, getting back to the subject, I say that things which are being and have been discovered in the heavens in our own time are such that they can give entire satisfaction to all philosophers, because just such events as we have been calling generations and corruptions have been seen and are being seen in particular bodies and in the whole expanse of heaven. Excellent astronomers have observed many comets generated and dissipated in places above the lunar orbit, besides the two new stars[†] of 1572 and 1604, which were indisputably beyond all the planets. And on the face of the sun itself, with the aid of the telescope, they have seen produced and dissolved dense and dark matter, appearing much like the clouds upon the earth; and many of these are so vast as to exceed not only the Mediterranean Sea, but all of Africa, with Asia thrown in. Now, if Aristotle had seen these things, what do you think he would have said and done, Simplicio?

Being certain of the conclusion assists in finding the proof by the analytical method.

Pythagoras sacrificed a hecatomb for a geometrical proof he discovered.

New stars have appeared in the heavens.

Spots generated and dissipated on the face of the sun.

Sunspots bigger than all Asia and Africa.

SIMP. I do not know what would have been done or said by
Aristotle, who was the master of all science, but I know to some
extent what his followers do and say, and what they ought to do
and say in order not to remain without a guide, a leader, and a
chief in philosophy.

As to the comets, have not these modern astronomers who
wanted to make them celestial been vanquished by the *Anti-
Tycho?* Vanquished, moreover, by their own weapons; that is, by
means of parallaxes and of calculations turned about every which
way, and finally concluding in favor of Aristotle that they are
all elemental. A thing so fundamental to the innovators having
been destroyed, what more remains to keep them on their feet?

SALV. Calm yourself, Simplicio. What does this modern author
of yours say about the new stars of 1572 and 1604, and of the
solar spots? As far as the comets are concerned I, for my part,
care little whether they are generated below or above the moon,
nor have I ever set much store by Tycho's verbosity. Neither do
I feel any reluctance to believe that their matter is elemental,
and that they may rise as they please without encountering any
obstacle from the impenetrability of the Peripatetic heavens,
which I hold to be far more tenuous, yielding, and subtle than
our air. And as to the calculation of parallaxes, in the first place
I doubt whether comets are subject to parallax; besides, the inconstancy of the observations upon which they have been computed renders me equally suspicious of both his opinions and his

adversary's — the more so because it seems to me that the *Anti-
Tycho* sometimes trims to its author's taste those observations
which do not suit his purposes, or else declares them to be
erroneous.

SIMP. With regard to the new stars, the *Anti-Tycho* thoroughly
disposes of them in a few words, saying that such recent new
stars are not positively known to be heavenly bodies, and that
if its adversaries wish to prove any alterations and generations
in the latter, they must show us mutations made in stars which
have already been described for a long time and which are celestial objects beyond doubt. And this can never possibly be done.

As to that material which some say is generated and dissolved
on the face of the sun, no mention is made of it at all, from which
I should gather that the author takes it for a fable, or for an
illusion of the telescope,† or at best for some phenomenon produced by the air; in a word, for anything but celestial matter.

SALV. But you, Simplicio, what have you thought of to reply to
the opposition of these importunate spots[†] which have come to
disturb the heavens, and worse still, the Peripatetic philosophy?
It must be that you, as its intrepid defender, have found a reply
and a solution which you should not deprive us of.

SIMP. I have heard different opinions on this matter. Some say,
"They are stars which, like Venus and Mercury, go about the sun
in their proper orbits, and in passing under it present themselves
to us as dark; and because there are many of them, they frequently happen to collect together, and then again to separate."
Others believe them to be figments of the air; still others, illusions of the lenses; and still others, other things. But I am most
inclined to believe — yes, I think it certain — that they are a
collection of various different opaque objects, coming together
almost accidentally; and therefore we often see that in one spot
there can be counted ten or more such tiny bodies of irregular
shape that look like snowflakes, or tufts of wool, or flying moths.
They change places with each other, now separating and now
congregating, but mostly right under the sun, about which, as
their center, they move. But it is not therefore necessary to say
that they are generated or decay. Rather, they are sometimes
hidden behind the body of the sun; at other times, though far
from it, they cannot be seen because of their proximity to its im-
measurable light. For in the sun's eccentric[†] sphere there is
established a sort of onion composed of various folds, one within
another, each being studded with certain little spots, and moving; and although their movements seem at first to be inconstant
and irregular, nonetheless it is said to be ultimately observed that
after a certain time the same spots are sure to return. This seems
to me to be the most appropriate expedient that has so far been
found to account for such phenomena, and at the same time to
maintain the incorruptibility and ingenerability of the heavens.
And if this is not enough, there are more brilliant intellects who
will find better answers.

SALV. If what we are discussing were a point of law or of the
humanities, in which neither true nor false exists, one might trust
in subtlety of mind and readiness of tongue and in the greater
experience of the writers, and expect him who excelled in those
things to make his reasoning most plausible, and one might judge
it to be the best. But in the natural sciences, whose conclusions
are true and necessary and have nothing to do with human will,

In the natural
sciences the art
of oratory is
ineffective.

one must take care not to place oneself in the defense of error;
for here a thousand Demostheneses and a thousand Aristotles
would be left in the lurch by every mediocre wit who happened
to hit upon the truth for himself. Therefore, Simplicio, give up
this idea and this hope of yours that there may be men so much
more learned, erudite, and well-read than the rest of us as to be
able to make that which is false become true in defiance of nature. And since among all opinions that have thus far been produced regarding the essence of sunspots, this one you have just
explained appears to you to be the correct one, it follows that
all the rest are false. Now to free you also from that one —
which is an utterly delusive chimera — I shall, disregarding the
many improbabilities in it, convey to you but two observed facts
against it.

One is that many of these spots are seen to originate in the
middle of the solar disc, and likewise many dissolve and vanish
far from the edge of the sun, a necessary argument that they
must be generated and dissolved. For without generation and
corruption, they could appear there only by way of local motion,
and they all ought to enter and leave by the very edge.

Conclusive proof
of the sunspots
being contiguous
to the body
of the sun.

The other observation, for those not in the rankest ignorance
of perspective, is that from the changes of shape observed in the
spots, and from their apparent changes in velocity, one must
infer that the spots are in contact with the sun's body, and that,
touching its surface, they are moved either with it or upon it,
and in no sense revolve in circles distant from it. Their motion

Motion of the
spots near the
sun's circumfer-
ence appears
slow.

proves this by appearing to be very slow around the edge of the
solar disc, and quite fast toward its center; the shapes of the
spots prove the same by appearing very narrow around the sun's

Shape of the
spots narrow
around the edge
of the solar disc,
and why.

edge in comparison with how they look in the vicinity of the
center. For around the center they are seen in their majesty and
as they really are; but around the edge, because of the curvature
of the spherical surface, they show themselves foreshortened.
These diminutions of both motion and shape, for anyone who
knows how to observe them and calculate diligently, correspond
exactly to what ought to appear if the spots are contiguous to
the sun, and hopelessly contradict their moving in distant circles,
or even at small intervals from the solar body. This has been
abundantly demonstrated by our mutual friend in his *Letters to
Mark Welser on the Solar Spots*. It may be inferred from the

same changes of shape that none of these are stars or other spherical bodies, because of all shapes only the sphere is never seen foreshortened, nor can it appear to be anything but perfectly round. So if any of the individual spots were a round body, as all stars are deemed to be, it would present the same roundness in the middle of the sun's disc as at the extreme edge, whereas they so much foreshorten and look so thin near that extremity, and are on the other hand so broad and long toward the center, as to make it certain that these are flakes of little thickness or depth with respect to their length and breadth.

Sunspots not spherical in shape, but spread out like thin flakes.

Then as to its being observed ultimately that the same spots are sure to return after a certain period, do not believe that, Simplicio; those who said that were trying to deceive you. That this is so, you may see from their having said nothing to you about those that are generated or dissolved on the face of the sun far from the edge; nor told you a word about those which foreshorten, this being a necessary proof of their contiguity to the sun. The truth about the same spots returning is merely what is written in the said *Letters;* namely, that some of them are occasionally of such long duration that they do not disappear in a single revolution around the sun, which takes place in less than a month.

SIMP. To tell the truth, I have not made such long and careful observations that I can qualify as an authority on the facts of this matter; but certainly I wish to do so, and then to see whether I can once more succeed in reconciling what experience presents to us with what Aristotle teaches. For obviously two truths cannot contradict one another.

SALV. Whenever you wish to reconcile what your senses show you with the soundest teachings of Aristotle, you will have no trouble at all. Does not Aristotle say that because of the great distance, celestial matters cannot be treated very definitely?
SIMP. He does say so, quite clearly.

Because of their great distance the heavens cannot be confidently discussed, according to Aristotle.

SALV. Does he not also declare that what sensible experience shows ought to be preferred over any argument,† even one that seems to be extremely well founded? And does he not say this positively and without a bit of hesitation?
SIMP. He does.

Senses prevail over reason for Aristotle.

SALV. Then of the two propositions, both of them Aristotelian doctrines, the second — which says it is necessary to prefer the

The heavens may
be said to be
alterable by a
doctrine more in
agreement with
Aristotle than
that in which
they are made
inalterable.

Thanks to the
telescope, we can
discuss celestial
matters better
than Aristotle
could.

Simplicio's
declamation.

senses over arguments — is a more solid and definite doctrine than the other, which holds the heavens to be inalterable. Therefore it is better Aristotelian philosophy to say, "Heaven is alterable because my senses tell me so," than to say, "Heaven is inalterable because Aristotle was so persuaded by reasoning." Add to this that we possess a better basis for reasoning about celestial things than Aristotle did. He admitted such perceptions to be very difficult for him by reason of the distance from his senses, and conceded that one whose senses could better represent them would be able to philosophize about them with more certainty. Now we, thanks to the telescope, have brought the heavens thirty or forty times closer to us than they were to Aristotle, so that we can discern many things in them that he could not see; among other things these sunspots, which were absolutely invisible to him. Therefore we can treat of the heavens and the sun more confidently than Aristotle could.

SAGR. I can put myself in Simplicio's place and see that he is deeply moved by the overwhelming force of these conclusive arguments. But seeing on the other hand the great authority that Aristotle has gained universally; considering the number of famous interpreters who have toiled to explain his meanings; and observing that the other sciences, so useful and necessary to mankind, base a large part of their value and reputation upon Aristotle's credit; Simplicio is confused and perplexed, and I seem to hear him say, "Who would there be to settle our controversies if Aristotle were to be deposed? What other author should we follow in the schools, the academies, the universities? What philosopher has written the whole of natural philosophy, so well arranged, without omitting a single conclusion? Ought we to desert that structure under which so many travelers have recuperated? Should we destroy that haven, that Prytaneum where so many scholars have taken refuge so comfortably; where, without exposing themselves to the inclemencies of the air, they can acquire a complete knowledge of the universe by merely turning over a few pages? Should that fort be leveled where one may abide in safety against all enemy assaults?"

I pity him no less than I should some fine gentleman who, having built a magnificent palace at great trouble and expense, employing hundreds and hundreds of artisans, and then beholding it threatened with ruin because of poor foundations, should

attempt, in order to avoid the grief of seeing the walls destroyed,
adorned as they are with so many lovely murals; or the columns
fall, which sustain the superb galleries, or the gilded beams; or
the doors spoiled, or the pediments and the marble cornices,
brought in at so much cost — should attempt, I say, to prevent
the collapse with chains, props, iron bars, buttresses, and shores.

SALV. Well, Simplicio need not yet fear any such collapse; I
undertake to insure him against damage at a much smaller cost.
There is no danger that such a multitude of great, subtle, and It is the Peripa-
tetic philosophy
that is inalter-
able.
wise philosophers will allow themselves to be overcome by one
or two who bluster a bit. Rather, without even directing their
pens against them, by means of silence alone, they place them
in universal scorn and derision. It is vanity to imagine that one
can introduce a new philosophy by refuting this or that author.
It is necessary first to teach the reform of the human mind and
to render it capable of distinguishing truth from falsehood, which
only God can do.

But where have we strayed, going from one argument to another? I shall not be able to get back to the path without guid-
ance from your memory.

SIMP. I remember quite well. We were dealing with the reply
of the *Anti-Tycho* to the objections against the immutability of
the heavens. Among these you brought in this matter of the sunspots, not mentioned by its author, and I believe you wished to
give consideration to his reply in the case of the new stars.

SALV. Now I remember the rest. Continuing this subject, it seems
to me that in the counterargument of the *Anti-Tycho* there are
some things that ought to be criticized. First of all, if the two new
stars, which that author can do no less than place in the highest
regions of heaven, and which existed a long time and finally
vanished, caused him no anxiety about insisting upon the inalterability of heaven simply because they were not unquestionably parts of heaven or mutations in the ancient stars, then to
what purpose does he make all this fuss and bother about getting
the comets away from the celestial regions at all costs? Would
it not have been enough for him to say that they are not unquestionably parts of heaven and not mutations in the ancient stars,
and hence that they do not prejudice in any way either the
heavens or the doctrines of Aristotle?

In the second place I am not satisfied about his state of mind when he admits that any alterations which might be made in the stars would be destructive of the celestial prerogatives of incorruptibility, etc., since the stars are celestial things, as is obvious and as everybody admits, and when on the other hand he is not the least perturbed if the same alterations take place elsewhere in the expanse of heaven outside the stars themselves. Does he perhaps mean to imply that heaven is not a celestial thing? I should think that the stars were called celestial things because of their being in the heavens, or because of their being made of heavenly material, and that therefore the heavens would be even more celestial than they; I could not say similarly that anything was more terrestrial than earth itself, or more igneous than fire.

Next, his not having made mention of the sunspots, which are conclusively proved to be produced and dissolved and to be situated next to the body of the sun and to revolve with it or in relation to it, gives me a good indication that this author may write more for the comforting of others than from his own convictions. I say this because he shows himself to be acquainted with mathematics, and it would be impossible for him not to be convinced by the proofs that such material is necessarily contiguous to the sun and undergoes generations and dissolutions so great that nothing of comparable size has ever occurred on earth. And if the generations and corruptions occurring on the very globe of the sun are so many, so great, and so frequent, while this can reasonably be called the noblest part of the heavens, then what argument remains that can dissuade us from believing that others take place on the other globes?

SAGR. I cannot without great astonishment—I might say without great insult to my intelligence—hear it attributed as a prime perfection and nobility of the natural and integral bodies of the universe that they are invariant, immutable, inalterable, etc., while on the other hand it is called a great imperfection to be alterable, generable, mutable, etc. For my part I consider the earth very noble and admirable precisely because of the diverse alterations, changes, generations, etc. that occur in it incessantly. If, not being subject to any changes, it were a vast desert of sand or a mountain of jasper, or if at the time of the flood the waters which covered it had frozen, and it had remained an enormous globe of ice where nothing was ever born or ever altered or changed, I should deem it a useless lump in the uni-

Generability and alteration are greater perfections in worldly bodies than the opposite conditions.

The earth most noble because so many mutations occur in it.

The earth useless and idle if alterations are removed.

verse, devoid of activity and, in a word, superfluous and essentially nonexistent. This is exactly the difference between a living animal and a dead one; and I say the same of the moon, of Jupiter, and of all other world globes.

The deeper I go in considering the vanities of popular reasoning, the lighter and more foolish I find them. What greater stupidity can be imagined than that of calling jewels, silver, and gold "precious," and earth and soil "base"? People who do this ought to remember that if there were as great a scarcity of soil as of jewels or precious metals, there would not be a prince who would not spend a bushel of diamonds and rubies and a cartload of gold just to have enough earth to plant a jasmine in a little pot, or to sow an orange seed and watch it sprout, grow, and produce its handsome leaves, its fragrant flowers, and fine fruit. It is scarcity and plenty that make the vulgar take things to be precious or worthless; they call a diamond very beautiful because it is like pure water, and then would not exchange one for ten barrels of water. Those who so greatly exalt incorruptibility, inalterability, etc. are reduced to talking this way, I believe, by their great desire to go on living, and by the terror they have of death. They do not reflect that if men were immortal, they themselves would never have come into the world. Such men really deserve to encounter a Medusa's head which would transmute them into statues of jasper or of diamond, and thus make them more perfect than they are.

SALV. Maybe such a metamorphosis would not be entirely to their disadvantage, for I think it would be better for them not to argue than to argue on the wrong side.

SIMP. Oh, there is no doubt whatever that the earth is more perfect the way it is, being alterable, changeable, etc., than it would be if it were a mass of stone or even a solid diamond, and extremely hard and invariant. But to the extent that these conditions bring nobility to the earth, they would render less perfect the celestial bodies, in which they would be superfluous. For the celestial bodies—that is, the sun, the moon, and the other stars, which are ordained to have no other use than that of service to the earth — need nothing more than motion and light to achieve their end.

SAGR. Has nature, then, produced and directed all these enormous, perfect, and most noble celestial bodies, invariant, eternal,

Earth nobler than gold and jewels.

Scarcity and plenty make things costly or cheap.

Incorruptibility extolled by the vulgar from fear of death.

Detractors of corruptibility deserve to be turned into statues.

Heavenly bodies, ordained to serve the earth, have no need of anything except motion and light.

and divine, for no other purpose than to serve the changeable, transitory, and mortal earth? To serve that which you call the dregs of the universe, the sink of all uncleanness? Now to what purpose would the celestial bodies be made eternal, etc. in order to serve something transitory, etc.? Take away this purpose of serving the earth, and the innumerable host of celestial bodies is left useless and superfluous, since they have not and cannot have any reciprocal activities among themselves, all of them being inalterable, immutable, and invariant. For instance, if the moon is invariant, how would you have the sun or any other star act upon it? The action would doubtless have no more effect than an attempt to melt a large mass of gold by looking at it or by thinking about it. Besides, it seems to me that at such times as the celestial bodies are contributing to the generations and alterations on the earth, they too must be alterable. Otherwise I do not see how the influence of the moon or sun in causing generations on the earth would differ from placing a marble statue beside a woman and expecting children from such a union.

Simp. Corruptibility, alteration, mutation, etc. do not pertain to the whole terrestrial globe, which as to its entirety is no less eternal than the sun or moon. But as to its external parts it is generable and corruptible, and it is certainly true that generations and corruptions are perpetual in those parts, and, as perpetual, that they require celestial and eternal operations. Therefore it is necessary that celestial bodies be eternal.

Sagr. This is all very well, but if there is nothing prejudicial to the immortality of the entire terrestrial globe in the corruptibility of its superficial parts, and if this generability, corruptibility, alterability, etc. give to it a great ornament and perfection, then why can you not and should you not likewise admit alterations, generations, etc. in the external parts of the celestial globes, adding these as an ornament without diminishing their perfection or depriving them of actions; even increasing those by making them operative not only upon the earth but reciprocally among themselves, and the earth also upon them?

Simp. This cannot be, because the generations, mutations, etc. which would occur, say, on the moon, would be vain and useless, and *natura nihil frustra facit.*

Sagr. And why should they be vain and useless?

Simp. Because we plainly see and feel that all generations,

Heavenly bodies
have no mutual
interactions.

Alterability not
of the whole
terrestrial globe,
but of some
parts.

Celestial bodies
alterable in their
external parts.

changes, etc. that occur on earth are either directly or indirectly designed for the use, comfort, and benefit of man. Horses are born to accommodate men; for the nutriment of horses, the earth produces hay and the clouds water it. For the comfort and nourishment of men are created herbs, cereals, fruits, beasts, birds, and fishes. In brief, if we proceed to examine and weigh carefully all these things, we shall find that the goal toward which all are directed is the need, the use, the comfort and the delight of men. Now of what use to the human race could generations ever be which might happen on the moon or other planets? Unless you mean that there are men also on the moon who enjoy their fruits; an idea which if not mythical is impious.

All generations and changes made on earth are for the benefit of mankind.

SAGR. I do not know nor do I suppose that herbs or plants or animals similar to ours are propagated on the moon, or that rains and winds and thunderstorms occur there as on the earth; much less that it is inhabited by men. Yet I still do not see that it necessarily follows that since things similar to ours are not generated there, no alterations at all take place, or that there cannot be things there that do change or are generated and dissolve; things not only different from ours, but so far from our conceptions as to be entirely unimaginable by us.

The moon has no species like ours and is not inhabited by men.

On the moon there may be species of things different from ours.

I am certain that a person born and raised in a huge forest among wild beasts and birds, and knowing nothing of the watery element, would never be able to frame in his imagination another world existing in nature differing from his, filled with animals which would travel without legs or fast-beating wings, and not upon its surface alone like beasts upon the earth, but everywhere within its depths; and not only moving, but stopping motionless wherever they pleased, a thing which birds in the air cannot do. And that men lived there too, and built palaces and cities, and traveled with such ease that without tiring themselves at all they could proceed to far countries with their families and households and whole cities. Now as I say, I am sure that such a man could not, even with the liveliest imagination, ever picture to himself fishes, the ocean, ships, fleets, and armadas. Thus, and more so, might it happen that in the moon, separated from us by so much greater an interval and made of materials perhaps much different from those on earth, substances exist and actions occur which are not merely remote from but completely beyond all our imaginings, lacking any resemblance to ours and therefore

One who knew nothing of the watery element could not imagine ships or fishes.

On the moon there may be substances different from ours.

being entirely unthinkable. For that which we imagine must be either something already seen or a composite of things and parts of things seen at different times; such are sphinxes, sirens, chimeras, centaurs, etc.

SALV. Many times have I given rein to my fancies about these things, and my conclusion is that it is indeed possible to discover some things that do not and cannot exist on the moon, but none which I believe can be and are there, except very generally; that is, things occupying it, acting and moving in it, perhaps in a very different way from ours, seeing and admiring the grandeur and beauty of the universe and of its Maker and Director and continually singing encomiums in His praise. I mean, in a word, doing what is so frequently decreed in the Holy Scriptures; namely, a perpetual occupation of all creatures in praising God.

SAGR. These are among the things which, speaking very generally, could be there. But I should like to hear you mention those which you believe cannot be there, as it must be possible for you to name them more specifically.

SALV. I warn you, Sagredo, that this will be the third time we have thus strayed imperceptibly, step by step, from our principal topic, and we shall get to the point of our argument but slowly if we make digressions. Therefore it will perhaps be good if we defer this matter, along with others we have agreed to put off until a special session.

SAGR. Please, now that we are on the moon, let us go on with things that pertain to it, so that we shall not have to make another trip over so long a road.

SALV. As you wish. To begin with the most general things, I believe that the lunar globe is very different from the terrestrial, although in some points conformity is to be seen. I shall speak first of their resemblances and then of differences.

First agreement between the moon and the earth, that of shape, proved by the way it is lighted by the sun.

The moon certainly agrees with the earth in its shape, which is indubitably spherical. This follows necessarily from its disc being seen perfectly circular, and from the manner of its receiving light from the sun. For if its surface were flat, it would all become covered with light at once, and likewise would all be deprived of light in an instant; not first the part directed toward the sun and then successively the following parts, so that the whole apparent disc is illuminated at opposition but not before. And on the other hand the contrary would occur if the visible surface were concave; that is, illumination would commence at the part opposite to the sun.

In the second place, it is itself dark and opaque like the earth, by which opacity it is fitted to receive and reflect the light of the sun; for if it were not so, it could not do this.

Third, I hold its material to be very dense and solid, no less than the earth's, of which a sufficiently clear proof to me is the unevenness of the major parts of its surface, evidenced by the many prominences and cavities revealed by the aid of the telescope. The prominences there are mainly very similar to our most rugged and steepest mountains, and some of them are seen to be drawn out in long tracts of hundreds of miles. Others are in more compact groups, and there are also many detached and solitary rocks, precipitous and craggy. But what occur most frequently there are certain ridges (*argini*) (I shall use this word because no more descriptive one occurs to me), somewhat raised, which surround and enclose plains of different sizes and various shapes, but for the most part circular. In the middle of many of these there is a mountain in sharp relief, and some few are filled with a rather dark substance similar to that of the large spots that are seen with the naked eye; these are the largest ones, and there are a very great number of smaller ones, almost all of them circular.

Fourth, just as the surface of our globe is divided into two chief parts—the land and the sea—so in the lunar disc we see a sharp distinction between the brighter areas and the less bright. I believe that the appearance of the earth illuminated by the sun would be very similar to this for one who could see it from the moon or from some similar distance, and that the surface of the seas would appear darker, and that of the land brighter.

Fifth, as from the earth we see the moon now completely lighted, now half, now more, now less, sometimes sickle-shaped and sometimes completely invisible (that is, when it is beneath the sun's rays so that the part which faces the earth remains darkened), just so would the illumination made by the sun on the face of the earth be seen from the moon, with precisely the same period and the same alterations of shape. Sixth, . . .

SAGR. Hold on a minute, Salviati. I understand perfectly well that for anyone on the moon the illumination of the earth would be similar, in its various shapes, to that which we discover in

Second agreement is that the moon is dark like the earth.

Third, the material of the moon is dense like the earth, and mountainous.

Fourth, the moon divided into two parts, differing by lightness and darkness, as the terrestrial globe into seas and land surfaces.

From a distance the surfaces of the oceans would appear darker than those of the land.

Fifth, the earth's changes of shape similar to those of the moon, and made with the same periodicity.

the moon. But I am not yet satisfied that this would appear to take place in the same period, seeing that what the sun's illumination does on the lunar surface in a month, it does on that of the earth in twenty-four hours.

SALV. It is true that the effect of the sun, as to the illumination of these two bodies and the touching of their surfaces with its splendor, hastens over the earth in a day and takes a month on the moon. But the variations of shape under which the illuminated parts of the earth's surface would be seen from the moon do not depend upon this alone, but upon the various changing relations that the moon has with the sun. Thus, for example, if the moon should exactly follow the motion of the sun, and always happen to stand in line between it and the earth in that relation which we call conjunction, forever looking toward the same hemisphere of the earth which the sun faced, all of this would be seen perpetually lighted. On the other hand, if the moon remained always in opposition to the sun, it would never see the earth, of which the part continually turned toward the moon would be dark and therefore invisible. But when the moon is in quadrature with the sun, that half of the earth's hemisphere exposed to the view of the moon which is toward the sun is luminous, and the other turned away from the sun is dark; and therefore the lighted part of the earth would show itself to the moon in a semicircular shape.

SAGR. I am completely convinced. I now understand very well that, as the moon leaves opposition to the sun, from which position it sees nothing of the lighted part of the earth's surface, and approaches day by day toward the sun, it commences little by little to discover some small portion of the illuminated face of the earth, which it sees in the shape of a thin sickle because of the earth being round. The moon, getting closer day by day to the sun in virtue of its motion, progressively discovers always more of the lighted hemisphere of the earth, so that at quadrature exactly half is revealed, just as we see the same amount of it. As it continues to approach conjunction, greater parts of the illuminated surface are revealed, and finally at conjunction the entire hemisphere appears luminous. To sum up, I understand quite well that whatever happens for the inhabitants of the earth in seeing the phases of the moon is what would happen for anyone seeing the earth from the moon, but in reverse order. That is, when for us the moon is full and in opposition to the sun, to him the earth would be in conjunction with the sun and completely dark and invisible; conversely, that state which to us is conjunction of the moon with the sun, and therefore new moon, would to him be opposition of the earth to the sun, and, so to speak, "full earth," that is, all lighted. And finally, whatever proportion of the moon's surface looks lighted to us at any time, just that proportion of the earth would look dark from the moon at the same time, and just as much of the moon would remain deprived of light for us as would appear lighted on the earth as seen from the moon. So that only in quadrature do we see a half circle of the moon lighted, and he that much of the earth. These reciprocal operations seem to me to differ in one respect, however. Assuming for the sake of the argument that there is someone on the moon who can see the earth, he will see the entire surface of the earth every day, by virtue of the moon's motion with respect to the earth every twenty-four or twenty-five hours. But we shall never see more than half the moon, since it makes no revolution of its own, as it would have to do for all of it to show itself.

SALV. Provided that the very opposite is not implied; namely, that its own rotation is the reason that we do not see the other side—for such would have to be the case if the moon should have an epicycle.† But why do you leave out a certain other difference, a counterpart to this one you put forward?

SAGR. And what is that? I have no other in mind at the moment.

SALV. It is that if the earth (as you have noted well) sees no more than half the moon, whereas from the moon the whole earth may be seen, still on the other hand all the earth sees the moon, while only one half of the moon sees the earth. For the inhabitants of the upper half of the moon, so to speak, which is invisible to us, are deprived of any view of the earth; maybe these are the Contraterrenes.† But here I happen to remember a specific event newly observed on the moon by our Academic friend, by means of which two necessary consequences may be inferred. One is that we do see somewhat more than half the moon, and the other is that the moon's motion bears an exact relation to the center of the earth. And what he observed was as follows.

If the moon did have a natural agreement and correspondence

All the earth sees only half the moon and only half the moon sees all the earth.

From the earth more than half the lunar globe is seen.

with the earth, facing it with some very definite part, then the straight line which joins their centers would always have to pass through the same point on the surface of the moon, so that anyone looking from the center of the earth would always see the same lunar disc bounded by exactly the same circumference. But for anyone located on the earth's surface, the rays passing from his eyes to the center of the moon's globe would not pass through that very point on its surface through which passes the line drawn from the center of the earth to the center of the moon, unless the latter were directly overhead. Hence when the moon is to the east or west, the point of incidence of the visual rays is above that of the line connecting the centers, and therefore some part of the edge of the moon's hemisphere is revealed, and a similar section hidden on the under side; I mean "revealed" and "hidden" with respect to that hemisphere which would be seen from the exact center of the earth. And since that part of the moon's circumference which is on top at rising is underneath at setting, the difference in appearance of these upper and lower parts ought to be noticeable enough because of various spots or markings on those parts being first revealed and then hidden. A similar variation should be observable also at the northern and southern extremities of the same disc, according as the moon is at its most southerly or most northerly point along the meridian.† When it is northerly, some of its northern parts are hidden and the southern revealed, and vice versa.

Now the telescope has made it certain that this conclusion is in fact verified. For there are two special markings on the moon, one of which is seen to the northwest when the moon is on the meridian, and the other almost diametrically opposite. The former is visible even without a telescope, but not the latter. The one toward the northwest is a small oval spot separated from three larger ones. The opposite one is smaller, and likewise stands apart from larger marks in a sufficiently clear field. In both of these the variation mentioned already is quite clearly observed; they are seen opposite to one another, now close to the edge of the lunar disc and now farther away. The difference is such that the distance between the northwesterly spot and the edge of the disc is at one time more than twice what it is at another. As to the other spot, being much closer to the edge of the disc, the change is more than threefold from one time to the

Two spots on the moon by which it is observed to point at the center of the earth in its motion.

other. From which it is obvious that the moon, as if drawn by a magnetic force, faces the earth constantly with one surface and never deviates in this regard.

SAGR. Will the new observations and discoveries made with this admirable instrument never cease?

SALV. If its progress follows the course of other great inventions, one may hope that in time things will be seen which we cannot even imagine at present.

To get back to our original discussion, I state that the sixth agreement between the moon and the earth is that just as the moon supplies us with the light we lack from the sun a great part of the time, and by reflection of its rays makes the nights fairly bright, so the earth repays it by reflecting the solar rays when the moon most needs them, giving a very strong illumination—as much greater than what the moon gives us, it would seem to me, as the surface of the earth is greater than that of the moon.

Sixth, the earth and moon reciprocally illuminate each other.

SAGR. Stop there, Salviati, and allow me the pleasure of showing you how from just this first hint I have seen through the cause of an event which I have thought about a thousand times without ever getting to the bottom of it.

You mean that a certain baffling light† which is seen on the moon, especially when it is horned, comes from the reflection of the sun's light from the surface of the earth and the sea; and this light is seen most clearly when the horns are the thinnest. For at that time the luminous part of the earth that is seen from the moon is greatest, in accordance with your conclusion a little while ago that the luminous part of the earth shown to the moon is always as great as the dark part of the moon which is turned toward the earth. Hence when the moon is thinly horned and consequently in large part shadowy, the illuminated part of the earth seen from the moon is large, and so much the more powerful is its reflection of light.

Light is reflected from the earth on the moon.

SALV. That is exactly what I meant. Really, it is a great pleasure to talk with discriminating and perceptive persons, especially when people are progressing and reasoning from one truth to another. For my part I more often encounter heads so thick that when I have repeated a thousand times what you have just seen immediately for yourself, I never manage to get it through them.

SIMP. If you mean being unable to show them so that they understand it, that is a great surprise to me; I am sure that if they did

not understand it from your explanation they would not understand it from anyone's, since yours seems to me very clear in
its expression. But if you mean not having persuaded them so
that they believe it, I am not at all surprised, for I must confess
myself one of those who understand your reasoning without
being satisfied by it. For me, there remain many difficulties in
this and in parts of others of your six analogies which I shall
propound when you are through presenting the rest.

SALV. I shall be brief, then, and hurry through the rest because
of my desire to discover any truth whatever, in which the objections of an intelligent man like yourself can assist me very much.

Seventh, the
earth and moon
are reciprocally
eclipsed.

The seventh resemblance, then, is their reciprocal reaction to
injuries as well as to favors. Just as the moon, at the height of
its illumination, is often deprived of light and eclipsed by the
interposition of the earth between it and the sun, so in retribution
it interposes itself between the earth and the sun, and with its
shadow darkens the earth. Though indeed the revenge is not
equal to the offense, for frequently the moon remains immersed
totally and for rather a long time in the shadow of the earth, but
the earth is never darkened by the moon completely or for long.
Still, considering the smallness in size of the moon in comparison
with the immensity of the sun, one may surely say that in a
certain sense the moon's valor and spirit are commendable.

So much for the resemblances. Discussion of the differences
should follow now, but since Simplicio wants to favor us with
his doubts against the above, it would be good to hear them and
consider them before going on.

SAGR. Yes indeed, because probably Simplicio will not have any
misgivings about the differences and disparities between the
earth and the moon, since he already considers their substances
quite different.

SIMP. Of the resemblances you have set forth in order to draw
a parallel between the earth and the moon, I find that I can
admit without misgivings only the first one and a couple of others.
I admit the first, that is, the spherical shape, though even in this
there is a difficulty; for I consider the moon's sphere to be as
smooth and polished as a mirror, whereas that of this earth that
we touch with our hands is very rough and rugged. But this
matter of the irregularity of the surface comes considerably into
one of the other correspondences you have set forth, and so I
reserve what I have to say until we get to that.

That the moon is opaque and dark in itself, as you say in your second analogy, I admit only as to the first attribute of opacity, which the solar eclipses assure me of. For if the moon were transparent, the sky would not become as dark as it does in a total eclipse of the sun. Transparency of the lunar globe would permit a refracted light to pass through as do the densest clouds. But as to the darkness, I do not believe that the moon is entirely without light, like the earth. On the contrary, that brightness which is observed on the balance of its disc outside of the thin horns lighted by the sun I take to be its own natural light; not a reflection from the earth, which is incapable of reflecting the sun's rays by reason of its extreme roughness and darkness.

Secondary light considered the moon's own.

Earth powerless to reflect the sun's rays.

In your third parallel, I agree with you in one part and disagree in another. I concur in judging the body of the moon to be very solid and hard like the earth's. Even more so, for if from Aristotle we take it that the heavens are of impenetrable hardness[†] and the stars are the denser parts of the heavens, then it must be that they are extremely solid and most impenetrable.

Celestial material impenetrable, according to Aristotle.

SAGR. What excellent stuff, the sky, for anyone who could get hold of it for building a palace! So hard, and yet so transparent!

SALV. Rather, what terrible stuff, being completely invisible because of its extreme transparency. One could not move about the rooms without grave danger of running into the doorposts and breaking one's head.

SAGR. There would be no such danger if, as some of the Peripatetics say, it is intangible; it cannot even be touched, let alone bumped into.

Celestial material intangible.

SALV. That would be no comfort, inasmuch as celestial material, though indeed it cannot be touched (on account of lacking the tangible quality), may very well touch elemental bodies; and by striking upon us it would injure us as much, and more, as it would if we had run against it.

But let us forsake these palaces, or more appropriately these castles in the air, and not hinder Simplicio.

SIMP. The question you have thus so casually raised has a place among the difficulties treated in philosophy, and I have heard upon this subject the very beautiful thoughts of a great professor at Padua.[†] But this is no time to go into that.

Back to our purpose. I reply that I consider the moon more solid than the earth, not for the reason you already gave, of the roughness and ruggedness of its surface, but on the contrary

Surface of the
moon smoother
than a mirror.

Prominences and
cavities on the
moon are illu-
sions of dark
and bright.

from its being suited to receive a polish and a lustre superior
to that of the smoothest mirror, as observed in the hardest stones
on earth. For thus must be its surface in order to make such a
vivid reflection of the sun's rays. The appearances you speak of,
the mountains, rocks, ridges, valleys, etc., are all illusions. I have
heard it strongly maintained in public debates against these innovators that such appearances belong merely to the unevenly
dark and light parts of which the moon is composed inside and
out. We see the same thing occur in crystal, amber, and many
perfectly polished precious stones, where, from the opacity of
some parts and the transparency of others, various concavities
and prominences appear to be present.

In the fourth analogy I concede that the surface of the terrestrial globe, seen from a distance, would have two different
appearances, one lighter and the other darker, but I consider that
the differences would fall out in reverse of what you say. I believe
that the surface of the water would appear shining because it is
smooth and transparent, while that of the land would remain
dark by reason of its opacity and roughness, these being badly
suited for the reflection of sunlight.

Concerning the fifth comparison, I admit it entirely, and am
convinced that if the earth did shine like the moon it would show
itself to anyone who saw it from there under a form similar to
that which we see in the moon. I understand also how the period
of its illumination and variation of shape would be one month,
although the sun circles it every twenty-four hours. And finally
I have no trouble granting that only half the moon sees all the
earth, while all the earth sees only half the moon.

In the sixth comparison, I think it most false that the moon
can receive light from the earth, which is completely dark,
opaque, and unfit to reflect sunlight as the moon reflects it so
well to us. And as I have said, I consider the light which is seen
over the rest of the face of the moon (outside the horns brightly
illuminated by the sun) to be the moon's own proper and natural
light, and it would be quite a feat to make me think otherwise.

The seventh, of mutual eclipses, I can also admit, although
properly speaking it is customary to call that an eclipse of the
sun which you want to call an eclipse of the earth.

This is all that occurs to me at present to say to you in refutation of the seven resemblances. If it pleases you to reply in any
way to these points, I shall be glad to listen.

SALV. If I have rightly understood so far as you have answered, it seems to me that there remain in dispute between you and me certain properties which I have made common to the moon and the earth, and they are these: You consider the moon to be as polished and smooth as a mirror and, as such, fitted to reflect the sunlight, and the earth, on the other hand, because of its roughness, as having no power to make a similar reflection. You concede the moon to be solid and hard; you deduce this from its being polished and smooth, and not from its being mountainous. As to its appearing mountainous, you assign as a cause its parts being more and less opaque and clear. And finally you believe that the secondary light of the moon is its own, and not reflected from the earth—although it seems that you do not deny some reflection from our seas, which are smooth of surface.

I have little hope of removing your error that the reflection of the moon is made in the manner of a mirror, seeing that for this purpose the reading of *Il Saggiatore* and the *Lettere Solari* of our mutual friend has had no effect upon your ideas; if, indeed, you have carefully read what has been written on the subject in those places.

SIMP. I have perused it rather superficially, as permitted by the little leisure left to me from more solid studies. So if you think my difficulties may be resolved by going over some of that reasoning or by adducing other proofs, I shall listen attentively.

SALV. I shall say what comes to my mind at the moment, possibly a mixture of my own ideas and what I read in those books; I remember that I was entirely convinced by them, although at first their conclusions seemed very paradoxical to me.

We are inquiring, Simplicio, whether in order to produce a reflection of light similar to that which comes to us from the moon, it is necessary that the surface from which the reflection comes shall be as smooth and polished as a mirror, or whether a rough and ill-polished surface, neither smooth nor shiny, may not be better suited. Now if two reflections should come to us, one brighter than the other, from two surfaces situated opposite to us, I ask you which of the two surfaces you believe would look the lighter to our eyes, and which the darker?

SIMP. I think without any doubt that the surface which reflected the light more brilliantly would look lighter to me, and the other darker.

Detailed proof
that the moon
has a rough
surface.

SALV. Now please take that mirror which is hanging on the wall,
and let us go out into that court; come with us, Sagredo. Hang
the mirror on that wall, there, where the sun strikes it. Now let
us withdraw into the shade. Now, there you see two surfaces
struck by the sun, the wall and the mirror. Which looks brighter
to you; the wall, or the mirror? What, no answer?

SAGR. I am going to let Simplicio answer; he is the one who is
experiencing the difficulty. For my part, from this small beginning of an experiment I am persuaded that the moon must indeed
have a very badly polished surface.

SALV. Tell me, Simplicio; if you had to paint a picture of that
wall with the mirror hanging on it, where would you use the
darkest colors? In depicting the wall or the mirror?

SIMP. Much darker in depicting the mirror.

SALV. Now if the most powerful reflection of light comes from
the surface that looks brightest, the wall here would be reflecting
the rays of the sun more vividly than the mirror.

SIMP. Very clever, my dear sir; and is this the best experiment
you have to offer? You have placed us where the reflection from
the mirror does not strike. But come with me a bit this way; no,
come along.

SAGR. Perhaps you are looking for the place where the mirror
throws its reflection?

SIMP. Yes, sir!

SAGR. Well, just look at it — there on the opposite wall, exactly
as large as the mirror, and little less bright than it would be if
the sun shone there directly.

SIMP. Come along, then, and look at the surface of the mirror
from there, and then tell me whether I should say it is darker
than that of the wall.

SAGR. Look at it yourself; I am not anxious to be blinded, and
I know perfectly well without looking that it looks as bright
and vivid as the sun itself, or little less so.

SIMP. Well, then, what do you say? Is the reflection from a
mirror less powerful than that from a wall? I notice that on this
opposite wall, which receives the reflection from the illuminated
wall along with that of the mirror, the reflection from the mirror
is much the brighter. And I see likewise that from here the mirror
itself looks very much brighter to me than the wall.

SALV. You have got ahead of me by your perspicacity, for this was the very observation which I needed for explaining the rest. You see the difference, then, between the reflections made by the surface of the wall and that of the mirror, which are struck in exactly the same way by the sun's rays. You see how the reflection that comes from the wall diffuses itself over all the points opposite to it, while that from the mirror goes to a single place no larger than the mirror itself. You see likewise how the surface of the wall always looks equally light in itself, no matter from what place you observe it, and somewhat lighter than that of the mirror from every place except that small area where the reflection from the mirror strikes; from there, the mirror appears very much brighter than the wall. From this sensible and palpable experiment it seems to me that you can very readily decide whether the reflection which comes here from the moon comes like that from a mirror, or like that from a wall; that is, whether from a smooth or a rough surface.

SAGR. If I were on the moon itself I do not believe that I could touch the roughness of its surface with my hand more definitely than I now perceive it by understanding your argument. The moon, seen in any position with respect to the sun and to us, always shows the surface exposed to the sun equally bright. This effect corresponds precisely with that of the wall, which seen from any place appears equally bright; it conflicts with that of the mirror, which from one place alone looks luminous and from all others dark. Besides, the light that comes to me from the reflection of the wall is weak and tolerable in comparison with that from the mirror, which is extremely strong and little less offensive to the eyes than the primary and direct rays of the sun. It is in just such a way that we can calmly contemplate the face of the moon. If that were like a mirror, appearing as large as the sun because of its closeness, it would be of an absolutely intolerable brilliance, and would seem to us almost as if we were looking at another sun.

SALV. Please, Sagredo, do not attribute to my demonstration more than belongs to it. I am about to confront you with a fact that I think you will find not so easy to explain. You take it as a great difference between the moon and the mirror that the former yields its reflections equally in all directions, as the wall does, while the mirror sends its reflection to one definite place alone. From this, you conclude that the moon is like the wall

The 74

First

Day

Plane mirrors
throw reflections
on a single place,
but spherical
mirrors on all
places.

and not like the mirror. But I tell you that this mirror sends
its reflection to one place alone because its surface is flat, and
since reflected rays must leave at equal angles with incident
rays, they have to leave a plane surface as a unit toward one
place. But the surface of the moon is not flat, it is spherical; and
the rays incident upon such a surface are found to be reflected
in all directions at angles equal to those of incidence, because of
the infinity of slopes which make up a spherical surface. Therefore the moon can send its reflections everywhere and need not
send them all to a single place like those of a plane mirror.

SIMP. This is exactly one of the objections which I wanted to
make.

SAGR. If it is one of them, then you must have others; but let
me tell you that so far as this first one is concerned, it seems to
me to be not so much for you as against you.

SIMP. You have called it obvious that the reflection made by
that wall is as bright and luminous as that of the moon, whereas
I think it trifling in comparison with the moon's. For "in this
matter of illumination, one must look for and define the sphere
of activity."† Who doubts that celestial bodies have greater
spheres of activity than our transitory mortal elements? And as
to that wall, is it after all anything more than a bit of earth;
dark, and unfit to illumine?

Sphere of activ-
ity greater for
celestial than for
elemental bodies.

SAGR. Here again I believe that you are quite mistaken. But I
return to the first point raised by Salviati, and I tell you that in
order to make an object appear luminous, it is not sufficient for
the rays of the illuminating body to fall upon it; it is also necessary for the reflected rays to get to our eyes. This is to be clearly
seen in the case of the mirror, upon which no doubt the rays of
the sun are falling, but which nevertheless does not appear to
be bright and illuminated unless we put our eyes in the particular
place where the reflection is going.

Let us consider this in regard to what would happen if the
mirror had a spherical surface. Unquestionably we should find
that of the whole reflection made by the illuminated surface,
only a small part would reach the eyes of a particular observer,
there being only the very least possible part of the entire surface
which would have the correct slope to reflect the rays to the
particular location of his eyes. Hence only the least part of the
spherical surface would shine for his eyes, all the rest looking

dark. If then the moon were smooth as a mirror, only a very small part would show itself to the eyes of a particular person as illuminated by the sun, although an entire hemisphere would be exposed to the sun's rays. The rest would remain, to this observer's eyes, unilluminated and therefore invisible.[†] To conclude, the whole moon would be invisible, since that particle which gave the reflection would be lost by reason of its smallness and great distance. And just as the moon would remain invisible to the eyes, so its illumination would remain nil; for it is indeed impossible that a luminous body should by its splendor take away our darkness, and we be unable to see it.

SALV. Wait a minute, Sagredo, for I see certain signs in Simplicio's face and actions which indicate to me that he is neither convinced nor satisfied by what you, with the best evidence and with perfect truth, have said. And now it occurs to me how to remove all doubt by another experiment. I have seen in a room upstairs a large spherical mirror; have it brought here. And while it is on its way, Simplicio, consider carefully the amount of light which comes from the reflection of the flat mirror to this wall here under the balcony.

SIMP. I see that it is little less lighted than if the sun were striking it directly.

SALV. So it is. Now tell me; if, taking away that little flat mirror, we were to put the large spherical one in its place, what result do you think that would have upon the reflection on this same wall?

SIMP. I think it would produce a much greater and broader light.

SALV. But what would you say if the illumination should be nil, or so small that you could hardly perceive it?

SIMP. When I have seen the effect, I shall think up a reply.

SALV. Here is the mirror, which I wish to have placed beside the other. But let us first go over there, near the reflection from the flat mirror, and note carefully its brightness. You see how bright it is here where it strikes, and how you can distinctly make out these details of the wall.

SIMP. I have looked and observed very closely; now place the other mirror beside the first.

SALV. That is where it is. It was placed there as soon as you began to look at the detail, and you did not perceive it because the increase of light over the rest of the wall was just as great.

If the moon were like a spherical mirror it would be invisible.

Now take away the flat mirror. See there, all the reflection is taken away, although the large convex mirror remains. Remove that also, and then replace it as you please; you will see no change whatever in the light upon the whole wall. Thus you see it shown to your senses how the reflection of the sun made from a spherical convex mirror does not noticeably illuminate the surrounding places. Now what have you to say to this experiment?

SIMP. I am afraid you have introduced some trickery. Yet I see, in looking at that mirror, that it gives out a dazzling light that almost blinds me; and what is more significant, I see it all the time, wherever I go, changing place on the surface of the mirror according as I look at it from this place or that; a conclusive proof that the light is reflected very vividly on all sides, and consequently upon the entire wall as upon my eyes.

SALV. Now you see how carefully and with what reserve one must proceed in giving assent to what is shown by argument alone. There is no doubt that what you say is plausible enough, and yet you can see that sensible experience refutes it.

SIMP. How, then, does one proceed in this business?

SALV. I shall tell you what I think about it, but I do not know how it will strike you. First of all, that brilliance which you see so vividly on the mirror, and which seems to you to occupy such a large part of it, is not such a big piece. It is really very tiny, but its extreme brightness causes an adventitious irradiation of your eyes through the reflection made in the moisture at the edges of your eyelids, which extends over the pupils. It is like the little hat that seems to be seen around the flame of a candle at some distance; or you may want to compare it with the apparent rays around a star. For example, if you match the little body of the Dog Star as seen in the daytime through the telescope, when it is without irradiations, with the same seen at night by the naked eye, you will perceive beyond all doubt that with its irradiations it appears thousands of times larger than the bare and real starlet. A similar or larger augmentation is made by the image of the sun which you see in that mirror; I say larger, because it is more vivid than that of the star, as is obvious from one's being able to look at the star with less injury to one's vision than at this reflection in the mirror.

Thus the reflection which has been imparted over this entire

Little irradiated
bodies of the
stars appear
thousands of
times larger than
when barren.

wall comes from a small part of the mirror, while that which was coming a little while ago from the flat mirror was imparted and confined to a very small part of the same wall. What is the marvel, then, that the first reflection shone very brightly, and this other remained almost imperceptible?

SIMP. I am more perplexed than ever; I must bring up the other difficulty. How can it be that the wall, being of so dark a material and so rough a surface, is able to reflect light more powerfully and vividly than a smooth and well-polished mirror?

SALV. Not more vividly,† but more diffusely. As to vividness, you see that the reflection of that little flat mirror, where it is thrown there under the balcony, shines strongly; and the rest of the wall, which receives a reflection from the wall to which the mirror is attached, is not lighted up to any great extent (as is the small part struck by the reflection from the mirror). If you wish to understand the whole matter, consider how the surface of this rough wall is composed of countless very small surfaces placed in an innumerable diversity of slopes, among which of necessity many happen to be arranged so as to send the rays they reflect to one place, and many others to another. In short, there is no place whatever which does not receive a multitude of rays reflected from very many little surfaces dispersed over the whole surface of the rough body upon which the luminous rays fall. From all this it necessarily follows that reflected rays fall upon every part of any surface opposite that which receives the primary incident rays, and it is accordingly illuminated.

Light reflected from rough bodies more diffused than that from smooth, and why.

It also follows that the same body on which the illuminating rays fall shows itself lighted and bright all over when looked at from any place. Therefore the moon, by being a rough surface rather than smooth, sends the sun's light in all directions, and looks equally light to all observers. If the surface, being spherical, were as smooth as a mirror, it would be entirely invisible, seeing that that very small part of it which can reflect the image of the sun to the eyes of any individual would remain invisible because of the great distance, as we have already remarked.

The moon, if smooth and polished, would be invisible.

SIMP. I thoroughly understand your entire argument. Still, it seems to me that one can explain it away with very little trouble, so as to be able to maintain that the moon is round and polished and that it reflects the sun's light to us in the way a mirror does. Nor need the image of the sun be seen in its center, for "Not in its own form may the sun be seen at such a great distance in a little image of the sun, but the illumination of the whole moon may be perceived by us through the light produced by the sun."

We may see such a thing in a gilded and polished plate which, being struck by luminous rays, shows itself resplendent all over to one who observes it from a distance; and only from nearby is the little image of the luminous body seen in the middle of it.

SALV. Naïvely confessing my incomprehension, I must say that I understand nothing of this argument of yours except that part about the gilded plate. If you will allow me to speak freely, I am strongly of the opinion that you do not understand it either, but have committed to memory words written by somebody out of a desire to contradict and to show himself more intelligent than his opponent. To show this, moreover, to those who, in order to appear intelligent themselves, would applaud what they did not understand, and form the better opinion of people according to the deficiency of their own understanding; if indeed the author himself is not one of those (and there are many) who write of what they do not understand, and whose writings are therefore not understood.

Some write of
what they do not
understand, and
therefore what
they write is
not understood.

But putting all this aside, I reply to you regarding the gilded plate that if it is flat and not very large it could appear from a distance to be lighted all over, if struck by a strong light. But it will be seen thus only when the eye is on a definite line, namely that of the reflected rays. And it would be seen more glittering than if it were made of silver, the color and density of the metal being better suited to receiving a perfect burnishing. And if its surface were well polished but not exactly flat, having various slopes, then its splendor could be seen from even more places— as many as could be reached by the rays from its various faces. That is why diamonds are worked into many facets, so that their delightful brilliance may be perceived from many places. But if the plate were very large, then despite the distance and even though it were perfectly flat, it would not be seen shining all over.

Diamonds are
fashioned with
many facets,
and why.

In order to explain better, let us take a very large gilded plate exposed to the sun; it will show to a distant eye the image of the sun occupying only a part of the plate, that from which the reflection of the incident solar rays comes. It is true that on account of the vividness of the light such an image would appear crowned with many rays, and would therefore seem to occupy a much larger part of the plate than it really did. To verify this, one might note the exact place on the plate from which the reflection came, and likewise figuring how large the shining space appears, cover the major part of this space leaving only the middle revealed; the size of the apparent brilliance would not be a whit diminished, but it would be seen widely spread over the cloth or material used for the covering. So if anyone, seeing from a distance a little gilded plate shining all over, should imagine that the same phenomenon would have to occur with a plate as large as the moon, he would be as much deceived as if he were to think that the moon is no larger than the bottom of a bathtub.

Then if the plate were spherical, the strong reflection would be seen only in a single point, though because of the brilliance it would indeed appear fringed with many vibrant rays. The rest of the ball would be seen as colored, and this only if it were not very highly polished; for if it were perfectly polished it would appear dark. We have an example of this daily before our eyes in silver vases, which, when they are merely bleached by boiling, are white as snow all over and cannot render images at all; but if any part of them is burnished, it quickly becomes dark and gives images like a mirror. The darkness comes merely from the leveling of a very fine grain that made the surface of the silver rough and therefore suited to reflect light in all directions, so that all places showed themselves equally lighted. When those minute inequalities are leveled by burnishing so that the reflection of the incident rays is directed toward a definite place, then from such a place the burnished part looks much clearer and lighter than the rest which is only bleached; but from any other place it is seen to be quite dark. And note that the diversity of what is seen upon looking at a burnished surface causes such a different appearance that to imitate or depict burnished armor, for example, one must combine pure black and white, one beside the other, in parts of the arms where the light falls equally.

Burnished silver appears darker than unburnished, and why.

Burnished steel appears very bright from some viewpoints and very dark from others.

SAGR. Then if these doctors of philosophy were content to grant that the moon, Venus, and the other planets had surfaces not as smooth and bright as a mirror, but were something short of that, like a silver plate merely bleached but not burnished, would this be sufficient to make it visible and fit to reflect the sun's light for us?

SALV. Partly, but it would not make a light as powerful as is made by its being mountainous and full of great prominences and cavities. However, these doctors of philosophy never do concede it to be less polished than a mirror; they want it more so, if that can be imagined, for they deem that only perfect shapes suit perfect bodies. Hence the sphericity of the heavenly globes must be absolute. Otherwise, if they were to concede me any inequality, even the slightest, I would grasp without scruple for some other, a little greater; for since such perfection consists in indivisibles, a hair spoils it as badly as a mountain.

SAGR. This gives rise to two questions on my part. One is to understand why the greater irregularity of the surface makes the reflection of the light more powerful, and the other is why these Peripatetic gentlemen want such a precise shape.

SALV. I shall answer the first and let Simplicio worry about replying to the second. You must know, then, that a given surface receives more or less illumination from the same light according as the rays of light fall upon it less or more obliquely; the greatest illumination occurs where the rays are perpendicular. And here I shall show you this by means of your senses. First I fold this sheet of paper so that one part makes an angle with the other, and now I expose it to the light reflected from that wall opposite to us. You see how this part that receives the rays obliquely is less light than this other where the rays fall at right angles. Note how the illumination becomes weaker as I make it receive them more and more obliquely.

SAGR. I see the effect, but I do not understand the cause.

SALV. If you thought about it a minute you would find it, but so as to save time, here is a sort of proof in this figure.

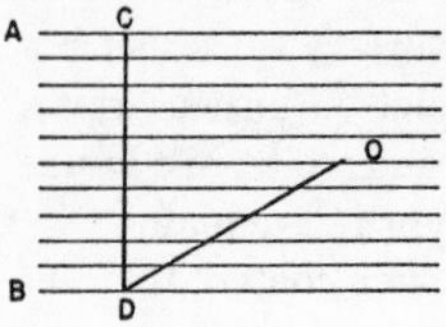

FIG. 7

SAGR. Just seeing the diagram has cleared the whole matter up, so go on.

SIMP. Please explain further for me, since I am not that quick-witted.

SALV. Imagine that all the parallel lines which you see leaving from between the points A and B are rays that strike the line CD at right angles. Now tilt CD so that it leans like DO. Do you not see that many of the rays which struck CD pass by without touching DO? And if DO is illuminated with less rays, it is surely reasonable that the light it receives from them is weaker.

A rougher surface gives more reflection of light than one less rough.

Perpendicular rays illuminate more than oblique ones, and why.

More oblique rays illuminate less, and why.

Now let us get back to the moon, which, being spherical in shape would, if its surface were as smooth as this paper, receive much less light near the edges of its lighted hemisphere than upon the central parts; for the rays would fall upon the former quite obliquely, and upon the latter at right angles. For that reason at full moon, when we see nearly all the hemisphere illuminated, the central parts ought to look brighter than those near the edges; but that is not what is seen. Now imagine the face of the moon to be full of high mountains. Do you not see that their peaks and ridges, being elevated above the convexity of a perfectly spherical surface, are exposed to the sun and accommodated to receive the rays much less obliquely, and therefore to look as much lighted as the rest?

SAGR. All right, but even if there are such mountains and it is true that the sun strikes them much straighter than it would the slopes of a smooth surface, still it is also true that the valleys among these mountains would remain dark because of the great shadows which the mountains would cast at such a time; whereas the central parts, though full of mountains and valleys, would remain without shadows through the elevation of the sun. Therefore they would be much lighter than the parts at the edge, those being spotted with shadows no less than with light. Yet no such difference is observed.

SIMP. I was turning over in my mind a like difficulty.

SALV. How much quicker Simplicio is to perceive difficulties that strengthen Aristotle's position than he is to see their solutions! But I suspect that sometimes he deliberately keeps those to himself; and having in the present instance been able to see the objection, which incidentally is quite ingenious, I cannot believe that he has not also discovered the answer. Hence I shall try to worm it out of him, as the saying goes. Now tell me, Simplicio, do you believe that there can be shadows where the rays of the sun are striking?

SIMP. I do not believe so; I am sure not. The sun being the strongest light, which scatters darkness with its rays, it is impossible that darkness could remain where it arrived. Besides, we know by definition that *tenebrae sunt privatio luminis*.

SALV. Then the sun, looking at the earth or moon or any other opaque body, never sees any of its shady parts, having no other eyes to see with than its light-bearing rays. Consequently anyone who was located on the sun would never see anything shady, because the rays of his vision would always travel in company with the illuminating sunshine.

SIMP. That is very true; it is beyond contradiction.

SALV. Now when the moon is in opposition to the sun, what difference is there between the path which the rays of your vision take and the way the rays of the sun go?

SIMP. Oh, now I understand you. You mean that since the rays of vision and those of the sun are going along the same lines, we can never see any of the shaded valleys of the moon. But please give up your opinion that I am a hypocrite or a dissembler; I give you my word as a gentleman that I did not perceive this reply, and I might never have discovered it without your help or without long study.

SAGR. The solution that you two have arrived at for this last question satisfies me also, yet at the same time this remark that the sun's rays and visual rays travel together has raised another difficulty elsewhere. I do not know whether I can express it, because it has just occurred to me and I have not yet assimilated it, but let us see whether among us we can clarify it.

Doubtless the outer parts of a smooth but unpolished hemisphere which is lighted by the sun, receiving its rays obliquely, receive many less rays than do the central parts, which catch them straight on. And it may be that a wide band, say of twenty degrees, near the edge of the hemisphere receives no more rays than another near the center no broader than four degrees, so that the former would be much darker than the latter and would appear so to anyone who saw them both head on or at their best, so to speak. But if the eye that looked at them were so located that the breadth of twenty degrees belonging to the dark band appeared to be no wider than that of the four-degree band located at the center of the hemisphere, I think it not impossible that the one might look just as light and luminous as the other. After all, the reflections of two equal quantities of rays would come to the eye within two equal angles — that is, of four degrees — for they are reflected from the center band four degrees in width and the other which is twenty degrees but which is seen as four degrees because of foreshortening. And such a place is occupied by the eye when it is located between the hemisphere and the body which illuminates it, because then the lines of sight and of the rays are the same. Therefore it seems possible that the moon could

have a very regular surface, and at the full still appear no less
luminous at the edges than in the center.
SALV. The problem is ingenious and deserves to be considered.
Since it has occurred to you offhand, I shall likewise answer it
with whatever comes to mind, though perhaps by thinking more
about it I might be able to hit upon a better answer.

But before I propose any solution it will be good for us to
make certain by an experiment whether there corresponds to
your objection any such fact as it appears to prove. Therefore
taking the same paper again and bending a small part over the
rest, we shall expose it to the light so that the rays fall straight
upon the smaller part and obliquely upon the rest, and test
whether the former, which receives the rays straight, looks
brighter. See here; the experiment already shows that it is
noticeably the more luminous.

Now, if your objection were valid, events would have to fall
out as follows. Lowering our eyes so as to see the larger, less
illuminated part in foreshortening, this part will be made to
appear no larger than the other more illuminated part, and consequently will subtend no larger a visual angle than that. Then
its light should, I say, increase so that it will appear as bright
as the other. Here I am looking at it, and I see it so obliquely
that it looks even narrower than the other; nevertheless it does
not brighten for me a bit. Look now and see whether the same
happens for you.

SAGR. I have done so, but however I lowered my eyes, I did not
see a bit of lightening or brightening over the surface in question.
It rather seemed to me to darken.

SALV. Then we are satisfied as to the futility of the objection.
Next, as to its solution, I believe that as the surface of this paper
is something less than smooth, few rays are reflected back in
the direction of incidence as compared with the many reflected in
other directions, and that of these few, more are lost the more the
visual rays approach the incident light rays. And since it is not
the incident rays but those which are reflected to the eye which
make the object appear luminous, more of these are lost by lowering the eye than are acquired; as you yourself said it appeared
to you, seeing the page become darker.

SAGR. I am satisfied with the experiment and with the reasoning.
It now remains for Simplicio to answer my other question,

Why perfect
sphericity in the
heavenly bodies
is assumed by
the Peripatetics.

telling me what impels the Peripatetics to desire such exact rotundity in the celestial bodies.

SIMP. Being ingenerable, incorruptible, inalterable, invariant, eternal, etc. implies that the celestial bodies are absolutely perfect; and being absolutely perfect entails their having all kinds of perfection. Therefore their shape is also perfect; that is to say, spherical; and absolutely and perfectly spherical, not approximately and irregularly.

SALV. And how do you derive this incorruptibility?

SIMP. Directly, from lacking contraries; indirectly, from simple circular motion.

SALV. So from what I gather from your argument, in establishing the essence of the celestial bodies as incorruptible, inalterable, etc., rotundity does not enter as a necessary cause or a requisite; for if rotundity could cause inalterability, we could at will make wood or wax or other elemental material incorruptible simply by reducing it to a spherical shape.

SIMP. And is it not obvious to you that a ball of wood maintains itself better and longer than does a steeple or some other angular form made of the same material?

Shape is a cause
not of incorrupt-
ibility but of
longer duration.

The corruptible
may be com-
pared, but not
the incorruptible.

Perfection of
shape operates
on corruptible
but not on
eternal bodies.

SALV. That is quite true, but the corruptible does not thus become incorruptible; it still remains corruptible, though indeed of longer duration. Hence it is to be noted that the corruptible is capable of being more or less so, and we can say, "This is less corruptible than that"; as, for example, jasper is less corruptible than sandstone. But the incorruptible does not admit of more or less; one cannot say, "This is more incorruptible than that other," if both are incorruptible and eternal. Difference of shape, then, cannot operate except on materials capable of greater or less duration; on the eternal, which cannot exist except in equal permanence, the operation of shape ceases.

Now since celestial material is incorruptible by reason not of shape but of something else, there is no need to be so solicitous about this perfect sphericity. For if material is incorruptible, it may have any shape you like and it will still be so.

If spherical shape
would confer
permanence, all
bodies would be
eternal.

SAGR. I will go further and say that, admitting spherical shape to have the faculty of conferring incorruptibility, all bodies, of any shape whatever, would be eternal and incorruptible. For if the round body is incorruptible, then corruptibility must subsist in the parts that depart from perfect roundness. For instance, there

exists within a cube a perfectly round ball, incorruptible as such;
corruptibility, then, resides in the corners which cover and conceal the roundness. The most that could happen, then, would be
that those corners—those excrescences, so to speak—produce
corruption.

And if we wish to go more deeply into the matter, then within
the parts toward the corners there are other smaller balls of the
same material. These are also incorruptible, being round; and
the same holds for the leftover parts surrounding these eight
lesser spheres — and in these, too, still other spheres may be
imagined. Thus in the end, resolving the whole cube into innumerable balls, you must admit it to be incorruptible. The same
argument and a similar analysis can be made for any other shape.

SALV. This line of reasoning works both ways. If, for example,
a crystal sphere were incorruptible on account of its shape (that
is, if it had the faculty of resisting any alteration from within
or without), then the addition of further crystal and the transformation of this sphere into, say, a cube, would be seen not to
have altered it either internally or externally. And it would surely
be less apt to resist new enclosures of the same material than
others of different materials — especially if it is true, as Aristotle
says, that corruption is accomplished through contraries. And
with what could this crystal ball be surrounded that would be
less contrary to it than crystal itself?

But we are not keeping track of the flight of time, and we
shall be late in finishing our discussion if such long arguments
are made about every detail. Besides which, a person's memory
becomes so confused with such a multitude of things that I can
scarcely recall the propositions that Simplicio, in his orderly
way, brought up for our consideration.

SIMP. I remember quite well; on this matter of the mountainousness of the moon there still remains the cause that I adduced for
such an appearance, maintaining that it is an illusion produced
by the moon's constituent parts being nonuniformly opaque and
transparent.

SAGR. A little while ago when Simplicio, in accordance with the
opinion of a certain Peripatetic friend of his, attributed the
apparent irregularities of the moon to its parts being unevenly
opaque and transparent, creating illusions similar to those seen
in crystals and gems of various kinds, something occurred to me

Mother of pearl
suitable for imi-
tating the appar-
ent unevenness
of the moon's
surface.

that would be much better adapted to the showing of such effects, and I believe that his philosopher would give anything for it. This is mother of pearl, which is worked into various shapes; even when brought to an extreme polish, it appears to the eye so pitted and raised in various places that even touching it can hardly make us believe in its smoothness.

SALV. This is really a very beautiful idea, and what has not yet been done may be done in good time, so if other gems and crystals have been brought up that have nothing to do with the illusions of mother of pearl, these may well be brought up also. Until then, so as not to deprive anyone of that opportunity, I shall withhold the answer which might go here, and merely attempt for the present to satisfy the objections brought up by Simplicio.

I say, then, that this argument of yours is too general, and since you do not apply it to all the appearances, one by one, which are seen on the moon and which incline me and others to hold it to be mountainous, I do not believe that you could find anyone who would be content with such a view. Nor do I believe that you or the author himself gets any more satisfaction from

The observed
unevenness of
the moon cannot
be imitated by
means of greater
and less opaque-
ness and
clearness.

Various views of
the moon may be
imitated by any
opaque material.

it than from anything else that is quite beside the point. Out of the countless different appearances that are revealed night after night during one lunation, you could not imitate a single one by arbitrarily fashioning a smooth ball out of more and less opaque and transparent pieces. On the other hand, balls may be made of any solid and opaque material which, merely by having prominences and cavities and by receiving varied illumination, will precisely demonstrate the very changes and scenes which are discovered from one time to another in the moon. On such balls

Various appear-
ances from which
are argued the
mountainousness
of the moon.

you may see the ridges of prominence exposed to the sun's light to be very bright, and behind them you may see the projections of very dark shadows; you will see them greater or less according as these prominences are found more or less distant from the boundary that separates the light part of the moon from the dark. You will see this edge and boundary not evenly spread, as it would be if the ball were smooth, but broken and jagged. Beyond this boundary you will see, in the darkened part, many illuminated summits separated from the already luminous portion. As the illumination becomes more elevated you will see the shadows mentioned before diminish until they vanish entirely, to be seen no more at all when the entire hemisphere is lighted; and then in reverse, as the light passes toward

the other lunar hemisphere, you will recognize the same prominences observed before and see the projections of their shadows
made in the opposite direction, and lengthening. Of all these
things, I say to you again, you cannot represent one for me with
your "opaque" and "transparent."

SAGR. Oh, yes, one can be imitated; namely, that of the full moon,
at which time all is illuminated and there are no longer discovered shadows or anything else that receives any alterations
from the prominences and cavities. But please, Salviati, waste
no more time on this particular, because anyone who has had the
patience to make observations of one or two lunations and is not
satisfied with this very sensible truth could well be adjudged to
have lost his wits; and on such people, why spend time and
words in vain?

SIMP. Really, I have not made such observations, having had
neither the curiosity nor the instruments suitable for making
them. But I wish by all means to do so, and for the present we
can leave this question pending and pass on to the point that
comes next, setting forth the reasons you have for believing that
the earth can reflect light of the sun no less strongly than the
moon can. For to me it seems so dark and opaque that such an
effect strikes me as quite impossible.

SALV. What you think is a cause making the earth unfit for
illuminations, Simplicio, is really not one at all. Would it not be
interesting if I should see into your reasoning better than yourself?

SIMP. Whether I reason well or badly, you might indeed know
better than I do; but whether I reason well or badly I shall never
believe that you can see into my reasoning better than I.

SALV. Even that I shall make you believe in due course. Tell me,
when the moon is nearly full, so that it can be seen by day and
also in the middle of the night, does it appear more brilliant in
the daytime or at night?

SIMP. Incomparably more at night. It seems to me that the moon
resembles those pillars of cloud and fire which guided the children of Israel; for in the presence of the sun it shows itself like a
little cloud, but then at night it is most splendid. Thus I have
observed the moon by day sometimes among small clouds, and it
looked like a little bleached one; but on the following night it
shone very splendidly.

The moon ap-
pears more re-
splendent by
night than by
day.

The moon seen
by day like a
little cloud.

SALV. So that if you had never happened to see the moon except by day, you would not have judged it brighter than one of those little clouds?

SIMP. I do believe you are right.

SALV. Now tell me, do you believe that the moon is really brighter at night than by day, or just that by some accident it looks that way?

SIMP. I believe that it shines intrinsically as much by day as by night, but that its light looks greater at night because we see it in the dark field of the sky. In the daytime, because everything around it is very bright, by its small addition of light it appears much less bright.

SALV. Now tell me, have you ever seen the terrestrial globe lit up by the sun in the middle of the night?

SIMP. That seems to me to be a question that is not asked except in sport, or only of some person notorious for his lack of wit.

SALV. No, no; I take you for a very sensible man, and ask the question in earnest. So answer just the same, and then if it seems to you that I am talking nonsense, I shall be taken for the brainless one; for he is a greater fool who asks a silly question than he to whom the question is put.

SIMP. Then if you do not take me for a complete simpleton, pretend that I have answered you by saying that it is impossible for anyone who is on earth, as we are, to see by night that part of the earth where it is day; that is to say, the part which is struck by the sun.

SALV. So you have never chanced to see the earth illuminated except by day, but you see the moon shining in the sky on the darkest night as well. And that, Simplicio, is the reason for your believing that the earth does not shine like the moon; for if you could see the earth illuminated while you were in a place as dark as night, it would look to you more splendid than the moon. Now if you want to proceed properly with the comparison, we must draw our parallel between the earth's light and that of the moon as seen in daytime; not the nocturnal moon, because there is no chance of our seeing the earth illuminated except by day. Is that satisfactory?

SIMP. So it must be.

SALV. Now you yourself have already admitted having seen the moon by day among little whitish clouds, and similar in appearance to one of them. This amounts to granting at the outset that these little clouds, though made of elemental matter, are just as fit to receive light as the moon is. More so, if you will recall in memory having seen some very large clouds at times, white as snow. It cannot be doubted that if such a one could remain equally luminous on the darkest night, it would light up the surrounding regions more than a hundred moons.

Clouds fit to be lighted by the sun no less than is the moon.

If we were sure, then, that the earth is as much lighted by the sun as one of these clouds, no question would remain about its being no less brilliant than the moon. Now all doubt upon this point ceases when we see those same clouds, in the absence of the sun, remaining as dark as the earth all night long. And what is more, there is not one of us who has not seen such a cloud low and far off, and wondered whether it was a cloud or a mountain; a clear indication that mountains are no less luminous than those clouds.

SAGR. But why any more arguments? Yonder is the moon, more than half full, and over there is a high wall where the sun beats down. Come this way so that the moon is seen beside the wall. Now look; which appears the brighter to you? Do you not see that if there is any advantage it belongs to the wall?

A wall lighted by the sun, when compared with the moon, is no less bright than the latter.

The sun hits that wall, and from there it is reflected to the walls of this room; thence it is reflected into that chamber, so that it arrives there on its third reflection; and I am absolutely certain that there is more light there than if the light were arriving directly from the moon.

Third reflection from a wall illuminates better than first one from the moon.

SIMP. Oh, I do not think so, for the light which the moon gives, especially at the full, is very great.

SAGR. It seems great from the darkness of the shadowy surroundings, but it is not much absolutely; less than that of the twilight a half hour after sunset. This is obvious, because earlier than that you do not see enough to distinguish upon the ground the shadows of things illuminated by the moon. You could tell whether this third reflection in that chamber gives more light than the moon by going in there and reading a book, and then testing whether it is easier to read by moonlight. I believe it would be harder.

Moonlight weaker than twilight.

SALV. If you are satisfied now, Simplicio, you can see how you yourself really knew that the earth shone no less than the moon, and that not my instruction but merely the recollection of certain things already known to you have made you sure of it. For I have not shown you that the moon shines more brilliantly by night than by day; you already knew it, as you also knew that a little cloud is brighter than the moon. Likewise you knew that the illumination of the earth is not seen at night, and in short you knew everything in question without being aware that you knew it. Hence there should be no reason that it should be hard for you to grant that reflection from the earth can illuminate the dark part of the moon with no less a light than that with which the moon lights up the darkness of the night. More, because the earth is forty times† the size of the moon.

SIMP. I really thought that the secondary light of the moon was its own.

SALV. Well, you knew about that, too, and did not perceive that you knew it. Tell me, did you not know yourself that the moon shows itself brighter by night than by day with respect to the darkness of the surroundings? And from that did you not know that in general every bright body looks brighter when the surroundings are darker?

Lighter bodies
look brighter in
surrounding
darkness.

SIMP. I knew that perfectly well.

SALV. When the moon is crescent and the secondary light looks bright to you, is it not always close to the sun and is it not consequently seen in twilight?

SIMP. So it is, and many times I have wished that the sky would darken so that I could see that light more clearly, but the moon has set before the night grew dark.

SALV. Oh, then you knew perfectly well that this light would have appeared greater in the dark of night?

SIMP. Yes indeed, and still greater if the bright light of the horns lit up by the sun could be removed, the presence of which much obscures the lesser light.

SALV. Does it not happen sometimes that one can see the whole disc of the moon in blackest night, without its being illuminated by the sun at all?

SIMP. I do not know that this ever happens except in a total eclipse of the moon.

SALV. Well then, at that time its light ought to look most vivid, being in a very dark field and not obscured by light from the luminous horns. How bright has it looked to you in that state?

SIMP. Sometimes I have seen it copper-colored and a little whit-

ish, but other times it remained quite dark so that I have lost
sight of it.

SALV. If what you could see so clearly in the twilight despite the
obstacle of the adjacent splendor of the horns were the moon's
own light, how could all other light be removed from it in the
darkest night, and its own light fail to appear?

SIMP. I understand that there have been those who believed this
light to be imparted by other stars; especially by its neighbor,
Venus.

SALV. This likewise is folly, because then at the time of a total
eclipse the secondary light ought to appear more clearly than
ever. For it cannot be said that the shadow of the earth hides
the moon from Venus or the other stars. Nevertheless, the moon
is totally deprived of light at that time, for the terrestrial hemisphere which is then turned toward the moon is the one where
it is night; that is, where there is complete absence of any sunlight. And if you were to observe it carefully, you would see quite
plainly that just as the moon illuminates the earth very little
when it is thinly crescent, and that, as it waxes, the splendor reflected from it to us grows likewise, so when the moon is thinly
crescent (and, being between the sun and the earth, sees a very
large part of the lighted terrestrial hemisphere), the light looks
rather bright to us. But as the moon moves away from the sun
and approaches quadrature, the light is seen to languish; at
quadrature it is seen quite weakly because the lighted part of
the earth is constantly being lost from view. Yet the contrary
should hold if the light were its own or were communicated from
the stars, because then we should be able to see it in deep night
and in very dark surroundings.

SIMP. Wait a moment, please, for I have just remembered reading in a recent booklet of theses,† which is full of novelties, that
"This secondary light is not caused by the stars nor by the moon's
own light, and still less is it communicated from the earth; it
derives from the illumination of the sun itself, which penetrates   Secondary light
its whole body because the substance of the lunar globe is some-   of the moon
what transparent. But this more vividly illuminates the surface   caused by the
of the hemisphere which is exposed to the sun's rays, and the in-   sun, according
terior, drinking in and soaking up this light so to speak, like a   to some.
cloud or crystal, transmits it and makes the moon visibly lighted." This, if I remember correctly, he proves by authority, ex-

perience, and reason, adducing Cleomedes, Vitellio, Macrobius, and some other author, a modern;† adding that experience shows this light to look rather bright in daytime when the moon is near conjunction (that is, when it is crescent), and shines most brightly along its limb. Moreover he writes that in solar eclipses, when the moon is under the sun's disc, it is seen to be translucent, especially around the extreme edges. As to his reasons, I believe he then says that since this phenomenon cannot be derived from the earth nor the stars nor the moon itself, it necessarily follows that it comes from the sun.

Besides, on this supposition one may give suitable reasons quite elegantly for every single thing that happens. Thus for the appearance of the secondary light more vividly along the extreme limb, there is the reason of the short space penetrated by the sun's rays — since of all the lines cutting a circle, the longest is that which passes through the center, and of the others those which are more distant from that one are always shorter than those which are closer to it. From the same principle, he says, one may deduce why the said light is little diminished. And, finally, a cause is assigned in this way for it happening that the brightest circle, along the extreme edge of the moon, is observed during a solar eclipse to be in that part which is under the sun's disc, and not in the part outside this disc. This comes about because the sun's rays penetrate directly to our eyes through the part placed under the sun, but fall beyond our line of sight when they go through the parts outside.

SALV. If this philosopher had been the first author to hold such an opinion, I should not wonder at his being so fond of it as to want it considered true; but he having received it from others, I cannot think of any sufficient reason to excuse him for not having perceived its errors. Especially after he had heard the true cause of the effect, and after having been able to assure himself by a thousand experiments and obvious evidences that it is produced by the earth's reflection and nothing else. And to the extent that in the estimation of this author (and of others who withhold their assent) the latter explanation leaves something to be desired, I can forgive the more ancient authors who had not heard of it nor hit upon it, but who, I am sure, would have accepted it with little hesitation if they had heard of it.

And if I may say frankly what I think, I cannot believe that this modern author rejects it himself; but I think that, being unable to pass himself off as its original author, it occurred to him to try his hand at suppressing it, or at least at belittling it for the simple-minded. We know the number of these to be enormous, and there are many men who enjoy the multitudinous applause of the people more than the approbation of the exceptional few.

SAGR. Just a minute, Salviati; it seems to me that you are not getting clear to the heart of the matter. Those who have nets to snare the common people know also how to be the authors of other men's inventions, so long as these are not ancient ones and have not been published in the schools and in the market places so that they are more than familiar to everyone.

SALV. Oh, I am more cynical than you. Why talk about publications and notoriety? Does it make any difference whether the opinions and inventions are new to people or the people new to them? If you would be content with the acclaim of the tyros in science who flourish now and then, you would be able to make yourself the inventor of even the alphabet and become admirable to them in this way. And if in the course of time your cunning were discovered, that would not prejudice your aims very much, for others would come along to fill in the gaps in the ranks of your supporters.

It is the same for opinions to be new to men as for men to be new to the opinions.

But let us get back to showing Simplicio the futility of the arguments of his modern author, in which there are falsehoods and fallacies and contradictions. First, it is false that this secondary light is brighter around the extreme margin than in the central parts, so that a sort of ring or circle is formed that is more brilliant than the rest of the field. It is true that the moon shows such a circle when observed in twilight at its first appearance after new moon, but that originates deceptively in differences between the boundaries which terminate the lunar disc over which this secondary light is spread. For on the side toward the sun, the light is bounded by the bright horn of the moon; on the other side, it has for its boundary the dark field of the twilight, in relation to which it appears lighter than the whiteness of the lunar disc — which on the other side is obscured by the greater brilliance of the horns. If only this modern author had tried placing between his eye and the primary brilliance some screen such as the roof of a house, or some other partition, so that

Secondary light of the moon appears in the form of a ring—bright at the extreme circumference and not in the center—and why.

Mode of observing the secondary light of the moon.

only the part of the moon outside the horn remained visible, he would have seen it all equally luminous.

SIMP. I seem to remember his writing about making use of some such device for hiding the bright crescent.

SALV. Well, if that is so, then what I have called an oversight of his is turned into a lie that borders upon rashness, since anybody can put it to the test as often as he likes.

Moon's disc in a solar eclipse can be seen only by default.

Next, I question very much whether in an eclipse of the sun the disc of the moon is seen at all except by deprivation of light, especially when the eclipse is partial, as it must have been for this author's observations. But even if it were perceived as lighted, this would not contradict our opinion. It would favor this, since at that time the moon is opposite the hemisphere of the earth that is illuminated by the sun; and though the shadow of the moon darkens a part of it, the darkened part is very small in comparison with what remains illuminated. Then he adds here that in this event the part of the margin which is underneath the sun looks very bright, but that which remains outside does not, deducing from this the direct arrival at the eye of the sun's rays through the former part but not through the latter. Here is one of those fabrications that reveal the other fictions of the person who recounts them. For if the rays of the sun had to pass directly to our eyes in order to make the secondary light of the lunar disc visible, doesn't this poor fellow see that we should never observe such a secondary light *except* in eclipses of the sun? And if the presence of a part of the moon only half a degree distant from the sun's disc can deviate the sun's rays so that they do not arrive at our eyes, what will happen when it is twenty or thirty degrees away, as it is right after the new moon? How will the rays of the sun which then have to pass through the body of the moon find their way to our eyes?

The author of the booklet adapts facts to his purposes, not his purposes to the facts.

This fellow goes about thinking up, one by one, things that would be required to serve his purposes, instead of adjusting his purposes step by step to things as they are. Look: To make the brightness of the sun capable of piercing the substance of the moon, he makes the latter translucent like the transparency of a cloud or a crystal, for example. But I think he never got round to deciding with regard to such transparency whether the sun's rays would penetrate a cloud more than two thousand miles thick. Now let us suppose that he would boldly respond that this could easily happen for celestial bodies, which are very differently constructed from our impure and filthy elemental bodies, and let us convict him of error by means that admit of no response — or rather no subterfuge. If he wants to maintain that the substance of the moon is diaphanous, he will have to say that this is so when the rays of the sun have to go through its entire thickness of two thousand miles, but that when they are opposed by only a mile or so of it they do not penetrate it any more than they do one of our own mountains.

SAGR. This reminds me of a man who wanted to sell me a secret method of communicating with a person two or three thousand miles away, by means of a certain sympathy of magnetic needles. I told him that I would gladly buy, but wanted to see by experiment and that it would be enough for me if he would stand in one room and I in another. He replied that its operation could not be detected at such a short distance. I sent him on his way, with the remark that I was not in the mood at that time to go to Cairo or Moscow for the experiment, but that if he wanted to go I would stay in Venice and take care of the other end.

Laughing-stock made of a person who tried to sell a certain secret for conversing with someone a thousand miles away.

But let us hear how the deductions of our author go, and why he has to admit the material of the moon to be permeable by solar rays at a thickness of two thousand miles, but as opaque as one of our mountains at a depth of only a mile.

SALV. The mountains of the moon themselves give evidence of it. Struck on one side by the sun, they cast very black shadows to the opposite side, more abrupt and definite than the shadows of our own mountains; whereas if they were diaphanous, we should never have been able to discern any roughness on the surface of the moon, nor to see those luminous separated peaks along the boundary that divides the lighted from the darkened part. Still less would we see that same boundary so distinctly if it were true that the sunlight penetrated the depths of the moon. Rather, according to this author's own words, the boundary between the parts touched and not touched by the sun would have to appear very vague and mixed light-and-dark. For any material that gives passage to the sun's rays through a thickness of two thousand miles would have to be so transparent that there would be little difference in a hundredth or less part of that magnitude. Yet the boundary that separates the lighted part from the dark is abrupt and as distinct as black is from white,

especially where the division passes over the part of the moon
that is naturally brightest and roughest. Where it cuts the classic
(*antiche*) spots, which are plains, they go in a spherical curve
so as to receive the sun's rays obliquely, and the boundary is
not so abrupt because the illumination is fainter.

Finally, what he says about the secondary light not diminishing and receding according as the moon waxes (but retaining
the same strength) is quite false. Little is to be seen of it even
at quadrature, when on the contrary it ought to be seen most
vividly, since at that time the moon can be seen after twilight in
the deepest night.

From all this we conclude that the reflection from the earth
is very powerful on the moon. What is more important is that
there follows from this another beautiful resemblance, which
is that if it is true that the planets act reciprocally upon the
earth by their motion and by their light, perhaps the earth is no
less potent in acting upon them by its own light — and possibly
by its motion, too. But even if it does not move, these actions
may remain the same. For as we have already seen, the action
of light (that is, the reflected light of the sun) is precisely the
same; and motion does no more than to make variations in appearances, which take place in the same way by making the
earth move and holding the sun still as they would by the opposite.

SIMP. No philosopher is to be found who ever said that these
inferior bodies act upon celestial ones, while Aristotle said clearly
the opposite.

SALV. Aristotle and the others who did not know that the earth
and moon reciprocally illuminate each other deserve to be excused. But they would equally deserve to be reprehended if, while
wanting us to give in to them and believe that the moon acts on
the earth by light, they should insist on denying us the action
of the earth on the moon when we had demonstrated to them that
the earth lights up the moon.

SIMP. All in all, I find in my heart a great reluctance to grant this
companionship between the earth and moon of which you want
to persuade me, placing the earth in the host of the stars, so to
speak. For even if there were nothing else, the immense separation and distance between the earth and the heavenly bodies
seems to me to imply necessarily a great dissimilarity.

The earth may
operate recipro-
cally upon the
heavenly bodies
by means of
light.

SALV. See what an inveterate affection and a deeply rooted opinion can do, Simplicio. It is so strong that you make the very things seem to favor your opinion which you yourself adduce against it. If separation and distance are valid facts for arguing a great difference in natures, it is necessary on the other hand that closeness and contiguity should mean similarity; and how much closer is the moon to the earth than it is to the other heavenly bodies! Confess then, by your own admission (and you will have plenty of other philosophers for company), the great affinity between the earth and the moon. And now let us get on; propose whatever else remains to be considered about the difficulties that you posed against the congruence of these two bodies.

SIMP. There remains my question regarding the solidity of the moon, which I deduced from its being highly polished and smooth, and you from its being mountainous. Another trouble originated in my believing that the reflection from the seas ought to be stronger, on account of the evenness of their surface, than that from the land, whose surface is so rough and dark.

SALV. As to the first question I say that it is just as with the parts of the earth, which because of their heaviness attempt to get as close as possible to its center, though some do remain farther away than others — mountains farther than plains, for instance — and this because of their solidity and hardness, for if they were of fluid material they would level out. Just so, to see parts of the moon remain raised above the sphericity of the parts beneath them implies their hardness, because it is plausible that the moon's material forms itself into a spherical shape from a coöperative tendency (*concorde conspirazione*) of all its parts toward their center.

Concerning the other question, it seems to me that from our having considered events that take place in mirrors we can understand quite well that the reflection of light coming from the seas would be less than that coming from the land. I mean here its general reflection, for as to the specific reflection from a quiet sea toward one certain place, I have no doubt that anyone located in that place would see from the water a very strong reflection. But from all other places, the surface of the water would be seen darker than the land. And to show this to your own senses, let us go into that hall and pour a little water on the pave-

Affinity between earth and moon with regard to proximity.

Solidity of the lunar globe is argued from its being mountainous.

Reflection of light weaker from the ocean than from the land.

Experiment
showing the re-
flection from
water to be less
bright than from
soil.

ment. Tell me, now, doesn't this wet brick look darker than those dry ones? Of course it does, and it looks so from every place but one; namely, where the reflection of light from that window strikes. Therefore move gently backward a bit.

SIMP. From here I see the wet part brighter than the rest of the pavement, and I perceive that this happens because the reflection of the light from that window is coming straight toward me.

SALV. All that this wetting has done is to fill the little pores in the brick and reduce its surface to a smooth plane, from which the reflected rays then come unitedly toward a single place. The rest of the pavement is dry, and keeps its roughness; that is, an innumerable variety of slopes in its minute particles, from which the reflections of light go out in every direction but are much weaker than if they were all to go united together. Therefore the appearance of this part varies little or none when observed from various directions, but looks the same from everywhere — and much less bright than that one particular reflection from the wet part.

I conclude therefore that just as the surface of the ocean seen from the moon would appear level (except for islands and rocks), so it would appear less bright than that of the land, which is uneven and mountainous. And if it were not that I do not wish to seem too eager, as they say, I should tell you of having observed the secondary light of the moon (which I say is a reflection from the terrestrial globe) to be appreciably brighter two or three days before conjunction than after. That is, when we see it before dawn in the east it is brighter than when we see it in the evening after the setting of the sun in the west. The reason for this difference is that the terrestrial hemisphere opposite to the moon when it is in the east has fewer seas and more land, containing all Asia. But when the moon is in the west, it faces great seas — the whole Atlantic clear to America — a very plausible argument for the surface of the water showing itself less brilliantly than that of the land.

Moon's second-
ary light
brighter before
conjunction than
after.

SIMP. [Therefore, in your opinion, the earth would make an appearance similar to that which we see in the moon, of at most two parts.] But do you believe then that those great spots which are seen on the face of the moon are seas, and the brighter balance land, or some such thing?†

SALV. What you are now asking me is the first of the differences that I think exist between the moon and the earth, which we had better hurry along with, as we are staying too long on the moon. I say then that if there were in nature only one way for two surfaces to be illuminated by the sun so that one appears lighter than the other, and that this were by having one made of land and the other of water, it would be necessary to say that the moon's surface was partly terrene and partly aqueous. But because there are more ways known to us that could produce the same effect, and perhaps others that we do not know of, I shall not make bold to affirm one rather than another to exist on the moon.

We have already seen that a bleached silver plate changes from white to dark by the touch of the burnisher; the watery part of the earth looks darker than the dry; on the ridges of mountains the wooded parts look much gloomier than the open and barren places because the plants cast a great deal of shadow while the clearings are lighted by the sun. Such a mixture of shadows is so effective that in sculptured velvet the color of the cut silk looks much darker than that of the uncut, because of shadows cast between one thread and another; and plain velvet is likewise much darker than taffeta made of the same silk. So if on the moon there were things resembling dense forests, their aspect would probably be like that of the spots we see; a like difference would be created if they were seas; and, finally, there is nothing to prevent these spots being really of a darker color than the rest, for it is in that way that snow makes mountains appear brighter.

What is clearly seen in the moon is that the darker parts are all plains, with few rocks and ridges in them, though there are some. The brighter remainder is all full of rocks, mountains, round ridges, and other shapes, and in particular there are great ranges of mountains around the spots. That the spots are flat surfaces we are certain, from observing that the boundary which separates the light and dark parts makes an even cut in traversing the spots, whereas in the bright parts it looks broken and jagged. But I do not know whether this evenness of surface is enough by itself to cause the apparent darkness, and I rather think not.

Quite apart from this, I consider the moon very different from the earth. Though I fancy to myself that its regions are not idle and dead, still I do not assert that life and motion exist there.

Darker parts of the moon are plains, brighter parts mountainous.

Around the spots on the moon are long chains of mountains.

Things like ours are not generated on the moon, but different—if indeed generations exist there.

Moon not com-
posed of land
and water.

Aspects of the
sun necessary for
our species are
not duplicated
on the moon.

An ordinary day
on the moon is a
month long.

To the moon, the
sun drops and
ascends with a
variation of 10
degrees; to the
earth, 47 degrees.

No rains on
the moon.

and much less that plants, animals, or other things similar to
ours are generated there. Even if they were, they would be extremely diverse, and far beyond all our imaginings. I am inclined
to believe this because in the first place I think that the material
of the lunar globe is not land and water, and this alone is enough
to prevent generations and alterations similar to ours. But even
supposing land and water on the moon, there are in any case
two reasons that plants and animals similar to ours would not
be produced there.

The first is that the varying aspects of the sun are so necessary for our various species that these could not exist at all without them. Now the behavior of the sun toward the earth is much
different from that which it displays toward the moon. As to
daily illumination, we on the earth have for the most part twenty-
four hours divided between day and night, but the same effect
takes a month on the moon. The annual sinking and rising by
which the sun causes the various seasons and the inequalities of
day and night are finished for the moon in a month. And whereas
for us the sun rises and sinks so much that between its maximum
and minimum altitudes there lie forty-seven degrees of difference (that is, as much as the distance between the tropics), for
the moon it varies no more than ten degrees or a little less, which
is the amount of the maximum latitudes of its orbit with respect
to the ecliptic.

Now think what the action of the sun would be in the torrid
zone if for fifteen days without pause it continued to beat down
with its rays. It goes without saying that all the plants and herbs
and animals would be destroyed; hence if any species existed
there, they would be plants and animals very different from
present ones.

In the second place, I am sure that there are no rains on the
moon, because if clouds collected in any part of it, as around the
earth, they would hide some of the things on the moon that we
see with the telescope. Briefly, the scene would alter in some
respect; an effect which I have never seen during long and dili-
gent observations, having always discovered a very pure and
uniform serenity.

SAGR. To this it might be replied that either there might be great
dews or that it rains there during its nights; that is, when the sun
does not light it up.

SALV. If from other appearances we had any signs that there were species similar to ours there, and only the occurrence of rains was lacking, we should be able to find this or some other condition to take their place, as happens in Egypt by the inundations of the Nile. But finding no event whatever like ours, of the many that would be required to produce similar effects, there is no point in troubling to introduce one only, and even that one not from sure observation but because of mere possibility. Besides, if I were asked what my basic knowledge and natural reason told me regarding the production there of things similar to or different from ours, I should always reply, "Very different and entirely unimaginable by us"; for this seems to me to fit with the richness of nature and the omnipotence of the Creator and Ruler.

SAGR. It always seems to me extreme rashness on the part of some when they want to make human abilities the measure of what nature can do. On the contrary, there is not a single effect in nature, even the least that exists, such that the most ingenious theorists can arrive at a complete understanding of it. This vain presumption of understanding everything can have no other basis than never understanding anything. For anyone who had experienced just once the perfect understanding of one single thing, and had truly tasted how knowledge is accomplished, would recognize that of the infinity of other truths he understands nothing.

Never having completely understood anything makes some people believe they understand everything.

SALV. Your argument is quite conclusive; in confirmation of it we have the evidence of those who do understand or have understood some thing; the more such men have known, the more they have recognized and freely confessed their little knowledge. And the wisest of the Greeks, so adjudged by the oracle, said openly that he recognized that he knew nothing.

SIMP. It must be said, then, that either the oracle or Socrates himself was a liar, the former declaring him the wisest, and the latter saying he knew himself the most ignorant.

SALV. Neither of your alternatives follows, since both pronouncements can be true. The oracle judges Socrates wisest above all other men, whose wisdom is limited; Socrates recognizes his knowing nothing relative to absolute wisdom, which is infinite. And since much is the same part of infinite as little, or as nothing (for to arrive at an infinite number it makes no difference whether

Truth of the
oracle's response
in judging Soc-
rates wisest.

we accumulate thousands, tens, or zeros), Socrates did well to
recognize his limited knowledge to be as nothing to the infinity
which he lacked. But since there is nevertheless some knowledge
to be found among men, and this is not equally distributed to
all, Socrates could have had a larger share than others and thus
have verified the response of the oracle.

SAGR. I think I understand this point quite well. Among men
there exists the power to act, Simplicio, but it is not equally
shared by all; and no doubt the power of an emperor is greater
than that of a private person, but both are nil in comparison to
Divine omnipotence. Among men there are some who understand agriculture better than others; but what has knowing how
to plant a grapevine in a ditch got to do with knowing how to
make it take root, draw nourishment, take from this some part
good for building leaves, some other for forming tendrils, this
for the bunches, that for the grapes, the other for the skins, all
this being the work of most wise Nature? This is one single particular example of the innumerable works of Nature, and in

Divine knowl-
edge infinitely
infinite.

this alone may be recognized an infinite wisdom; hence one may
conclude that Divine wisdom is infinitely infinite.

Michelangelo's
sublime genius.

SALV. Here is another example. Do we not say that the art of discovering a beautiful statue in a block of marble has elevated the
genius of Michelangelo far, far above the ordinary minds of
other men? Yet this work is nothing but the copying of a single
attitude and position of the external and superficial members of
one motionless man. Then what is it in comparison with a man
made by Nature, composed of so many members, external and
internal, of so many muscles, tendons, nerves, bones, that serve
so many and such diverse motions? And what shall we say of
the senses, of spiritual power, and finally of the understanding?
May we not rightly say that the making of a statue yields by an
infinite amount to the formation of a live man, even to the formation of the lowest worm?

SAGR. And what difference do you think there was between the
dove of Archytas and a natural dove?

SIMP. Either I am without understanding or there is a manifest
contradiction in this argument of yours. Among your greatest
encomiums, if not indeed the greatest of all, is your praise for
the understanding which you attribute to natural man. A little
while ago you agreed with Socrates that his understanding was

nil. Then you must say that not even Nature understood how
to make an intellect that could understand.

SALV. You put the point very sharply, and to answer the objection
it is best to have recourse to a philosophical distinction and to say
that the human understanding can be taken in two modes, the
intensive or the *extensive*. *Extensively*, that is, with regard to *Intensively*, man
the multitude of intelligibles, which are infinite, the human un- understands
derstanding is as nothing even if it understands a thousand prop- much; *exten-*
ositions; for a thousand in relation to infinity is zero. But taking *sively*, little.
man's understanding *intensively*, in so far as this term denotes
understanding some proposition perfectly, I say that the human
intellect does understand some of them perfectly, and thus in
these it has as much absolute certainty as Nature itself has. Of
such are the mathematical sciences alone; that is, geometry and
arithmetic, in which the Divine intellect indeed knows infinitely
more propositions, since it knows all. But with regard to those
few which the human intellect does understand, I believe that its
knowledge equals the Divine in objective certainty, for here it
succeeds in understanding necessity, beyond which there can be
no greater sureness.

SIMP. This speech strikes me as very bold and daring.

SALV. These are very ordinary propositions† and far from any
shade of temerity or boldness. They do not detract in the least
from the majesty of Divine wisdom, just as saying that God
cannot undo what is done does not in the least diminish His omnipotence. But I question, Simplicio, whether your suspicion
does not arise from your having taken my words equivocally.
So in order to explain myself better, I say that as to the truth of
the knowledge which is given by mathematical proofs, this is the
same that Divine wisdom recognizes; but I shall concede to you God's mode of
indeed that the way in which God knows the infinite proposi- knowing is dif-
tions of which we know some few is exceedingly more excellent ferent from
than ours. Our method proceeds with reasoning by steps from man's.
one conclusion to another, while His is one of simple intuition. Human
We, for example, in order to win a knowledge of some proper- understanding
ties of the circle (which has an infinity of them), begin with one accomplished by
of the simplest, and, taking this for the definition of circle, pro- reasoning.
ceed by reasoning to another property, and from this to a third,
and then a fourth, and so on; but the Divine intellect, by a simple apprehension of the circle's essence, knows without time-

consuming reasoning all the infinity of its properties. Next, all these properties are in effect virtually included in the definitions of all things; and ultimately, through being infinite, are perhaps but one in their essence and in the Divine mind. Nor is all the above entirely unknown to the human mind either, but it is clouded with deep and thick mists, which become partly dispersed and clarified when we master some conclusions and get them so firmly established and so readily in our possession that we can run over them very rapidly. For, after all, what more is there to the square on the hypotenuse being equal to the squares on the other two sides, than the equality of two parallelograms on equal bases and between parallel lines? And is this not ultimately the same as the equality of two surfaces which when superimposed are not increased, but are enclosed within the same boundaries? Now these advances, which our intellect makes laboriously and step by step, run through the Divine mind like light in an instant; which is the same as saying that everything is always present to it.

I conclude from this that our understanding, as well in the manner as in the number of things understood, is infinitely surpassed by the Divine; but I do not thereby abase it so much as to consider it absolutely null. No, when I consider what marvelous things and how many of them men have understood, inquired into, and contrived, I recognize and understand only too clearly that the human mind is a work of God's, and one of the most excellent.

SAGR. I myself have many times considered in the same vein what you are now saying, and how great may be the acuteness of the human mind. And when I run over the many and marvelous inventions men have discovered in the arts as in letters, and then reflect upon my own knowledge, I count myself little better than miserable. I am so far from being able to promise myself, not indeed the finding out of anything new, but even the learning of what has already been discovered, that I feel stupid and confused, and am goaded by despair. If I look at some excellent statue, I say within my heart: "When will you be able to remove the excess from a block of marble and reveal so lovely a figure hidden therein? When will you know how to mix different colors and spread them over a canvas or a wall and represent all visible objects by their means, like a Michelangelo, a Raphael, or a

Titian?" Looking at what men have found out about arranging the musical intervals and forming precepts and rules in order to control them for the wonderful delight of the ear, when shall I be able to cease my amazement? What shall I say of so many and such diverse instruments? With what admiration the reading of excellent poets fills anyone who attentively studies the invention and interpretation of concepts! And what shall I say of architecture? What of the art of navigation?

But surpassing all stupendous inventions, what sublimity of mind was his who dreamed of finding means to communicate his deepest thoughts to any other person, though distant by mighty intervals of place and time! Of talking with those who are in India; of speaking to those who are not yet born and will not be born for a thousand or ten thousand years; and with what facility, by the different arrangements of twenty characters upon a page!

Invention of writing stupendous above all.

Let this be the seal of all the admirable inventions of mankind and the close of our discussions for this day. The hottest hours now being past, I think that Salviati might like to enjoy our cool ones in a gondola; and tomorrow I shall expect you both so that we may continue the discussions now begun.

End of the First Day

THE SECOND DAY

SALVIATI. Yesterday took us into so many and such great digressions twisting away from the main thread of our principal argument that I do not know whether I shall be able to go ahead without your assistance in putting me back on the track.

SAGR. I am not surprised that you should find yourself in some confusion, for your mind is as much filled and encumbered with what remains to be said as with what has been said. But I am simply a listener and have in my mind only the things I have heard, so perhaps I can put your discourse back on its path by briefly outlining these for you.

As I recall it, yesterday's discourse may be summarized as a preliminary examination of the two following opinions as to which is the more probable and reasonable. The first holds the substance of the heavenly bodies to be ingenerable, incorruptible, inalterable, invariant, and in a word free from all mutations except those of situation, and accordingly to be a quintessence† most different from our generable, corruptible, alterable bodies. The other opinion, removing this disparity from the world's parts, considers the earth to enjoy the same perfection as other integral bodies of the universe; in short, to be a movable and a moving body no less than the moon, Jupiter, Venus, or any other planet. Later many detailed parallels were drawn between the earth and the moon. More comparisons were made with the moon than with other planets, perhaps from our having more and better sensible evidence about the former by reason of its lesser distance. And having finally concluded this second opinion to have more likelihood than the other, it seems to me that our next step should be to examine whether the earth must be considered immovable, as most people have believed up to the present, or mobile, as many ancient philosophers believed and as others of more recent times consider it; and, if movable, what its motion may be.

SALV. Now I know and recognize the signposts along our road. But before starting in again and going ahead, I ought to tell you that I question this last thing you have said, about our having concluded in favor of the opinion that the earth is endowed with the same properties as the heavenly bodies. For I did not conclude this, just as I am not deciding upon any other controversial proposition. My intention was only to adduce those arguments and replies, as much on one side as on the other — those questions and solutions which others have thought of up to the present time (together with a few which have occurred to me after long thought) — and then to leave the decision to the judgment of others.

SAGR. I allowed myself to be carried away by my own sentiments, and believing that what I felt in my heart ought to be felt by others too, I made that conclusion universal which should have been kept particular. This really was an error on my part, especially as I do not know the views of Simplicio, here present.

SIMP. I confess that all last night I was meditating on yesterday's material, and truly I find it to contain many beautiful considerations which are novel and forceful. Still, I am much more impressed by the authority of so many great authors, and in particular . . . You shake your head, Sagredo, and smile, as if I had uttered some absurdity.

SAGR. I merely smile, but believe me, I am hardly able to keep from laughing, because I am reminded of a situation that I witnessed not many years ago together with some friends of mine, whom I could name to you for that matter.

SALV. Perhaps you had better tell us about it so that Simplicio will not go on thinking your mirth was directed at him.

SAGR. I'll be glad to. One day I was at the home of a very famous doctor in Venice, where many persons came on account of their studies, and others occasionally came out of curiosity to see some anatomical dissection performed by a man who was truly no less learned than he was a careful and expert anatomist. It happened

A philosopher's ridiculous answer as to where the nerves originate.

Origin of the
nerves according
to Aristotle and
according to the
doctors.

on this day that he was investigating the source and origin of the nerves, about which there exists a notorious controversy between the Galenist and Peripatetic doctors. The anatomist showed that the great trunk of nerves, leaving the brain and passing through the nape, extended on down the spine and then branched out through the whole body, and that only a single strand as fine as a thread arrived at the heart. Turning to a gentleman whom he knew to be a Peripatetic philosopher, and on whose account he had been exhibiting and demonstrating everything with unusual care, he asked this man whether he was at last satisfied and convinced that the nerves originated in the brain and not in the heart. The philosopher, after considering for awhile, answered: "You have made me see this matter so plainly and palpably that if Aristotle's text were not contrary to it, stating clearly that the nerves originate in the heart, I should be forced to admit it to be true."

SIMP. Sir, I want you to know that this dispute as to the source of the nerves is by no means as settled and decided as perhaps some people like to think.

SAGR. Doubtless it never will be, in the minds of such opponents. But what you say does not in the least diminish the absurdity of this Peripatetic's reply; who, as a counter to sensible experience, adduced no experiment or argument of Aristotle's, but just the authority of his bare *ipse dixit*.

SIMP. Aristotle acquired his great authority only because of the strength of his proofs and the profundity of his arguments. Yet one must understand him, and not merely understand him, but have such thorough familiarity with his books that the most complete idea of them may be formed, in such a manner that every saying of his is always before the mind. He did not write for the common people, nor was he obliged to thread his syllogisms together by the trivial ordinary method; rather, making use of the permuted method,† he has sometimes put the proof of a proposition among texts that seem to deal with other things.

Requirements
for philosophiz-
ing well in Aris-
totle's method.

Therefore one must have a grasp of the whole grand scheme, and be able to combine this passage with that, collecting together one text here and another very distant from it. There is no doubt that whoever has this skill will be able to draw from his books demonstrations of all that can be known; for every single thing is in them.

SAGR. My dear Simplicio, since having things scattered all over the place does not disgust you, and since you believe by the collection and combination of the various pieces you can draw the juice out of them, then what you and the other brave philosophers will do with Aristotle's texts, I shall do with the verses of Virgil and Ovid, making centos of them and explaining by means of these all the affairs of men and the secrets of nature. But why do I speak of Virgil, or any other poet? I have a little book, much briefer than Aristotle or Ovid, in which is contained the whole of science, and with very little study one may form from it the most complete ideas. It is the alphabet, and no doubt anyone who can properly join and order this or that vowel and these or those consonants with one another can dig out of it the truest answers to every question, and draw from it instruction in all the arts and sciences. Just so does a painter, from the various simple colors placed separately upon his palette, by gathering a little of this with a bit of that and a trifle of the other, depict men, plants, buildings, birds, fishes, and in a word represent every visible object, without any eyes or feathers or scales or leaves or stones being on his palette. Indeed, it is necessary that none of the things imitated nor parts of them should actually be among the colors, if you want to be able to represent everything; if there were feathers, for instance, these would not do to depict anything but birds or feather dusters.

A clever device for learning philosophy from any book one pleases.

SALV. And certain gentlemen still living and active were present when a doctor lecturing in a famous Academy, upon hearing the telescope described but not yet having seen it, said that the invention was taken from Aristotle. Having a text fetched, he found a certain place† where the reason is given why stars in the sky can be seen during daytime from the bottom of a very deep well. At this point the doctor said: "Here you have the well, which represents the tube; here the gross vapors, from whence the invention of glass lenses is taken; and finally here is the strengthening of the sight by the rays passing through a diaphanous medium which is denser and darker."

Invention of the telescope dug out of Aristotle.

SAGR. This manner of "containing" everything that can be known is similar to the sense in which a block of marble contains a beautiful statue, or rather thousands of them; but the whole point lies in being able to reveal them. Even better we might say that it is like the prophecies of Joachim or the answers of the

Alchemists inter-
pret the fables
of poets into
secrets for
making gold.

heathen oracles, which are understood only after the events they
forecast have occurred.

SALV. And why do you leave out the prophecies of the astrologers,
which are so clearly seen in horoscopes (or should we say in the
configurations of the heavens) after their fulfillment?

SAGR. It is in this way that the alchemists, led on by their madness, find that the greatest geniuses of the world never really
wrote about anything except how to make gold; but in order to
tell this without revealing it to the vulgar, this fellow in one
manner and that one in another have whimsically concealed it
under various disguises. And a very amusing thing it is to hear
their comments upon the ancient poets, revealing the important
mysteries hidden behind their stories — what the loves of the
moon mean, and her descent to the earth for Endymion; her dis-
pleasure with Acteon; the significance of Jupiter's turning himself into a rain of gold, or into a fiery flame; what great secrets
of the art there are in Mercury the interpreter, in Pluto's kid-
napings, and in golden boughs.

SIMP. I believe, and to some extent I know, that the world does
not lack certain giddy brains, but their folly should not redound
to the discredit of Aristotle, of whom it seems to me you sometimes speak with too little respect. His antiquity alone, and the
mighty name he has acquired among so many men of distinguished mind, should be enough to earn him respect among all
the learned.

SALV. That is not quite how matters stand, Simplicio. Some of
his followers are so excessively timid that they give us occasion
(or more correctly would give us occasion if we credited their
triflings) to think less of him. Tell me, are you so credulous as
not to understand that if Aristotle had been present and heard
this doctor who wanted to make him inventor of the telescope,
he would have been much angrier with him than with those who
laughed at this doctor and his interpretations? Is it possible for
you to doubt that if Aristotle should see the new discoveries in
the sky he would change his opinions and correct his books and
embrace the most sensible doctrines, casting away from himself
those people so weak-minded as to be induced to go on abjectly
maintaining everything he had ever said? Why, if Aristotle had
been such a man as they imagine, he would have been a man of
intractable mind, of obstinate spirit, and barbarous soul; a man

Some of Aris-
totle's followers
injure his repu-
tation by trying
too hard to
increase it.

of tyrannical will who, regarding all others as silly sheep, wished to have his decrees preferred over the senses, experience, and nature itself. It is the followers of Aristotle who have crowned him with authority, not he who has usurped or appropriated it to himself. And since it is handier to conceal oneself under the cloak of another than to show one's face in open court, they dare not in their timidity get a single step away from him, and rather than put any alterations into the heavens of Aristotle, they want to deny out of hand those that they see in nature's heaven.

SAGR. Such people remind me of that sculptor who, having transformed a huge block of marble into the image of a Hercules or a thundering Jove, I forget which, and having with consummate art made it so lifelike and fierce that it moved everyone with terror who beheld it, he himself began to be afraid, though all its vivacity and power were the work of his own hands; and his terror was such that he no longer dared affront it with his mallet and chisel.

Amusing case of a certain sculptor.

SALV. I often wonder how it can be that these strict supporters of Aristotle's every word fail to perceive how great a hindrance to his credit and reputation they are, and how the more they desire to increase his authority, the more they actually detract from it. For when I see them being obstinate about sustaining propositions which I personally know to be obviously false, and wanting to persuade me that what they are doing is truly philosophical and would be done by Aristotle himself, it much weakens my opinion that he philosophized correctly about other matters more recondite to me. If I saw them give in and change their opinions about obvious truths, I should believe that they might have sound proofs for those in which they persisted and which I did not understand or had not heard.

SAGR. Or truly, if it seemed to them that they staked too much of their own reputation and of Aristotle's in confessing that they did not know this or that conclusion discovered by someone else, would it not be a lesser evil for them to seek it among his texts by the collection of various of these according to the practice recommended by Simplicio? For if all things that can be known are in these texts, then it must follow that they can be discovered there.

SALV. Sagredo, do not sneer at this prudent scheme, which it seems to me you propose sarcastically. For it is not long since a

famous philosopher composed a book on the soul in which, discussing Aristotle's opinion as to its mortality or immortality, he
adduced many texts beyond those already quoted by Alexander.
As to those, he asserted that Aristotle was not even dealing with
such matters there, let alone deciding anything about them, and
he gave others which he himself had discovered in various remote
places and which tended to the damaging side. Being advised
that this would make trouble for him in getting a license to pub-

Convenient
decision of a
Peripatetic
philosopher.

lish it, he wrote back to his friend that he would nevertheless get
one quickly, since if no other obstacle came up he would have
no difficulty altering the doctrine of Aristotle; for with other
texts and other expositions he could maintain the contrary opinion, and it would still agree with the sense of Aristotle.

SAGR. Oh, what a doctor this is! I am his to command; for he will
not let himself be imposed upon by Aristotle, but will lead him
by the nose and make him speak to his own purpose! See how
important it is to know how to take time by the forelock! One
ought not to get into the position of doing business with Hercules
when he is under the Furies and enraged, but rather when he is
telling stories among the Lydian maids.

Cowardice of
some followers
of Aristotle.

Oh, the inexpressible baseness of abject minds! To make
themselves slaves willingly; to accept decrees as inviolable; to
place themselves under obligation and to call themselves persuaded and convinced by arguments that are so "powerful" and
"clearly conclusive" that they themselves cannot tell the purpose
for which they were written, or what conclusion they serve to
prove! But let us call it a greater madness that among themselves
they are even in doubt whether this very author held to the
affirmative or the negative side. Now what is this but to make
an oracle out of a log of wood, and run to it for answers; to fear
it, revere it, and adore it?

SIMP. But if Aristotle is to be abandoned, whom shall we have
for a guide in philosophy? Suppose you name some author.

SALV. We need guides in forests and in unknown lands, but on
plains and in open places only the blind need guides. It is better
for such people to stay at home, but anyone with eyes in his head
and his wits about him could serve as a guide for them. In saying

Too much adora-
tion of Aristotle
is blasphemous.

this, I do not mean that a person should not listen to Aristotle;
indeed, I applaud the reading and careful study of his works,
and I reproach only those who give themselves up as slaves to
him in such a way as to subscribe blindly to everything he says

and take it as an inviolable decree without looking for any other reasons. This abuse carries with it another profound disorder, that other people do not try harder to comprehend the strength of his demonstrations. And what is more revolting in a public dispute, when someone is dealing with demonstrable conclusions, than to hear him interrupted by a text (often written to some quite different purpose) thrown into his teeth by an opponent? If, indeed, you wish to continue in this method of studying, then put aside the name of philosophers and call yourselves historians, or memory experts; for it is not proper that those who never philosophize should usurp the honorable title of philosopher.

It is not fitting for those who never philosophize to usurp the title "philosopher."

But we had better get back to shore, lest we enter into a boundless ocean and not get out of it all day. So put forward the arguments and demonstrations, Simplicio — either yours or Aristotle's — but not just texts and bare authorities, because our discourses must relate to the sensible world and not to one on paper. And since in yesterday's argument the earth was lifted up out of darkness and exposed to the open sky, and the attempt to number it among the bodies we call heavenly was shown to be not so hopeless and prostrate a proposition that it remained without a spark of life, we should follow this up by examining that other proposition which holds it to be probable that the earth is fixed and utterly immovable as to its entire globe, and see what chance there is of making it movable, and with what motion.

The sensible world.

Now because I am undecided about this question, whereas Simplicio has his mind made up with Aristotle on the side of immovability, he shall give the reasons for his opinion step by step, and I the answers and the arguments of the other side, while Sagredo shall tell us the workings of his mind and the side toward which he feels it drawn.

SAGR. That suits me very well, provided that I retain the freedom to bring up whatever common sense may dictate to me from time to time.

SALV. Indeed, I particularly beg you to do so; for I believe that writers on the subject have left out few of the easier and, so to speak, more material considerations, so that only those are lacking and may be wished for which are subtler and more recondite. And to look into these, what ingenuity can be more fitting than that of Sagredo's acute and penetrating wit?

SAGR. Describe me as you like, Salviati, but please let us not get

Motions of the
earth are imper-
ceptible to its
inhabitants.

The earth can
have no other
motions than
those which ap-
pear to us to be
common to the
entire universe
excepting the
earth.

Diurnal motion
is seen to be most
general to the
whole universe
excepting the
earth.

into another kind of digression — the ceremonial. For now I am a philosopher, and am at school and not at court (*al Broio*).

SALV. Then let the beginning of our reflections be the considera-tion that whatever motion comes to be attributed to the earth must necessarily remain imperceptible to us and as if nonexistent, so long as we look only at terrestrial objects; for as inhabitants of the earth, we consequently participate in the same motion. But on the other hand it is indeed just as necessary that it display itself very generally in all other visible bodies and objects which, being separated from the earth, do not take part in this move-ment. So the true method of investigating whether any motion can be attributed to the earth, and if so what it may be, is to observe and consider whether bodies separated from the earth exhibit some appearance of motion which belongs equally to all. For a motion which is perceived only, for example, in the moon, and which does not affect Venus or Jupiter or the other stars, cannot in any way be the earth's or anything but the moon's.

Now there is one motion which is most general and supreme over all, and it is that by which the sun, moon, and all other planets and fixed stars — in a word, the whole universe, the earth alone excepted — appear to be moved as a unit from east to west in the space of twenty-four hours. This, in so far as first appearances are concerned, may just as logically belong to the earth alone as to the rest of the universe, since the same appear-ances would prevail as much in the one situation as in the other. Thus it is that Aristotle and Ptolemy, who thoroughly understood this consideration, in their attempt to prove the earth immovable do not argue against any other motion than this diurnal one, though Aristotle does drop a hint against another motion ascribed to it by an ancient writer,† of which we shall speak in the proper place.

SAGR. I am quite convinced of the force of your argument, but it raises a question for me from which I do not know how to free myself, and it is this: Copernicus attributed to the earth another motion than the diurnal. By the rule just affirmed, this ought to remain imperceptible to all observations on the earth, but be visible in the rest of the universe. It seems to me that one may deduce as a necessary consequence either that he was grossly mistaken in assigning to the earth a motion corresponding to no appearance in the heavens generally, or that if the correspond-

ent motion does exist, then Ptolemy was equally at fault in not
explaining it away, as he explained away the other.

SALV. This is very reasonably questioned, and when we come to
treat of the other movement you will see how greatly Copernicus
surpassed Ptolemy in acuteness and penetration of mind by
seeing what the latter did not — I mean the wonderful correspondence with which such a movement is reιected in all the
other heavenly bodies. But let us postpone this for the present
and return to the first consideration, with respect to which I shall
set forth, commencing with the most general things, those reasons
which seem to favor the earth's motion, so that we may then
hear their refutation from Simplicio.

First, let us consider only the immense bulk of the starry
sphere in contrast with the smallness of the terrestrial globe,
which is contained in the former so many millions of times. Now
if we think of the velocity of motion required to make a complete
rotation in a single day and night, I cannot persuade myself that
anyone could be found who would think it the more reasonable
and credible thing that it was the celestial sphere which did the
turning, and the terrestrial globe which remained fixed.

Why the diurnal
motion must
more probably
belong to the
earth than to the
rest of the
universe.

SAGR. If, throughout the whole variety of effects that could exist
in nature as dependent upon these motions, all the same consequences followed indifferently to a hairsbreadth from both positions, still my first general impression of them would be this: I
should think that anyone who considered it more reasonable for
the whole universe to move in order to let the earth remain fixed
would be more irrational than one who should climb to the top
of your cupola just to get a view of the city and its environs, and
then demand that the whole countryside should revolve around
him so that he would not have to take the trouble to turn his head.
Doubtless there are many and great advantages to be drawn
from the new theory and not from the previous one (which to
my mind is comparable with or even surpasses the above in
absurdity), making the former more credible than the latter. But
perhaps Aristotle, Ptolemy, and Simplicio ought to marshal their
advantages against us and set them forth, too, if such there are;
otherwise it will be clear to me that there are none and cannot
be any.

SALV. Despite much thinking about it, I have not been able to
find any difference, so it seems to me I have found that there can

Motion is non-
existent for
things equally
moved; so far as
it acts, it relates
to things
lacking it.

Proposition
taken by Aris-
totle from the
ancients but
altered by him.

First argument
for proving the
diurnal motion
to be the earth's.

be no difference; hence I think it vain to seek one further. For consider: Motion, in so far as it is and acts as motion, to that extent exists relatively to things that lack it; and among things which all share equally in any motion, it does not act, and is as if it did not exist. Thus the goods with which a ship is laden leaving Venice, pass by Corfu, by Crete, by Cyprus and go to Aleppo. Venice, Corfu, Crete, etc. stand still and do not move with the ship; but as to the sacks, boxes, and bundles with which the boat is laden and with respect to the ship itself, the motion from Venice to Syria is as nothing, and in no way alters their relation among themselves. This is so because it is common to all of them and all share equally in it. If, from the cargo in the ship, a sack were shifted from a chest one single inch, this alone would be more of a movement for it than the two-thousand-mile journey made by all of them together.

SIMP. This is good, sound doctrine, and entirely Peripatetic.

SALV. I should have thought it somewhat older. And I question whether Aristotle entirely understood it when selecting it from some good school of thought, and whether he has not, by altering it in his writings, made it a source of confusion among those who wish to maintain everything he said. When he wrote that everything which is moved is moved upon something immovable, I think he only made equivocal the saying that whatever moves, moves with respect to something motionless. This proposition suffers no difficulties at all, whereas the other has many.

SAGR. Please do not break the thread, but continue with the argument already begun.

SALV. It is obvious, then, that motion which is common to many moving things is idle and inconsequential to the relation of these movables among themselves, nothing being changed among them, and that it is operative only in the relation that they have with other bodies lacking that motion, among which their location is changed. Now, having divided the universe into two parts, one of which is necessarily movable and the other motionless, it is the same thing to make the earth alone move, and to move all the rest of the universe, so far as concerns any result which may depend upon such movement. For the action of such a movement is only in the relation between the celestial bodies and the earth, which relation alone is changed. Now if precisely the same effect follows whether the earth is made to move and the rest of the universe stay still, or the earth alone remains fixed while the whole universe shares one motion, who is going to believe that nature (which by general agreement does not act by means of many things when it can do so by means of few) has chosen to make an immense number of extremely large bodies move with inconceivable velocities, to achieve what could have been done by a moderate movement of one single body around its own center?

Nature does not act by means of many things when she can act by means of few.

SIMP. I do not quite understand how this very great motion is as nothing for the sun, the moon, the other planets, and the innumerable host of the fixed stars. Why do you say it is nothing for the sun to pass from one meridian to the other, rise above this horizon and sink beneath that, causing now the day and now the night; and for the moon, the other planets, and the fixed stars to vary similarly?

SALV. Every one of these variations which you recite to me is nothing except in relation to the earth. To see that this is true, remove the earth; nothing remains in the universe of rising and setting of the sun and moon, nor of horizons and meridians, nor day and night, and in a word from this movement there will never originate any changes in the moon or sun or any stars you please, fixed or moving. All these changes are in relation to the earth, all of them meaning nothing except that the sun shows itself now over China, then to Persia, afterward to Egypt, to Greece, to France, to Spain, to America, etc. And the same holds for the moon and the rest of the heavenly bodies, this effect taking place in exactly the same way if, without embroiling the biggest part of the universe, the terrestrial globe is made to revolve upon itself.

From the diurnal motion, no change originates in all the heavenly bodies; all changes may be referred to the earth.

And let us redouble the difficulty with another very great one, which is this. If this great motion is attributed to the heavens, it has to be made in the opposite direction from the specific motion of all the planetary orbs, of which each one incontrovertibly has its own motion from west to east, this being very gentle and moderate, and must then be made to rush the other way; that is, from east to west, with this very rapid diurnal motion. Whereas by making the earth itself move, the contrariety of motions is removed, and the single motion from west to east accommodates all the observations and satisfies them all completely.

Second confirmation that the diurnal motion is the earth's.

SIMP. As to the contrariety of motions, that would matter little, since Aristotle demonstrates that circular motions are not contrary to one another, and their opposition cannot be called true contrariety.

SALV. Does Aristotle demonstrate that, or does he just say it because it suits certain designs of his? If, as he himself declares, contraries are those things which mutually destroy each other, I cannot see how two movable bodies meeting each other along a circular line conflict any less than if they had met along a straight line.

SAGR. Please stop a moment. Tell me, Simplicio, when two knights meet tilting in an open field, or two whole squadrons, or two fleets at sea go to attack and smash and sink each other, would you call their encounters contrary to one another?

SIMP. I should say they were contrary.

SAGR. Then why are two circular motions not contrary? Being made upon the surface of the land or sea, which as you know is spherical, these motions become circular. Do you know what circular motions are† not contrary to each other, Simplicio? They are those of two circles which touch from the outside; one being turned, the other naturally moves the opposite way. But if one circle should be inside the other, it is impossible that their motions should be made in opposite directions without their resisting each other.

SALV. "Contrary" or "not contrary," these are quibbles about words, but I know that with facts it is a much simpler and more natural thing to keep everything with a single motion than to introduce two, whether one wants to call them contrary or opposite. But I do not assume the introduction of two to be impossible, nor do I pretend to draw a necessary proof from this; merely a greater probability. The improbability is shown for a

third time in the relative disruption of the order which we surely see existing among those heavenly bodies whose circulation is not doubtful, but most certain. This order is such that the greater orbits complete their revolutions in longer times, and the lesser

in shorter; thus Saturn, describing a greater circle than the other planets, completes it in thirty years; Jupiter revolves in its smaller one in twelve years, Mars in two; the moon covers its much smaller circle in a single month. And we see no less sensibly that of the satellites of Jupiter (*stelle Medicee*),† the closest one to that planet makes its revolution in a very short time, that is in about forty-two hours; the next, in three and a half days; the third in seven days and the most distant in sixteen. And this very harmonious trend will not be a bit altered if the earth is made to move on itself in twenty-four hours. But if the earth is desired to remain motionless, it is necessary, after passing from the brief period of the moon to the other consecutively larger ones, and ultimately to that of Mars in two years, and the greater one of Jupiter in twelve, and from this to the still larger one of Saturn, whose period is thirty years — it is necessary, I say, to pass on beyond to another incomparably larger sphere, and make this one finish an entire revolution in twenty-four hours. Now this is the minimum disorder that can be introduced, for if one wished to pass from Saturn's sphere to the stellar, and make the latter so much greater than Saturn's that it would proportionally be suited to a very slow motion of many thousands of years,† a much greater leap would be required to pass beyond that to a still larger one and then make that revolve in twenty-hour hours. But by giving mobility to the earth, order becomes very well observed among the periods; from the very slow sphere of Saturn one passes on to the entirely immovable fixed stars, and manages to escape a fourth difficulty necessitated by supposing the stellar sphere to be movable. This difficulty is the immense disparity between the motions of the stars, some of which would be moving very rapidly in vast circles, and others very slowly in little tiny circles, according as they are located farther from or closer to the poles. This is indeed a nuisance, for just as we see that all those bodies whose motion is undoubted move in large circles, so it would not seem to have been good judgment to arrange bodies in such a way that they must move circularly at immense distances from the center, and then make them move in little tiny circles.

Not only will the size of the circles and consequently the velocities of motion of these stars be very diverse from the orbits and motions of some others, but (and this shall be the fifth difficulty) the same stars will keep changing their circles and their velocities, since those which two thousand years ago were on the celestial equator, and which consequently described great circles with their motion, are found in our time to be many degrees distant, and must be made slower in motion and reduced to moving in smaller circles. Indeed, it is not impossible that a time will come when some of the stars which in the past have always been

Times of circulation of Jupiter's satellites.

A twenty-four-hour motion attributed to the highest sphere disorders the periods of the lower ones.

Fourth confirmation.

Great inequality between the movements of the fixed stars when their sphere is made movable.

Motions of the fixed stars now accelerated and again retarded if the stellar sphere is movable.

moving will be reduced, by reaching the pole, to holding fast, and then after that time will start moving once more; whereas all those stars which certainly do move describe, as I said, very large circles in their orbits and are unchangeably preserved in them.

For anyone who reasons soundly, the unlikelihood is increased — and this is the sixth difficulty — by the incomprehensibility of what is called the "solidity" of that very vast sphere in whose depths are firmly fixed so many stars which, without changing place in the least among themselves, come to be carried around so harmoniously with such a disparity of motions. If, however, the heavens are fluid (as may much more reasonably be believed) so that each star roves around in it by itself, what law will regulate their motion so that as seen from the earth they shall appear as if made into a single sphere? For this to happen, it seems to me that it is as much more effective and convenient to make them immovable than to have them roam around, as it is easier to count the myriad tiles set in a courtyard than to number the troop of children running around on them.

Finally, for the seventh objection, if we attribute the diurnal rotation to the highest heaven, then this has to be made of such strength and power as to carry with it the innumerable host of fixed stars, all of them vast bodies and much larger than the earth, as well as to carry along the planetary orbs despite the fact that the two move naturally in opposite ways. Besides this, one must grant that the element of fire and the greater part of the air are likewise hurried along, and that only the little body of the earth remains defiant and resistant to such power. This seems to

The earth, pen-
dant and bal-
anced in a fluid
medium, seems
unable to resist
the urgency of
the diurnal
motion.

me to be most difficult; I do not understand why the earth, a suspended body balanced on its center and indifferent to motion or to rest, placed in and surrounded by an enclosing fluid, should not give in to such force and be carried around too. We encounter no such objections if we give the motion to the earth, a small and trifling body in comparison with the universe, and hence unable to do it any violence.

SAGR. I am aware of some ideas whirling around in my own imagination which have been confusedly roused in me by these arguments. If I wish to keep my attention on the things about to be said, I shall have to try to get them in better order and to place the proper construction upon them, if possible. Perhaps it

will help me to express myself more easily if I proceed by interrogation. Therefore I ask Simplicio, first, whether he believes
that the same simple movable body can naturally partake of
diverse movements, or whether only a single motion suits it, this
being its own natural one?

SIMP. For a simple movable body there can be but a single mo- For a simple
tion, and no more, which suits it naturally; any others it can movable body
there is one nat-
possess only incidentally and by participation. Thus when a man ural motion
walks along the deck of a ship, his own motion is that of walking, alone; all others
are by participa-
while the motion which takes him to port is his by participation; tion.
for he could never arrive there by walking if the ship did not take
him there by means of its motion.

SAGR. Second, tell me about this motion which is communicated
to a movable body by participation, when it itself is moved by
some other motion different from that in which it participates.
Must this shared motion in turn reside in some subject, or can it
indeed exist in nature without other support?

SIMP. Aristotle answers all these questions for you. He tells you Motion does not
that just as there is only one motion for one movable body, so exist without a
movable subject.
there is but one movable body for that motion. Consequently no
motion can either exist or even be imagined except as inhering in
its subject.

SAGR. Now in the third place I should like you to tell me whether
you believe that the moon and the other planets and celestial
bodies have their own motions, and what these are.

SIMP. They have, and they are those motions in accordance with
which they run through the zodiac — the moon in a month, the
sun in a year, Mars in two, the stellar sphere in so many thousands. These are their own natural motions.

SAGR. Now as to that motion with which the fixed stars, and with
them all the planets, are seen rising and setting and returning to
the east every twenty-four hours. How does that belong to them?

SIMP. They have that by participation.

SAGR. Then it does not reside in them; and neither residing in
them, nor being able to exist without some subject to reside in,
it must be made the proper and natural motion of some other
sphere.

SIMP. As to this, astronomers and philosophers have discovered
another very high sphere, devoid of stars, to which the diurnal
rotation naturally belongs. To this they have given the name

primum mobile; this speeds along with it all the inferior spheres, contributing to and sharing with them its motion.

SAGR. But when all things can proceed in most perfect harmony without introducing other huge and unknown spheres; without other movements or imparted speedings; with every sphere having only its simple motion, unmixed with contrary movements, and with everything taking place in the same direction, as must be the case if all depend upon a single principle, why reject the means of doing this, and give assent to such outlandish things and such labored conditions?

SIMP. The point is to find a simple and ready means.

SAGR. This seems to me to be found, and quite elegantly. Make the earth the *primum mobile;* that is, make it revolve upon itself in twenty-four hours in the same way as all the other spheres. Then, without its imparting such a motion to any other planet or star, all of them will have their risings, settings, and in a word all their other appearances.

SIMP. The crucial thing is being able to move the earth without causing a thousand inconveniences.

SALV. All inconveniences will be removed as you propound them. Up to this point, only the first and most general reasons have been mentioned which render it not entirely improbable that the daily rotation belongs to the earth rather than to the rest of the universe. Nor do I set these forth to you as inviolable laws, but merely as plausible reasons. For I understand very well that one single experiment or conclusive proof to the contrary would suffice to overthrow both these and a great many other probable arguments. So there is no need to stop here; rather let us proceed ahead and hear what Simplicio answers, and what greater probabilities or firmer arguments he adduces on the other side.

SIMP. First I shall say some things in general about all these considerations taken together, and then get down to certain particulars.

It seems to me that you base your case throughout upon the greater ease and simplicity of producing the same effects. As to their causation, you consider the moving of the earth alone equal to the moving of all the rest of the universe except the earth, while from the standpoint of action, you consider the former much easier than the latter. To this I answer that it seems that way to me also when I consider my own powers, which are

One experiment or established proof vanquishes all probable reasons.

not finite merely, but very feeble. But with respect to the power of the Mover, which is infinite, it is just as easy to move the universe as the earth, or for that matter a straw. And when the power is infinite, why should not a great part of it be exercised rather than a small? From this it appears to me that the general argument is ineffective.

SALV. If I had ever said that the universe does not move because of any lack of power in the Mover, I should have been mistaken, and your correction would be opportune; I grant you that it is as easy for an infinite force to move a hundred thousand things as to move one. But what I have been saying was with regard not to the Mover, but only the movables; and not with regard to their resistance alone, which is certainly less for the earth than for the universe, but with regard to other particulars considered just now.

Next, as to your saying that a great part of an infinite power may better be exercised than a small part, I reply to you that one part of the infinite is not greater than another, when both are finite; nor can it be said of an infinite number that a hundred thousand is a greater part than two is, though the former is fifty thousand times as great as the latter. And if what is required in order to move the universe is a finite power, then even though this would be very large in comparison with what would be required to move the earth alone, nevertheless a greater part of the infinite power would not thereby be employed, nor would that which remained idle be less than infinite. Hence to apply a little more or less power for a particular effect is insignificant. Besides, the operations of such power do not have for their end and goal the diurnal movement alone, for there are many other motions of the universe that we know of, and there may be very many more unknown to us.

Giving our attention, then, to the movable bodies, and not questioning that it is a shorter and readier operation to move the earth than the universe, and paying attention to the many other simplifications and conveniences that follow from merely this one, it is much more probable that the diurnal motion belongs to the earth alone than to the rest of the universe excepting the earth. This is supported by a very true maxim of Aristotle's which teaches that *frustra fit per plura quod potest fieri per pauciora.*

SIMP. In referring to this axiom you have left out one little clause that means everything, especially for our present purposes. The detail left out is *aeque bene;* hence it is necessary to examine whether both assumptions can satisfy us *equally well* in every respect.

SALV. Finding out whether both positions satisfy us equally well will be included in the detailed examination of the appearances which they have to satisfy. For we have argued *ex hypothesi* up to now, and will continue to argue so, assuming that both positions are equally adapted to the fulfillment of all the appearances. So I suspect that this detail which you declare to have been omitted by me was rather superfluously added by you. Saying "equally well" names a relation, which necessarily requires at least two terms, one thing not being capable of being related to itself; one cannot say, for example, that quiet is equally good with quiet. Therefore to say: "It is pointless to use many to accomplish what may be done with fewer" implies that what is to be done must be the same thing, and not two different things. And because the same thing cannot be said to be equally well done with itself, the addition of the phrase "equally well" is superfluous, and a relation with only one term.

In the axiom
*Frustra fit per
plura* etc., it is
superfluous to
add *aeque bene.*

SAGR. If we do not want to repeat what happened yesterday, please get back to the point; and you, Simplicio, begin producing those difficulties that seem to you to contradict this new arrangement of the universe.

SIMP. The arrangement is not new; rather, it is most ancient, as is shown by Aristotle refuting it, the following being his refutations:†

Aristotle's rea-
sons for the earth
being at rest.

"First, whether the earth is moved either in itself, being placed in the center, or in a circle, being removed from the center, it must be moved with such motion by force, for this is not its natural motion. Because if it were, it would belong also to all its particles. But every one of them is moved along a straight line toward the center. Being thus forced and preternatural, it cannot be everlasting. But the world order is eternal; therefore, etc.

"Second, it appears that all other bodies which move circularly lag behind, and are moved with more than one motion, except the *primum mobile.* Hence it would be necessary that the earth be moved also with two motions; and if that were so, there would have to be variations in the fixed stars. But such are not to be seen; rather, the same stars always rise and set in the same place without any variations.

"Third, the natural motion of the parts and of the whole is toward the center of the universe, and for that reason also it rests therein." He then discusses the question whether the motion of the parts is toward the center of the universe or merely toward that of the earth, concluding that their own tendency is to go toward the former, and that only accidentally do they go toward the latter, which question was argued at length yesterday.

Finally he strengthens this with a fourth argument taken from experiments with heavy bodies which, falling from a height, go perpendicularly to the surface of the earth. Similarly, projectiles thrown vertically upward come down again perpendicularly by the same line, even though they have been thrown to immense height. These arguments are necessary proofs that their motion is toward the center of the earth, which, without moving in the least, awaits and receives them.

He then hints at the end that astronomers adduce other reasons in confirmation of the same conclusions — that the earth is in the center of the universe and immovable. A single one of these is that all the appearances seen in the movements of the stars correspond with this central position of the earth, which correspondence they would not otherwise possess. The others, adduced by Ptolemy and other astronomers, I can give you now if you like; or after you have said as much as you want to in reply to these of Aristotle.

SALV. The arguments produced on this matter are of two kinds. Some pertain to terrestrial events without relation to the stars, and others are drawn from the appearances and observations of celestial things. Aristotle's arguments are drawn mostly from the things around us, and he leaves the others to the astronomers. Hence it will be good, if it seems so to you, to examine those taken from earthly experiments, and thereafter we shall see to the other sort. And since some such arguments are adduced by Ptolemy, Tycho, and other astronomers and philosophers, in addition to their accepting, confirming, and supporting those of Aristotle, these may all be taken together in order not to have to give the same or similar answers twice. Therefore, Simplicio, present them, if you will; or, if you want me to relieve you of that burden, I am at your service.

SIMP. It will be better for you to bring them up, for having given

Arguments of two kinds on the question of motion or rest for the earth.

Arguments of Ptolemy, Tycho, and others besides Aristotle's.

First argument,
taken from
heavy bodies
falling from on
high.

Confirmation by
the example of a
body falling
from the top of
a ship's mast.

Second argu-
ment, taken from
a projectile
thrown to a
great height.

Third argument,
taken from can-
non shots to east
and west.

them greater study you will have them readier at hand, and in great number too.

SALV. As the strongest reason of all is adduced that of heavy bodies, which, falling down from on high, go by a straight and vertical line to the surface of the earth. This is considered an irrefutable argument for the earth being motionless. For if it made the diurnal rotation, a tower from whose top a rock was let fall, being carried by the whirling of the earth, would travel many hundreds of yards to the east in the time the rock would consume in its fall, and the rock ought to strike the earth that distance away from the base of the tower. This effect they support with another experiment, which is to drop a lead ball from the top of the mast of a boat at rest, noting the place where it hits, which is close to the foot of the mast; but if the same ball is dropped from the same place when the boat is moving, it will strike at that distance from the foot of the mast which the boat will have run during the time of fall of the lead, and for no other reason than that the natural movement of the ball when set free is in a straight line toward the center of the earth. This argument is fortified with the experiment of a projectile sent a very great distance upward; this might be a ball shot from a cannon aimed perpendicular to the horizon. In its flight and return this consumes so much time that in our latitude the cannon and we would be carried together many miles eastward by the earth, so that the ball, falling, could never come back near the gun, but would fall as far to the west as the earth had run on ahead.

They add moreover the third and very effective experiment of shooting a cannon ball point-blank† to the east, and then another one with equal charge at the same elevation to the west; the shot toward the west ought to range a great deal farther out than the other one to the east. For when the ball goes toward the west, and the cannon, carried by the earth, goes east, the ball ought to strike the earth at a distance from the cannon equal to the sum of the two motions, one made by itself to the west, and the other by the gun, carried by the earth, toward the east. On the other hand, from the trip made by the ball shot toward the east it would be necessary to subtract that which was made by the cannon following it. Suppose, for example, that the journey made by the ball in itself was five miles and that the earth in that latitude traveled three miles during the flight of the ball; in the shot toward the west, the ball would fall to earth eight miles distant from the gun — that is, its own five toward the west and the gun's three to the east. But the shot toward the east would range no further than two miles, which is all that remains after subtracting from the five of the shot the three of the gun's motion toward the same place. Now experiment shows the shots to fall equally; therefore the cannon is motionless, and consequently the earth is, too. Not only this, but shots to the south or north likewise confirm the stability of the earth; for they would never hit the mark that one had aimed at, but would always slant toward the west because of the travel that would be made toward the east by the target, carried by the earth while the ball was in the air. And not merely shots along the meridians, but even those made to the east or west would not range truly; for the easterly shots would carry high and the westerly low whenever they were aimed point-blank. Since the shots in both directions take the path of a tangent — that is, a line parallel to the horizon — and the horizon is always falling away to the east and rising in the west if the diurnal motion belongs to the earth (which is why the eastern stars appear to rise and the western stars to set), it follows that the target to the east would drop away under the shot, wherefore the shot would range high, and the rising of the western target would make the shot to the west low. Hence in no direction would shooting ever be accurate; and since experience is contrary to this, it must be said that the earth is immovable.

SIMP. Oh, these are excellent arguments, to which it will be impossible to find a valid answer.

SALV. Perhaps they are new to you?

SIMP. Yes, indeed, and now I see with how many elegant experiments nature graciously wishes to aid us in coming to the recognition of the truth. Oh, how well one truth accords with another, and how all coöperate to make themselves indomitable!

SAGR. What a shame there were no cannons in Aristotle's time! With them he would indeed have battered down ignorance, and spoken without the least hesitation concerning the universe.

SALV. It suits me very well that these arguments are new to you, for now you will not remain of the same opinion as most Peripatetics, who believe that anyone who departs from Aristotle's doctrine must therefore have failed to understand his proofs. But you will certainly see further novelties; you will hear the

The argument confirmed by shots toward north and south.

The same confirmed by shots to east and west.

Followers of
Copernicus not
motivated by ig-
norance of the
opposing reasons.

followers of the new system producing observations, experiments, and arguments against it more forcible than those adduced
by Aristotle and Ptolemy and the other opponents of the same
conclusions. Thus you will become assured that it is not through
ignorance or inexperience that they have learned to adhere to
such opinions.

SAGR. This is the time for me to tell you a few of the things that
happened to me when I first began to hear these opinions spoken
of. I was then a youth who had scarcely finished the course in
philosophy, giving this up in order to apply myself to other activities. It happened that a certain foreigner from Rostock,

Christian
Wursteisen lec-
tured about the
opinions of Co-
pernicus, and
what happened.

whose name I believe was Christian Wursteisen, a supporter of
the Copernican opinion, arrived in these parts and gave two or
three lectures in an academy on this subject. He had a throng
of hearers, more from the novelty of the subject than for any
other reason, I think. I did not attend them, having formed a
definite impression that this opinion could be nothing but solemn
foolery. Later, asking about it from some who had gone, I heard
them all making a joke of it except one, who told me that the
matter was not entirely ridiculous. Since I considered this person
an intelligent man and rather conservative, I was sorry that I
had not gone; and from then on, as I happened from time to time
to meet anyone who held the Copernican opinion, I asked him
whether he had always believed in it. Among all the many whom
I questioned, I found not a single one who did not tell me that
he had long been of the contrary opinion, but had come over to
this one, moved and persuaded by the force of its arguments.
Examining them one by one then, to see how well they had
mastered the arguments on the other side, I found them all to
have these ready at hand, so that I could not truly say that they
had forsaken that position out of ignorance or vanity or, so to

All followers of
Copernicus had
previously been
against this opin-
ion; followers of
Aristotle and
Ptolemy never
held the
contrary.

speak, to show off their cleverness. On the other hand, so far as
I questioned the Peripatetics and the Ptolemaics (for out of
curiosity I asked many of them) how much they had studied
Copernicus's book, I found very few who had so much as seen
it, and none who I believed understood it. Moreover I tried to
find out from the followers of the Peripatetic doctrine whether
any of them had ever held the other opinion, and likewise found
none.

From this, considering that everyone who followed the opinion of Copernicus had at first held the opposite, and was very well informed concerning the arguments of Aristotle and Ptolemy, and that on the other hand none of the followers of Ptolemy and Aristotle had been formerly of the Copernican opinion and had left that to come round to Aristotle's view — considering these things, I say, I commenced to believe that one who forsakes an opinion which he imbibed with his milk and which is supported by multitudes, to take up another that has few followers and is rejected by all the schools and that truly seems to be a gigantic paradox, must of necessity be moved, not to say compelled, by the most effective arguments. This made me very curious to get to the bottom of this matter. And I consider it great fortune to have met you two, from whom without any trouble I can hear everything that has been said — perhaps all that can be said — on this subject, certain that I ought by virtue of your reasonings to be lifted out of doubt and put into a position of certainty.

SIMP. Nevertheless your opinion and your hope may be mistaken, for in the end you may find yourself more confused than ever.

SAGR. It seems to me impossible for that to happen.

SIMP. Why not? I am good evidence myself; for the farther on this goes, the more confused I become.

SAGR. This is a sign that those arguments which have until now seemed conclusive to you, and which seemed to give you assurance of the correctness of your opinion, are beginning to change their aspect in your mind; by degrees they are allowing you to incline, if not pass over, to the contrary one. But I, who up to the present have been quite undecided, am very confident that I shall arrive at satisfaction and assurance, nor will you yourself contradict me in this if you will but hear what it is that gives me hope.

SIMP. I shall be glad to hear it, and it would please me no less to have it work upon me in the same way.

SAGR. Then be so good as to answer a few questions. Tell me first, Simplicio: Is not this the conclusion we are seeking to understand — whether it should be held with Aristotle and Ptolemy that the earth alone remains fixed in the center of the universe while all the celestial bodies move, or on the other hand that the stellar sphere remains fixed with the sun in its center, the earth being located elsewhere and having the motions which appear to be those of the sun and the fixed stars?

SIMP. These are the conclusions about which we are debating.

SAGR. Are not these two conclusions such that one must needs be true, and the other false?

SIMP. Such they are; we are in a dilemma, one side of which must necessarily be true and the other false. For between motion and rest, which are contradictories, there is no middle ground (as if one might say the earth neither moves nor stands still; the sun and the stars do not move and do not stand still).

SAGR. What kind of things are the earth, the sun, and the stars in nature? Are they trifling things, or important?

SIMP. They are principal bodies; most noble, integral parts of the universe; very vast, and most important.

Motion and rest are principal events in nature.

SAGR. And what kind of natural events are motion and rest?

SIMP. So great and basic that nature itself is defined by them.

SAGR. So that moving eternally and being completely immovable are two very important conditions in nature, show the very greatest dissimilarity, and are the main attributes of the chief bodies in the universe. Consequently from them only the most different results can follow.

SIMP. This is surely so.

SAGR. Now answer me on one other point. Do you believe that in dialectics, in rhetoric, in physics, metaphysics, mathematics, or finally in the generality of reasonings, there are arguments sufficiently powerful and demonstrative to persuade anyone of false no less than true conclusions?

Errors cannot be demonstrable as are truths.

SIMP. By no means. Rather, I take it to be definite and certain that for the proof of a true and necessary conclusion there are in nature not merely one but very many powerful demonstrations, and that such a proposition can be discussed and turned about and subjected to thousands of comparisons without ever falling into any absurdity, and that the more any sophist wants to becloud it, the clearer its certainty will always become. I believe on the other hand that to make a false proposition appear true and convincing, nothing can be adduced but fallacies, sophisms, paralogisms, quibbles, and silly inconsistent arguments full of pitfalls and contradictions.

For proving true conclusions there may be many conclusive arguments, but not for false ones.

SAGR. Very well. Eternal motion and permanent rest are such important events in nature and so very different from each other that only the most diverse consequences can depend upon them, especially when applied to such vast and significant bodies in the universe as the sun and the earth. And it is impossible that one of two contradictory propositions should not be true and the other false. Now if it is further impossible to adduce in proof of the false proposition anything but fallacies, while the true one may be proved by all manner of conclusive and demonstrative arguments, how could you suppose that whichever one of you approaches me in support of the true proposition would not have me convinced? I should have to be stupid indeed, warped in judgment, thick-witted, and blind to reason, not to distinguish light from darkness, jewels from coals, truth from falsity.

SIMP. I tell you, as I have told you on other occasions, that the greatest master there has been for teaching the recognition of sophisms, paralogisms, and other fallacies is Aristotle, who in this particular can never be mistaken.

SAGR. You are only angry that Aristotle cannot speak; yet I tell you that if Aristotle were here he would either be convinced by us or he would pick our arguments to pieces and persuade us with better ones. For look: Did not you yourself, upon hearing the experiments with cannons described, understand and admire them, and confess them more conclusive than Aristotle's arguments? Yet I do not hear Salviati, who put them forward and who has surely examined them and explored them minutely, confess himself persuaded by them, nor even by others of still greater force which he intimates that he is about to deliver to us. And I do not know upon what basis you accuse Nature of having been for many ages in her second childhood, having forgotten how to produce any reflective thinkers except those who make themselves slaves of Aristotle and have to think with his brain and see with his eyes.

But let us hear the rest of the arguments favorable to his opinion so that we may proceed with their testing, refining them in the crucible and weighing them in the assayer's balance.

SALV. Before going further I must tell Sagredo that I act the part of Copernicus in our arguments and wear his mask. As to the internal effects upon me of the arguments which I produce in his favor, I want you to be guided not by what I say when we are in the heat of acting out our play, but after I have put off the costume, for perhaps then you shall find me different from what you saw of me on the stage.

Now let us proceed. Ptolemy and his followers produce another experiment like that of the projectiles, and it pertains to things

Aristotle would either unravel the arguments or change his opinion.

which, separated from the earth, remain in the air a long time, such as clouds and birds in flight. Since of these it cannot be said that they are carried by the earth, as they do not adhere to it, it does not seem possible that they could keep up with its swiftness; rather, it ought to look to us as if they were being moved very rapidly westward. If we, carried by the earth, pass along our parallel (which is at least sixteen thousand miles long) in twenty-four hours, how could the birds keep up on such a course? Whereas we see them fly east just as much as west or any other direction, without any detectable difference.

Argument taken
from the wind
that seems to
strike us when
we are riding
horseback.

Besides this, if, when we travel on horseback, we feel the air strike rather strongly upon our faces, then what an east wind should we not perpetually feel when being borne in such a rapid course against the air! Yet no such effect is felt.

Argument taken
from the fact
that whirling has
power to extrude
and scatter
things.

Here is another very ingenious argument taken from certain experiences. Circular motion has the property of casting off, scattering, and driving away from its center the parts of the moving body, whenever the motion is not sufficiently slow or the parts not too solidly attached together. If, for example, we should very rapidly spin one of those great treadmills with which massive weights are moved by one or more men walking within them (such as huge stones used in mangles, or barges being dragged across the land from one waterway to another), then if the parts of this rapidly turned wheel were not very solidly joined, it would all come apart. Or, if many rocks or other heavy materials were strongly attached to its external surface, they would not be able to resist the impetus, and it would scatter them with great force to various places far from the wheel, and accordingly from its center. If, then, the earth were to be moved with so much greater a velocity, what weight, what tenacity of lime or mortar would hold rocks, buildings, and whole cities so that they would not be hurled into the sky by such precipitous whirling? And men and beasts, none of which are attached to the earth; how would they resist such an impetus? Whereas on the contrary, we see these and the much less resistant pebbles, sand, and leaves reposing quietly upon the earth, and even falling back upon it with very slow motion.

Here, Simplicio, are the very potent arguments taken, so to speak, from terrestrial things. There remain those of the other kind; that is, those with relation to celestial appearances, which arguments tend still more to show that the earth is in the center of the universe, and consequently deprive it of the annual motion around that center as attributed to it by Copernicus. These being of rather a different nature, they can be brought forth after we have judged the strength of those already propounded.

SAGR. Well, what do you say, Simplicio? Does it seem to you that Salviati understands and knows how to explain the Ptolemaic and the Aristotelian arguments? Do you think any Peripatetic understands the Copernican proofs so well?

SIMP. Had I not formed from previous arguments such a high opinion of Salviati's soundness of learning and Sagredo's sharpness of wit, with their kind permission I should wish to leave without hearing any more, as it would appear to me an impossible feat to contradict such palpable experiences. And without hearing any more, I should like to cling to my old opinion; for it seems to me that if, indeed, it is false, it may be excused on the grounds of its being supported by so many arguments of such great probability. If these are fallacies, what true demonstrations were ever more elegant?

SAGR. Yet we had better listen to Salviati's answers, which if true must be even more beautiful; infinitely more beautiful, and the others extremely ugly, if that metaphysical proposition is correct which says that the true and the beautiful are one and the same, as are likewise the false and the ugly. Therefore, Salviati, let us not delay a moment more.

The true and the beautiful are the same, and so are the false and the ugly.

SALV. If I remember correctly, Simplicio's first argument was this: The earth cannot move circularly, because such a motion would be a forced one and therefore not perpetual. The reason that it would be forced was that if it were natural, the earth's parts would also naturally move in rotation, which is impossible because the nature of these parts is to be moved downward in a straight line.

To this I reply that I should have liked it better if Aristotle had made himself clearer when he said, "The parts would also be moved circularly," since this "being moved circularly" can be understood in two ways. One is that every particle separated from the whole would move circularly around its own center, describing its tiny circlets. The other is that the whole globe being moved around its center in twenty-four hours, the parts would also revolve around the same center in twenty-four hours.

Reply to the first argument of Aristotle.

The first would be a piece of nonsense no less than if one were to say that every part of the circumference of a circle had to be a circle, or even that since the earth is spherical every part of the earth must be a ball, because that is required by the maxim *eadem est ratio totius et partium.* But if he meant the other—that in imitation of the whole the parts would naturally move around the center of the whole globe in twenty-four hours — I say that that is precisely what they do, and that it is up to you, as Aristotle's representative, to prove that they do not.

SIMP. This is proved by Aristotle in the same place, when he says that the natural motion of the parts is along straight lines toward the center of the universe; therefore circular motion cannot naturally belong to them.

SALV. But do you not see that in the same words there is also the refutation of this reply?

SIMP. How? Where?

SALV. Does he not say that a circular motion for the earth would be forced, and therefore not eternal? And that this is absurd, since the world order is eternal?

SIMP. That is what he says.

SALV. But if that which is forced cannot be eternal, then by the converse that which cannot be eternal cannot be natural;[†] but there is no way for the earth's downward motion to be eternal, and so much the less can it be natural, nor can any motion be natural to it which cannot be eternal to it. But if we make the earth circularly movable, this can be eternal to it and to its parts, and therefore natural.

SIMP. Straight motion is most natural to the parts of the earth, and to them it is eternal, nor will it ever happen that they are not moved straight, always understanding that impediments are removed.

SALV. You are quibbling, Simplicio, and I should like to see you freed from the equivocation. So tell me, do you believe that a ship which is going to Palestine from the Straits of Gibraltar could eternally navigate toward that country, running always the same course?

SIMP. Certainly not.

SALV. And why not?

SIMP. Because that voyage is restricted and bounded by the Gates of Hercules and the shore of Palestine; and the distance

What is forced cannot be eternal, and what cannot be eternal cannot be natural.

being bounded, it is covered in a finite time, unless one wishes by turning back in the opposite direction to return and repeat the same voyage. But that would be an interrupted and not a continuous motion.

SALV. A perfectly correct reply. But how about the trip from the Straits of Magellan through the Pacific Ocean, the Straits of Molucca, around the Cape of Good Hope, from there to the original straits and again through the Pacific, and so on? Do you believe that this could be perpetual?

SIMP. It could be, because this is a circulation which returns upon itself; by repeating it an infinite number of times it could be perpetuated without any interruption.

SALV. Then on this voyage a ship could keep on navigating for all eternity.

SIMP. It could if the ship were indestructible; but the ship being dissolved, the journey would necessarily be terminated.

SALV. But in the Mediterranean, even if the ship were indestructible it could not on that account be sailed forever toward Palestine, such a voyage being bounded. So two things are required for a body moving without interruption to be moved eternally; one is that the motion shall by its nature be unbounded and infinite, and the other is that the moving body be likewise indestructible and eternal.

Two things required for motion to be perpetual: unbounded space and an indestructible movable body.

SIMP. All this is necessary.

SALV. Therefore of your own accord you have already confessed it to be impossible that any movable body is eternally moved in a straight line. For straight motion, whether you will have it be upward or downward, you yourself make bounded by the circumference and the center; hence although the movable body (that is, the earth) is eternal, yet straight motion being by its nature not eternal but bounded, the earth cannot naturally partake of it. Rather, as was said yesterday, Aristotle himself was obliged to make the earth's globe eternally fixed. When you say, then, that the parts of the earth would always be moved downward (all impediments removed), you equivocate egregiously; for on the contrary you must impede them, oppose them, and force them if you want them to be moved, since once they have fallen they have to be forcibly thrown up on high in order to fall again. And as to the impediments, these merely prevent them from getting to the center. If a tunnel were made that went past the cen-

Straight motion cannot be eternal, and hence cannot be natural to the earth.

ter, a clod would not pass beyond this center except in so far as it was carried by an impetus pushing it further, to return there afterward and finally come to rest there.

Hence as to maintaining that movement by a straight line suits or could suit naturally either the earth or any other movable body while the rest of the universe preserved its perfect order, give up this whole idea; if you will not grant the earth circular motion, exert your strength in upholding and defending its immobility.

SIMP. Regarding immobility, Aristotle's arguments (and even better, those others brought forward by you) seem to me so far to prove it conclusively. In my judgment prodigies will be needed to refute them.

Reply to the second argument.

SALV. Let us get on to the second argument, then, which was that those bodies of whose circular motion we are sure, excepting only the *primum mobile,* have more than one motion. Hence if the earth moves circularly, it must have two motions, from which there would follow alterations in the rising and setting of the fixed stars; but such are not seen to occur; therefore, etc. The simplest and most appropriate answer to this objection is in the argument itself, and it is Aristotle who puts it into our mouths. You cannot have failed to see this, Simplicio.

SIMP. I did not see this, and I do not see it now.

SALV. Astonishing! For it is there, and quite plain.

SIMP. By your leave, I shall have a look at the text.

SAGR. Let us have the text brought at once.

SIMP. I keep it always in my pocket. Here it is, and I know the exact place, which is in Book II of *De Caelo,* at chapter 14. Here, paragraph 97: *Praeterea, omnia quae feruntur latione circulari, subdeficere videntur, ac moveri pluribus una latione, praeter primam sphaeram: quare et Terram necessarium est, sive circa medium, sive in medio posita feratur, duabus moveri lationibus. Si autem hoc acciderit, necessarium est fieri mutationes ac conversiones fixorum astrorum. Hoc autem non videtur fieri, sed semper eadem apud eadem loca ipsius et oriuntur, et occidunt.* ("Again, everything that moves with the circular movement, except the first sphere, is observed to be passed, and to move with more than one motion. The earth, then, also, whether it move about the center or as stationary at it, must necessarily move with two motions. But if this were so, there would have to be passings and turnings of the fixed stars. Yet no such thing is observed. The same stars always rise and set in the same parts of the earth.") Now here I see no fallacy whatever, and it looks to me as if the argument is quite conclusive.

SALV. And for my part, this rereading has confirmed the fallacy in the argument and has in addition revealed another falsity. For look: Aristotle wants to reject two positions, or I should say two conclusions; one is that of those who, placing the earth in the center, make it move upon itself about its own center, while the other belongs to those who, placing the earth distant from the center, would make it move circularly around that center. Both these positions jointly he opposes with the same argument. Now I say that he errs in both the first and the second opposition, the error in the first being an equivocation or paralogism, and in the second a false inference.

Let us take the first position, which places the earth in the center and makes it movable upon itself about its own center. Let us confront this with Aristotle's objection, saying: "All the circularly moving bodies seem to lag behind and to move with more motions than one, except the first sphere (that is, the *primum mobile*); therefore the earth, moving around its own center and being placed at the center, must be moved with two motions and must fall behind; but if this were the case, the risings and settings of the fixed stars would have to vary, which is not seen to happen; therefore the earth is not moved etc." Here is the paralogism; in order to reveal it, argue with Aristotle in the following way: "You say, O Aristotle, that the earth, placed in the center, cannot move upon itself, because it would be necessary to attribute to it two motions. Therefore if it were not necessary to attribute more than a single motion to the earth, you would not hold it impossible that it might move with such a single motion. For you would have been restricting yourself to no purpose by resting the impossibility upon the plurality of motions if it also could not be moved with even a single one. Now of all the movables in the universe, you make only one move with just one motion, and all others with more than one. This movable you declare to be the first sphere; that is, the one by which all the stars, fixed and wandering, appear to be moved in unison from east to west. Then if the earth could be this prime sphere which, by moving with one motion alone, makes the stars appear to be moved from east to west, you would not deny it this motion. But those who say that the earth revolves upon itself at the center do not attribute to it any motion except that one by which all the stars appear to be moving from east to west, which amounts to the earth's being that first sphere which you yourself concede moves with but a single motion. Therefore, O Aristotle, if you want to prove anything, you must show that the earth, placed in the center, cannot move with even a single motion — or else that not even the first sphere can have but a single motion. Otherwise you commit the fallacy in your own syllogism, where it is obvious, at once denying and granting the same thing."

I come now to the second position, which is that of those who place the earth at a distance from the center and make it movable about the center; that is, who make it a planet or a wandering star. Aristotle's argument is directed against this position, and is conclusive as to form, but it errs as to content. For granting that the earth moves in such a way, with two motions (*lazione*), it does not necessarily follow that alterations must occur in the risings and settings of the fixed stars, as I shall explain in the proper place. And here I wish indeed to excuse Aristotle's error, and even to praise him for having hit upon the most subtle argument against the Copernican position which can be found. And if the objection is an acute and apparently cogent one, you shall see how much more subtle and ingenious is its solution; one not to be discovered by a mind less penetrating than that of Copernicus. From the difficulty of understanding it you will be able to infer how much greater was the difficulty of finding it in the first place. In the meantime, let us postpone the reply, which you will hear in due course after this same objection of Aristotle's has been repeated and moreover greatly strengthened for him.

Reply to the
third argument.

We pass on now to the third argument, also Aristotle's, to which there is no need to reply further, it having been adequately answered between yesterday and today. In this he objects that the natural motion of heavy bodies is in straight lines toward the center, and then he inquires whether it is toward the center of the earth or the center of the universe, concluding that it is naturally toward the center of the universe and only accidentally toward that of the earth.

We may go on therefore to the fourth, with which it will be proper to deal at length, this being founded upon that experience

from which most of the remaining arguments derive their force.
Aristotle says, then, that a most certain proof of the earth's
being motionless is that things projected perpendicularly upward
are seen to return by the same line to the same place from which
they were thrown, even though the movement is extremely high.
This, he argues, could not happen if the earth moved, since in the
time during which the projectile is moving upward and then
downward it is separated from the earth, and the place from
which the projectile began its motion would go a long way toward the east, thanks to the revolving of the earth, and the
falling projectile would strike the earth that distance away from
the place in question. Thus we can accommodate here the argument of the cannon ball as well as the other argument, used by
Aristotle and Ptolemy, of seeing heavy bodies falling from great
heights along a straight line perpendicular to the surface of the
earth. Now, in order to begin to untie these knots, I ask Simplicio
by what means he would prove that freely falling bodies go along
straight and perpendicular lines directed toward the center,
should anyone refuse to grant this to Aristotle and Ptolemy.

SIMP. By means of the senses, which assure us that the tower is
straight and perpendicular, and which show us that a falling
stone goes along grazing it, without deviating a hairsbreadth to
one side or the other, and strikes at the foot of the tower exactly
under the place from which it was dropped.

SALV. But if it happened that the earth rotated, and consequently
carried along the tower, and if the falling stone were seen to
graze the side of the tower just the same, what would its motion
then have to be?

SIMP. In that case one would have to say "its motions," for there
would be one with which it went from top to bottom, and another
one needed for following the path of the tower.

SALV. The motion would then be a compound of two motions;
the one with which it measures the tower, and the other with
which it follows it. From this compounding it would follow that
the rock would no longer describe that simple straight perpendicular line, but a slanting one, and perhaps not straight.

SIMP. I don't know about its not being straight, but I understand
well enough that it would have to be slanting, and different from
the straight perpendicular line it would describe with the earth
motionless.

SALV. Hence just from seeing the falling stone graze the tower,

Paralogism of
Aristotle and
Ptolemy in as-
suming as known
the very thing
in question.

you could not say for sure that it described a straight and perpendicular line, unless you first assumed the earth to stand still.

SIMP. Exactly so; for if the earth were moving, the motion of the
stone would be slanting and not perpendicular.

SALV. Then here, clear and evident, is the paralogism of Aristotle
and of Ptolemy, discovered by you yourself. They take as known
that which is intended to be proved.

SIMP. In what way? It looks to me like a syllogism in proper
form, and not a *petitio principii.*

SALV. In this way: Does he not, in his proof, take the conclusion
as unknown?

SIMP. Unknown, for otherwise it would be superfluous to prove it.

SALV. And the middle term; does he not require that to be known?

SIMP. Of course; otherwise it would be an attempt to prove
ignotum per aeque ignotum.

SALV. Our conclusion, which is unknown and is to be proved; is
this not the motionlessness of the earth?

SIMP. That is what it is.

SALV. Is not the middle term,† which must be known, the straight
and perpendicular fall of the stone?

SIMP. That is the middle term.

SALV. But wasn't it concluded a little while ago that we could
not have any knowledge of this fall being straight and perpendicular unless it was first known that the earth stood still? Therefore in your syllogism, the certainty of the middle term is drawn
from the uncertainty of the conclusion. Thus you see how, and
how badly, it is a paralogism.

SAGR. On behalf of Simplicio I should like, if possible, to defend
Aristotle, or at least to be better persuaded as to the force of your
deduction. You say that seeing the stone graze the tower is not
enough to assure us that the motion of the rock is perpendicular
(and this is the middle term of the syllogism) unless one assumes
the earth to stand still (which is the conclusion to be proved).
For if the tower moved along with the earth and the rock grazed
it, the motion of the rock would be slanting, and not perpendicular. But I reply that if the tower were moving, it would be impossible for the rock to fall grazing it; therefore, from the scrap-
ing fall is inferred the stability of the earth.

SIMP. So it is. For to expect the rock to go grazing the tower if
that were carried along by the earth would be requiring the rock
to have two natural motions; that is, a straight one toward the

center, and a circular one about the center, which is impossible. SALV. So Aristotle's defense consists in its being impossible, or at least in his having considered it impossible, that the rock might move with a motion mixed of straight and circular. For if he had not held it to be impossible that the stone might move both toward and around the center at the same time, he would have understood how it could happen that the falling rock might go grazing the tower whether that was moving or was standing still, and consequently he would have been able to perceive that this grazing could imply nothing as to the motion or rest of the earth.

Nevertheless this does not excuse Aristotle, not only because if he did have this idea he ought to have said so, it being such an important point in the argument, but also, and more so, because it cannot be said either that such an effect is impossible or that Aristotle considered it impossible. The former cannot be said because, as I shall shortly prove to you, this is not only possible but necessary; and the latter cannot be said either, because Aristotle himself admits that fire moves naturally upward in a straight line and also turns in the diurnal motion which is imparted by the sky to all the element of fire and to the greater part of the air. Therefore if he saw no impossibility in the mixing of straight-upward with circular motion, as communicated to fire and to the air up as far as the moon's orbit, no more should he deem this impossible with regard to the rock's straight-downward motion and the circular motion natural to the entire globe of the earth, of which the rock is a part.

SIMP. It does not look that way to me at all. If the element of fire goes around together with the air, this is a very easy and even a necessary thing for a particle of fire, which, rising high from the earth, receives that very motion in passing through the moving air, being so tenuous and light a body and so easily moved. But it is quite incredible that a very heavy rock or a cannon ball which is dropping without restraint should let itself be budged by the air or by anything else. Besides which, there is the very appropriate experiment of the stone dropped from the top of the mast of a ship, which falls to the foot of the mast when the ship is standing still, but falls as far from that same point when the ship is sailing as the ship is perceived to have advanced during the time of the fall, this being several yards when the ship's course is rapid.

SALV. There is a considerable difference between the matter of

Disparity be-
tween the fall of
a rock from the
top of a ship's
mast and from
the top of a
tower.

the ship and that of the earth under the assumption that the diurnal motion belongs to the terrestrial globe. For it is quite obvious that just as the motion of the ship is not its natural one, so the motion of all the things in it is accidental; hence it is no wonder that this stone which was held at the top of the mast falls down when it is set free, without any compulsion to follow the motion of the ship. But the diurnal rotation is being taken as the terrestrial globe's own and natural motion, and hence that of all its parts, as a thing indelibly impressed upon them by nature. Therefore the rock at the top of the tower has as its primary tendency a revolution about the center of the whole in twenty-four hours, and it eternally exercises this natural propensity no matter where it is placed. To be convinced of this, you have only to alter an outmoded impression made upon your mind, saying: "Having thought until now that it is a property of the earth's globe to remain motionless with respect to its center, I have never had any difficulty in or resistance to understanding that each of its particles also rests naturally in the same quiescence. Just so, it ought to be that if the natural tendency of the earth were to go around its center in twenty-four hours, each of its particles would also have an inherent and natural inclination not to stand still but to follow that same course."

And thus without encountering any obstacle you would be able to conclude that since the motion conferred by the force of the oars upon a boat, and through the boat upon all things contained in it, is not natural but foreign to them, then it might well be that this rock, once separated from the boat, is restored to its natural state and resumes its exercise of the simple tendency natural to it.

I might add that at least that part of the air which is lower than the highest mountains must be swept along and carried around by the roughness of the earth's surface, or must naturally follow the diurnal motion because of being a mixture of various terrestrial vapors and exhalations. But the air around a boat propelled by oars is not moved by them. So arguing from the boat to the tower has no inferential force. The rock coming from the top of the mast enters a medium which does not have the motion of the boat; but that which leaves the top of the tower finds itself in a medium which has the same motion as the entire terrestrial globe, so that far from being impeded by the air, it

The lower air, as
far as the top of
the highest
mountains, fol-
lows the motion
of the earth.

rather follows the general course of the earth with assistance from the air.

SIMP. I am not convinced that the air could impress its own motion upon a huge stone or a large ball of iron or lead weighing, say, two hundred pounds, as it might upon feathers, snow, and other very light bodies. In fact, I can see that a weight of that sort does not move a single inch from its place even when exposed to the wildest wind you please; now judge whether the air alone would carry it along.

Motion of the air is able to carry very light things with it, but not very heavy things.

SALV. There is an enormous difference between this experience of yours and our example. You make the wind arrive upon this rock placed at rest, and we are exposing to the already moving air a rock which is also moving with the same speed, so that the air need not confer upon it some new motion, but merely maintain — or rather, not impede — what it already has. You want to drive the rock with a motion foreign and unnatural to it; we, to conserve its natural motion in it. If you want to present a more suitable experiment, you ought to say what would be observed (if not with one's actual eyes, at least with those of the mind) if an eagle, carried by the force of the wind, were to drop a rock from its talons. Since this rock was already flying equally with the wind, and thereafter entered into a medium moving with the same velocity, I am pretty sure that it would not be seen to fall perpendicularly, but, following the course of the wind and adding to this that of its own weight, would move in a slanting path.

SIMP. It would be necessary to be able to make such an experiment and then to decide according to the result. Meanwhile, the result on shipboard confirms my opinion up to this point.

SALV. You may well say "up to this point," since perhaps in a very short time it will look different. And to keep you no longer on tenterhooks, as the saying goes, tell me, Simplicio: Do you feel convinced that the experiment on the ship squares so well with our purpose that one may reasonably believe that whatever is seen to occur there must also take place on the terrestrial globe?

SIMP. So far, yes; and though you have brought up some trivial disparities, they do not seem to me of such moment as to suffice to shake my conviction.

SALV. Rather, I hope that you will stick to it, and firmly insist that the result on the earth must correspond to that on the ship, so that when the latter is perceived to be prejudicial to your case you will not be tempted to change your mind.

You say, then, that since when the ship stands still the rock falls to the foot of the mast, and when the ship is in motion it falls apart from there, then conversely, from the falling of the rock at the foot it is inferred that the ship stands still, and from its falling away it may be deduced that the ship is moving. And since what happens on the ship must likewise happen on the land, from the falling of the rock at the foot of the tower one necessarily infers the immobility of the terrestrial globe. Is that your argument?

SIMP. That is exactly it, briefly stated, which makes it easy to understand.

SALV. Now tell me: If the stone dropped from the top of the mast when the ship was sailing rapidly fell in exactly the same place on the ship to which it fell when the ship was standing still, what use could you make of this falling with regard to determining whether the vessel stood still or moved?

SIMP. Absolutely none; just as by the beating of the pulse, for instance, you cannot know whether a person is asleep or awake, since the pulse beats in the same manner in sleeping as in waking.

SALV. Very good. Now, have you ever made this experiment of the ship?

SIMP. I have never made it, but I certainly believe that the authorities who adduced it had carefully observed it. Besides, the cause of the difference is so exactly known that there is no room for doubt.

SALV. You yourself are sufficient evidence that those authorities may have offered it without having performed it, for you take it as certain without having done it, and commit yourself to the good faith of their dictum. Similarly it not only may be, but must be that they did the same thing too — I mean, put faith in their predecessors, right on back without ever arriving at anyone who had performed it. For anyone who does will find that the experiment shows exactly the opposite of what is written; that is, it will show that the stone always falls in the same place on the ship, whether the ship is standing still or moving with any speed you please. Therefore, the same cause holding good on the earth as on the ship, nothing can be inferred about the earth's

The stone falling from the ship's mast strikes in the same place whether the ship moves or stands still.

motion or rest from the stone falling always perpendicularly to the foot of the tower.

SIMP. If you had referred me to any other agency than experiment, I think that our dispute would not soon come to an end; for this appears to me to be a thing so remote from human reason that there is no place in it for credulity or probability.

SALV. For me there is, just the same.

SIMP. So you have not made a hundred tests, or even one? And yet you so freely declare it to be certain? I shall retain my incredulity, and my own confidence that the experiment has been made by the most important authors who make use of it, and that it shows what they say it does.

SALV. Without experiment, I am sure that the effect will happen as I tell you, because it must happen that way; and I might add that you yourself also know that it cannot happen otherwise, no matter how you may pretend not to know it — or give that impression. But I am so handy at picking people's brains that I shall make you confess this in spite of yourself.

Sagredo is very quiet; it seemed to me that I saw him move as though he were about to say something.

SAGR. I was about to say something or other, but the interest aroused in me by hearing you threaten Simplicio with this sort of violence in order the reveal the knowledge he is trying to hide has deprived me of any other desire; I beg you to make good your boast.

SALV. If only Simplicio is willing to reply to my interrogation, I cannot fail.

SIMP. I shall reply as best I can, certain that I shall be put to little trouble; for of the things I hold to be false, I believe I can know nothing, seeing that knowledge is of the true and not of the false.

SALV. I do not want you to declare or reply anything that you do not know for certain. Now tell me: Suppose you have a plane surface as smooth as a mirror and made of some hard material like steel. This is not parallel to the horizon, but somewhat inclined, and upon it you have placed a ball which is perfectly spherical and of some hard and heavy material like bronze. What do you believe this will do when released? Do you not think, as I do, that it will remain still?

SIMP. If that surface is tilted?

SALV. Yes, that is what was assumed.

SIMP. I do not believe that it would stay still at all; rather, I am sure that it would spontaneously roll down.

SALV. Pay careful attention to what you are saying, Simplicio, for I am certain that it would stay wherever you placed it.

SIMP. Well, Salviati, so long as you make use of assumptions of this sort I shall cease to be surprised that you deduce such false conclusions.

SALV. Then you are quite sure that it would spontaneously move downward?

SIMP. What doubt is there about this?

SALV. And you take this for granted not because I have taught it to you — indeed, I have tried to persuade you to the contrary — but all by yourself, by means of your own common sense.

SIMP. Oh, now I see your trick; you spoke as you did in order to get me out on a limb, as the common people say, and not because you really believed what you said.

SALV. That was it. Now how long would the ball continue to roll, and how fast? Remember that I said a perfectly round ball and a highly polished surface, in order to remove all external and accidental impediments. Similarly I want you to take away any impediment of the air caused by its resistance to separation, and all other accidental obstacles, if there are any.

SIMP. I completely understood you, and to your question I reply that the ball would continue to move indefinitely, as far as the slope of the surface extended, and with a continually accelerated motion. For such is the nature of heavy bodies, which *vires acquirunt eundo;* and the greater the slope, the greater would be the velocity.

SALV. But if one wanted the ball to move upward on this same surface, do you think it would go?

SIMP. Not spontaneously, no; but drawn or thrown forcibly, it would.

SALV. And if it were thrust along with some impetus impressed forcibly upon it, what would its motion be, and how great?

SIMP. The motion would constantly slow down and be retarded, being contrary to nature, and would be of longer or shorter duration according to the greater or lesser impulse and the lesser or greater slope upward.

SALV. Very well; up to this point you have explained to me the events of motion upon two different planes. On the downward inclined plane, the heavy moving body spontaneously descends and continually accelerates, and to keep it at rest requires the use of force. On the upward slope, force is needed to thrust it along or even to hold it still, and motion which is impressed upon it continually diminishes until it is entirely annihilated. You say also that a difference in the two instances arises from the greater or lesser upward or downward slope of the plane, so that from a greater slope downward there follows a greater speed, while on the contrary upon the upward slope a given movable body thrown with a given force moves farther according as the slope is less.

Now tell me what would happen to the same movable body placed upon a surface with no slope upward or downward.

SIMP. Here I must think a moment about my reply. There being no downward slope, there can be no natural tendency toward motion; and there being no upward slope, there can be no resistance to being moved, so there would be an indifference between the propensity and the resistance to motion. Therefore it seems to me that it ought naturally to remain stable. But I forgot; it was not so very long ago that Sagredo gave me to understand that that is what would happen.†

SALV. I believe it would do so if one set the ball down firmly. But what would happen if it were given an impetus in any direction?

SIMP. It must follow that it would move in that direction.

SALV. But with what sort of movement? One continually accelerated, as on the downward plane, or increasingly retarded as on the upward one?

SIMP. I cannot see any cause for acceleration or deceleration, there being no slope upward or downward.

SALV. Exactly so. But if there is no cause for the ball's retardation, there ought to be still less for its coming to rest; so how far would you have the ball continue to move?

SIMP. As far as the extension of the surface continued without rising or falling.

SALV. Then if such a space were unbounded, the motion on it would likewise be boundless?† That is, perpetual?

SIMP. It seems so to me, if the movable body were of durable material.

SALV. That is of course assumed, since we said that all external and accidental impediments were to be removed, and any fragility on the part of the moving body would in this case be one of the accidental impediments.

Now tell me, what do you consider to be the cause of the ball moving spontaneously† on the downward inclined plane, but only by force on the one tilted upward?

SIMP. That the tendency of heavy bodies is to move toward the center of the earth, and to move upward from its circumference only with force; now the downward surface is that which gets closer to the center, while the upward one gets farther away.

SALV. Then in order for a surface to be neither downward nor upward, all its parts must be equally distant from the center. Are there any such surfaces in the world?

SIMP. Plenty of them; such would be the surface of our terrestrial globe if it were smooth, and not rough and mountainous as it is. But there is that of the water, when it is placid and tranquil.

SALV. Then a ship, when it moves over a calm sea, is one of these movables which courses over a surface that is tilted neither up nor down, and if all external and accidental obstacles were removed, it would thus be disposed to move incessantly and uniformly from an impulse once received?

SIMP. It seems that it ought to be.

SALV. Now as to that stone which is on top of the mast; does it not move, carried by the ship, both of them going along the circumference of a circle about its center? And consequently is there not in it an ineradicable motion, all external impediments being removed? And is not this motion as fast as that of the ship?

SIMP. All this is true, but what next?

SALV. Go on and draw the final consequence by yourself, if by yourself you have known all the premises.

SIMP. By the final conclusion you mean that the stone, moving with an indelibly impressed motion, is not going to leave the ship, but will follow it, and finally will fall at the same place where it fell when the ship remained motionless. And I, too, say that this would follow if there were no external impediments to disturb the motion of the stone after it was set free. But there are two such impediments; one is the inability of the movable body to split the air with its own impetus alone, once it has lost the force from the oars which it shared as part of the ship while it was on the mast; the other is the new motion of falling downward, which must impede its other, forward, motion.

SALV. As for the impediment of the air, I do not deny that to you, and if the falling body were of very light material, like a feather or a tuft of wool, the retardation would be quite considerable. But in a heavy stone it is insignificant, and if, as you yourself just said a little while ago, the force of the wildest wind is not enough to move a large stone from its place, just imagine how much the quiet air could accomplish upon meeting a rock which moved no faster than the ship! All the same, as I said, I concede to you the small effect which may depend upon such an impediment, just as I know you will concede to me that if the air were moving at the same speed as the ship and the rock, this impediment would be absolutely nil.

As for the other, the supervening motion downward, in the first place it is obvious that these two motions (I mean the circular around the center and the straight motion toward the center) are not contraries, nor are they destructive of one another, nor incompatible. As to the moving body, it has no resistance whatever to such a motion, for you yourself have already granted the resistance to be against motion which increases the distance from the center, and the tendency to be toward motion which approaches the center. From this it follows necessarily that the moving body has neither a resistance nor a propensity to motion which does not approach toward or depart from the center, and in consequence no cause for diminution in the property impressed upon it. Hence the cause of motion is not a single one which must be weakened by the new action, but there exist two distinct causes. Of these, heaviness attends only to the drawing of the movable body toward the center, and impressed force only to its being led around the center, so no occasion remains for any impediment.

SIMP. This argument is really very plausible in appearance, but actually it is offset by a difficulty which is hard to overcome. You have made an assumption throughout which will not lightly be granted by the Peripatetic school, being directly contrary to Aristotle. You take it as well known and evident that the projectile when separated from its origin retains the motion which was forcibly impressed upon it there. Now this impressed force is as detestable to the Peripatetic philosophy as is any transfer of an accidental property from one subject to another. In their philosophy it is held, as I believe you know, that the projectile

According to
Aristotle the
projectile is
moved not by
impressed force
but by the
medium.

is carried by the medium, which in the present instance is the air. Therefore if that rock which was dropped from the top of the mast were to follow the motion of the ship, this effect would have to be attributed to the air, and not to the impressed force; but you assume that the air does not follow the motion of the ship, and is quiet. Furthermore, the person letting the stone fall does not need to fling it or give it any impetus with his arm, but has only to open his hand and let it go. So the rock cannot follow the motion of the boat either through any force impressed upon it by its thrower or by means of any assistance from the air, and therefore it will remain behind.

SALV. It seems to me that what you are saying is that there is no way for the stone to be projected, not being thrown by anybody's arm.

SIMP. This motion cannot properly be called one of projection.

SALV. Then what Aristotle says about the motion of projectiles, the things moved by projection, and their movers is quite beside our purpose; and if it has nothing to do with us, why do you bring it up?

SIMP. I adduce it for the sake of that impressed force which you introduced and gave a name to, but which, since it does not exist in the world, cannot act at all, since *non entium nullae sunt operationes*. Hence the cause of motion must be attributed to the medium, not only for projectiles, but for all other motions that are not natural ones. Due consideration has not been given to this, so what has been said up to this point remains ineffective.

SALV. Patience; all in good time. Tell me: Seeing that your objection is based entirely upon the nonexistence of impressed force, then if I were to show you that the medium plays no part in the continuation of motion in projectiles after they are separated from their throwers, would you allow impressed force to exist? Or would you merely move on to some other attack directed toward its destruction?

SIMP. If the action of the medium were removed, I do not see how recourse could be had to anything else than the property impressed by the motive force.

SALV. It will be best, so as to get as far away as possible from any reason for arguing about it forever, to have you explain as clearly as you can just what the action of the medium is in maintaining the motion of the projectile.

SIMP. Whoever throws the stone has it in his hand; he moves his arm with speed and force; by its motion not only the rock but the surrounding air is moved; the rock, upon being deserted by the hand, finds itself in air which is already moving with impetus, and by that it is carried. For if the air did not act, the stone would fall from the thrower's hand to his feet.

SALV. And you are so credulous as to let yourself be persuaded of this nonsense, when you have your own senses to refute it and to learn the truth? Look here: A big stone or a cannon ball would remain motionless on a table in the strongest wind, according to what you affirmed a little while ago. Now do you believe that if instead this had been a ball of cork or cotton, the wind would have moved it?

SIMP. I am quite sure the wind would have carried it away, and would have done this the faster, the lighter the material was. For we see this in clouds being borne with a speed equal to that of the wind which drives them.

SALV. And what sort of thing is the wind?

SIMP. The wind is defined as merely air in motion.

SALV. So then the air in motion carries light materials much faster and farther than it does heavy ones?

SIMP. Certainly.

SALV. But if with your arm you had to throw first a stone and then a wisp of cotton, which would move the faster and the farther?

SIMP. The stone, by a good deal; the cotton would merely fall at my feet.

SALV. Well, if that which moves the thrown thing after it leaves your hand is only the air moved by your arm, and if moving air pushes light material more easily than heavy, why doesn't the cotton projectile go farther and faster than the stone one? There must be something conserved in the stone, in addition to any motion of the air. Besides, if two strings of equal length were suspended from that rafter, with a lead ball attached to the end of one and a cotton ball to the other, and then if both were drawn an equal distance from the perpendicular and set free, there is no doubt that each would move toward the perpendicular and, propelled by its own impetus, would go beyond that by a certain interval and afterward return. Which of these pendulums do you believe would continue to move the longer before stopping vertically?

Action of the medium in continuing the motion of the projectile.

Many experiments and arguments refute the cause assumed by Aristotle for the motion of projectiles.

SIMP. The lead ball would go back and forth a great many times; the cotton ball, two or three at most.

SALV. So that whatever the cause of that impetus and mobility, it is conserved longer in the heavy material than in the light. Now I come to the next point, and ask you why the air does not carry away the citron on that table right now?

SIMP. Because the air itself is not moving.

SALV. So the person who does the throwing must give the air that motion with which it subsequently moves the thing thrown. But since this force is incapable of being impressed (for you said that an accidental property cannot be made to pass from one subject to another), how can it go from the arm to the air? Or perhaps the arm and the air are not different subjects?

SIMP. The answer is that the air, being neither heavy nor light in its own domain, is so disposed as to receive every impulse very readily, and to conserve it, too.

SALV. Well, if the pendulums have just shown us that the less a moving body partakes of weight, the less apt it is to conserve motion, how can it be that the air, which has no weight at all in air, is the only thing that does conserve the motion acquired? I believe, and I know that you also believe at this moment, that no sooner does the arm stop than the air around it stops. Let us go into that room and agitate the air as much as possible with a towel; then, stopping the cloth, have a little candle flame brought immediately into the room, or set flying a bit of gold leaf in it, and you will see from the quiet wandering of either one that the air has been instantly restored to tranquillity. I could give you many experiments, but if one of these is not enough, the case is quite hopeless.

SAGR. What an incredible stroke of luck it is that when an arrow is shot against the wind, the slender thread of air driven by the bowstring goes along with the arrow! But there is another point of Aristotle's which I should like to understand, and I beg Simplicio to oblige me with an answer.

If two arrows were shot with the same bow, one in the usual way and one sideways — that is, putting the arrow lengthwise along the cord and shooting it that way — I should like to know which one would go the farther? Please reply, even though the question may seem to you more ridiculous than otherwise; forgive me for being, as you see, something of a blockhead, so that my speculations do not soar very high.

SIMP. I have never seen an arrow shot sideways, but I think it would not go even one-twentieth the distance of one shot point first.

SAGR. Since that is just what I thought, it gives me occasion to raise a question between Aristotle's dictum and experience. For as to experience, if I were to place two arrows upon that table when a strong wind was blowing, one in the direction of the wind and the other across it, the wind would quickly carry away the latter and leave the former. Now apparently the same ought to happen with two shots from a bow, if Aristotle's doctrine were true, because the one going sideways would be spurred on by a great quantity of air moved by the bowstring — as much as the whole length of the arrow — whereas the other arrow would receive the impulse from only as much air as there is in the tiny circle of its thickness. I cannot imagine the cause of such a disparity, and should like very much to know it.

SIMP. The cause is obvious to me; it is because the arrow shot point foremost has to penetrate only a small quantity of air, and the other has to cleave as much as its whole length.

SAGR. Oh, so when arrows are shot they have to penetrate the air? If the air goes with them, or rather if it is the very thing which conducts them, what penetration can there be? Do you not see that in such a manner the arrow would be moving faster than the air? Now what conferred this greater velocity upon the arrow? Do you mean to say that the air gives it a greater speed than its own?

You know perfectly well, Simplicio, that this whole thing takes place just exactly opposite to what Aristotle says, and that it is as false that the medium confers motion upon the projectile as it is true that it is this alone which impedes it. Once you understand this, you will recognize without any difficulty that when the air really does move, it carries the arrow along with it much better sideways than point first, because there is lots of air driving it in the former case and little in the latter. But when shot from the bow, since the air stands still, the sidewise arrow strikes against much air and is much impeded, while the other easily overcomes the obstacles of the tiny amount of air that opposes it.

SALV. How many propositions I have noted in Aristotle (meaning always in his science) that are not only wrong, but wrong in

The medium impedes rather than confers the motion of projectiles.

such a way that their diametrical opposites are true, as happens in this instance! But keeping to our purpose, I believe that Simplicio is convinced that from seeing the rock always fall in the same place, nothing can be guessed about the motion or stability of the ship. If what had been said previously was not enough, this experience concerning the medium can make the whole thing certain. From this experience it may be seen that, at most, the falling body might drop behind if it were made of light material and the air did not follow the ship's motion; but if the air were moving with equal speed, no imaginable difference could be found in this or in any other experiment you please, as I shall soon explain to you. Now if in this example no difference whatever appears, what is it that you claim to see in the stone falling from the top of the tower, where the rotational movement is not adventitious and accidental to the stone, but natural and eternal, and where the air as punctiliously follows the motion of the earth as the tower does that of the terrestrial globe? Do you have anything else to say, Simplicio, on this particular?

SIMP. No more, except that so far I do not see the mobility of the earth to be proved.

SALV. I have not claimed to prove it yet, but only to show that nothing can be deduced from the experiments offered by its adversaries as one argument for its motionlessness, as I believe I shall show of the others.

SAGR. Excuse me, Salviati, but before going on to the others let me bring up a certain difficulty that has been going round in my head while you were so patiently going into such detail with Simplicio on this ship experiment.

SALV. What we are here for is to discuss things, and it is good for everyone to raise his objections as they occur to him, for that is the road to knowledge. So speak up.

SAGR. If it is true that the impetus of the ship's motion remains indelibly impressed on the stone after it has separated from the mast, and that furthermore this motion occasions no hindrance or slowing in the straight-downward motion which is natural to the stone, then an effect of a remarkable nature must take place.

Let the ship be motionless and the fall of the stone from the mast take two pulse beats. Then cause the ship to move, and drop the same stone from the same place; from what has been said, it will still take two pulse beats to arrive at the deck. In

Remarkable phe-
nomenon in the
motion of
projectiles.

these two pulse beats the ship will have gone, say, twenty yards,
so that the actual motion of the stone will have been a diagonal
line much longer than the first straight and perpendicular one,
which was merely the length of the mast; nevertheless, it will
have traversed this distance in the same time. Now, assuming the
ship to be speeded up still more, so that the stone in falling must
follow a diagonal line very much longer still than the other,
eventually the velocity of the ship may be increased by any
amount, while the falling rock will describe always longer and
longer diagonals, and still pass over them in the same two pulse
beats. Similarly, if a perfectly level cannon on a tower were fired
parallel to the horizon, it would not matter whether a small
charge or a great one was put in, so that the ball would fall a
thousand yards away, or four thousand, or six thousand, or ten
thousand, or more; all these shots would require equal times,
and each time would be equal to that which the ball would have
taken in going from the mouth of the cannon to the ground if it
were allowed to fall straight down without any other impulse.

Now it seems a marvelous thing that in the same short time
of a straight fall from a height of, say, a hundred yards to the
ground, the same ball driven by powder could go now four hundred, now a thousand, again four thousand, or even ten thousand
yards, so that all shots fired point-blank would stay in the air for
an equal time.

SALV. This reflection is very beautiful by reason of its novelty,
and if the effect is true it is most remarkable. And I have no doubt
as to its correctness. Barring the accidental impediment from the
air, I consider it certain that if, when one ball left the cannon,
another one were allowed to fall straight down from the same
height, they would both arrive on the ground at the same instant,
even though the former would have traveled ten thousand yards
and the latter a mere hundred. Of course we are assuming the
surface of the earth to be perfectly level; to guarantee this, the
shots might be made over some lake. The impediment due to the
air would then be one of retarding the very great speed of
the shot.

Now if you are satisfied with this, let us get to the solutions
of the other arguments, since, so far as I know, Simplicio is persuaded of the uselessness of this first one taken from bodies
falling from heights.

SIMP. I do not feel that all my doubts are removed, but perhaps the fault is mine for not being as alert and quick-witted as Sagredo. It seems to me that if this motion which the stone shares while on top of the ship's mast were, as you said, conserved in it also after it is separated from the ship, then it would likewise be necessary for a ball dropped to earth by the rider of a galloping horse to continue to follow the horse's path without lagging behind. I do not believe that this effect is seen except when the rider throws the ball forcibly in the direction in which he is riding. Outside of that, I believe that it will remain where it strikes the ground.

SALV. I think you are much deceived, and I am sure that experience will show you on the contrary that the ball, having hit the ground, does run along with the horse and does not drop behind, except as the roughness and unevenness of the path impedes it. And the reason seems clear to me, too. For if you, standing still, were to throw the same ball along the ground, would it not continue the motion also after it was out of your hand? And the distance would be the longer according as the surface was the more even; on ice, for example, it would go a long way.

SIMP. No doubt it would, if I gave it an impetus with my arm; but in the other example it was assumed that the horseman merely let it fall.

SALV. That is what I want to have happen. When you throw it with your arm, what is it that stays with the ball when it has left your hand, except the motion received from your arm which is conserved in it and continues to urge it on? And what difference is there whether that impetus is conferred upon the ball by your hand or by the horse? While you are on horseback, doesn't your hand, and consequently the ball which is in it, move as fast as the horse itself? Of course it does. Hence upon the mere opening of your hand, the ball leaves it with just that much motion already received; not from your own motion of your arm, but from motion dependent upon the horse, communicated first to you, then to your arm, thence to your hand, and finally to the ball.

I should add that if the rider threw the ball in the direction opposite to his course, when it struck it would sometimes still follow the horse's route, and sometimes it would lie still on the ground; it would move away from him only if the motion received from the arm exceeded that of the rider in velocity. And it is folly to say, as some do, that a cavalryman can cast his javelin before him, pursue it on his horse, overtake it and recapture it. I say this is folly because in order to have the projectile return to his hand he would have to throw it straight up, in the same way as if he were standing still. Let his course be what you will, provided only that it is uniform; then unless the thing thrown is extremely light, it will always fall back into the thrower's hand no matter how high it is thrown.

SAGR. By this doctrine I am reminded of some curious problems about projectiles, the first of which must seem very strange to Simplicio. It is this: I say that it is possible for a ball merely dropped by someone moving very rapidly in any way, when it has arrived at the ground, not only to follow his course but to anticipate this somewhat. This problem is connected with the fact that a movable body thrown along the plane of the horizon may acquire a new velocity rather greater than that conferred upon it by the thrower.

Various curious problems concerning the motion of projectiles.

I have often observed this effect with astonishment when I have been watching people play with hoops (*ruzzole*).[†] These, after they have left the hand, are seen to go in the air with a certain velocity which is afterward greatly increased upon their arrival on the ground; and if in rolling they bump into some obstacle which makes them jump into the air, they are seen to go much more slowly; falling back to the ground, they are once more moved with greater speed. But what is strangest of all is that I have also seen that they not only always go faster on the ground than in the air, but that of two stretches both passed on the ground, the motion in the second is sometimes faster than that in the first. Now what would Simplicio say to that?

SIMP. I should say in the first place that I have not observed any such things; second, that I do not believe them; and then, in the third place, if you should assure me of them and show me proofs of them, that you would be a veritable demon.

SAGR. One like Socrates's,[†] though; not one from hell. But the showing depends upon you; I say to you that if one does not know the truth by himself, it is impossible for anyone to make him know it. I can indeed point out things to you, things being neither true nor false; but as for the true — that is, the necessary; that which cannot possibly be otherwise — every man of ordinary intelligence either knows this by himself or it is impossible for him ever to know it. And I am sure that Salviati holds this opinion too. Therefore I tell you that the causes in the present problem are known to you, but are perhaps not recognized as such.

SIMP. Let us not argue about that now; allow me to tell you that I neither know nor understand the things in question. Therefore see if you can satisfy me as to these problems.

SAGR. This first one depends upon another, which is this: Why does a hoop when rotated with a cord go much farther, and consequently much more forcibly, than one merely spun by hand?

SIMP. Aristotle also makes perplexing problems about such toys (*questi proietti*).

SALV. Yes, indeed, and very ingenious ones; especially that one about why round wheels roll better than square ones.†

SAGR. Now as to the reason for that, Simplicio, can't you make up your own mind about it without somebody else teaching it to you?

SIMP. Of course, of course; stop your sneering.

SAGR. You know the reason for this other one, too, just as well. Tell me, do you know that a moving thing stops when it is impeded?

SIMP. I know that it does if the impediment is sufficiently great.

SAGR. Do you know that it is a greater impediment for a moving body to have to move on the ground than in the air, the ground being rough and hard, and the air soft and yielding?

SIMP. Since I do know this, I know that the hoop will go faster in the air than on the ground, so that my knowledge is just the opposite of what you thought it was.

SAGR. Not so fast, Simplicio. Do you know that among the parts of a moving body which is turning round its center, movements in every direction are to be found? So that some parts go up, some go down, some go forward, and some backward?

SIMP. I know it, and Aristotle taught it to me.

SAGR. Please tell me by what kind of proof.

SIMP. Proofs from the senses.

SAGR. Then has Aristotle made you see what you would not have seen without him? Did he even lend you his eyes? You mean that Aristotle said it to you, made you notice it, reminded you of it; not that he taught it to you.

Well, then, when a hoop turns on itself without changing place, in a direction not parallel but vertical to the horizon, some of its parts go up and the opposite parts go down; the upper parts go in one direction and the lower parts in the other. Now picture to yourself a hoop which without changing place turns rapidly on itself and stays suspended in the air, and while rotating in this way, is dropped to the earth perpendicularly. Do you think that when it gets to the ground it will continue to revolve on itself without changing place, as at first?

SIMP. By no means.

SAGR. Well, what will it do?

SIMP. It will run quickly along the ground.

SAGR. In what direction?

SIMP. In that toward which its whirling carries it.

SAGR. There are two parts to its whirling; namely, the upper and the lower, which move contrary to one another; therefore you must say which it will obey. As to the ascending and descending parts, one will not give in to the other; the whole will neither go down, being impeded by the earth, nor up, because of its weight.

SIMP. The hoop will go rolling along the ground in the direction toward which its upper parts tend.

SAGR. And why not where the contrary parts tend; that is, those which touch the earth?

SIMP. Because the earth impedes those by the roughness of the contact; that is, by the very harshness of the ground. But the upper parts, which are in the thin and yielding air, are impeded little or not at all, and therefore the hoop will go in their direction.

SAGR. So that those parts underneath attach themselves, so to speak, to the earth, which holds them back, and only those parts above push on.

SALV. And accordingly if the hoop should fall on ice or some other polished surface, it would not run on so well, but might perhaps continue to turn on itself without acquiring any other forward motion.

SAGR. It is easily possible that this might follow; at least the hoop would not go rolling as fast as it would after falling upon a somewhat rough surface. But tell me, Simplicio, when the hoop is let fall spinning rapidly upon itself, why does it not go forward in the air also, as it does afterward when it is on the ground?

SIMP. Because having the air above and beneath it, neither of its parts have anything to attach themselves to; and with no more reason to go forward than backward, it falls plumb.

SAGR. So that just this whirling about itself, without other impetus, can propel the hoop very rapidly when it gets to the ground.

Now we come to the remainder. What is the effect upon the hoop of that cord which the hoop spinner has tied to his arm, and with which he drives the hoop after having wrapped this cord around it?

SIMP. This forces it to turn upon itself in order to get free of the cord.

SAGR. So then when the hoop arrives on the ground it is spinning upon itself thanks to the cord. Then is this not a cause in itself for the hoop being moved more rapidly on the ground than it could be in the air?

SIMP. Certainly it is, for in the air it had no other impulse than that of the spinner's arm; although it did also have its whirling, this, as has been said, does not propel it at all in the air. But upon its arrival on the ground, the progression due to whirling is added to the motion of the arm and its speed is redoubled. And already I understand quite well that the speed of the hoop will diminish when it skips into the air, because it lacks the aid of the rotation; but on falling back to the ground it recovers this, and resumes moving faster than it did in the air. It remains only for me to understand how in this second trip on the ground it goes faster than during the first, for thus it would move perpetually, always accelerating.

SAGR. I did not say without qualification that this second motion would be faster than the first, but that it could sometimes happen to be faster.

SIMP. That is what I am not satisfied about, and wish to hear.

SAGR. You know this also by yourself. Tell me, if you were to drop the hoop from your hand without spinning it, what would happen when it struck the ground?

SIMP. Nothing; it would remain there.

SAGR. Might it not happen that it would acquire motion upon hitting the ground? Think about it.

SIMP. Not unless we let it fall on some steep stone, as children do in playing a kind of marbles (*chiose*),[†] and falling slantwise on this it should acquire a turning motion with which it could continue to roll along the earth; otherwise I do not know how it could do anything but stay where it landed.

SAGR. That is just how it can acquire an added whirling. When the hoop, then, having skipped high up, falls back down, why may it not happen to hit on the slant of some rock stuck in the ground and tilted in the direction of its motion? Acquiring more rotation from such a landing, its motion may be redoubled and it may be made faster than on its first striking the earth.

SIMP. I see now that this could easily happen. And come to think of it, if the hoop were made to turn the other way upon its arrival at the ground, this would have the opposite effect; that is, the twist given to it would retard that which it had from the player.

SAGR. It would retard it, and sometimes it would stop it entirely if this twist were fast enough. And herein lies the solution of the effect achieved by expert tennis players† to their own advantage when they deceive their opponents by cutting the ball, as it is called. This consists in returning it with the racket slanted, in such a way that the ball takes on a spin contrary to its forward motion. It then follows that when it comes to earth, the rebound which would give the adversary the usual time to return it — for if it were not spinning, it would go toward him — seems dead, and the ball squashes itself to the ground, or bounces much less than usual, and breaks the timing of the return. This also explains what we see bowlers† do who play to get a wooden ball closest to a given mark. When they are playing on a rocky road full of obstacles which would make the ball deviate countless ways and not go toward the mark at all, in order to avoid all these they send the ball through the air as if they were playing quoits, instead of rolling it along the ground. But since in throwing the ball some spin is conferred upon it as it leaves the hand when the hand is held under the ball in the usual way, they make use of a trick of gripping the ball, holding the hand above and the ball below. Otherwise when the ball hit the ground near the mark it would run far beyond it, because of both the motion of throwing and that of spinning; but in this way a contrary spin is imparted to it upon its release, and it stops, or runs only a little further upon hitting the ground near the mark.

But to return to the main problem which was the occasion for these others arising, I say that it is possible for a person in very rapid motion to drop from his hand a ball which, alighting on the ground, will not merely follow his motion but will run ahead of it still faster. In order to see such an effect, I would have the course be that of a wagon, outside of which there would be fastened to one side a tilted board, with the lower part toward the horses and the higher part toward the back wheels. Now with the wagon going full speed, if someone in it lets a ball fall down the slope of this board, this will acquire its own spin by coming down rolling; and this, added to the motion received from the wagon, will carry the ball on the ground rather faster than the wagon. And if another board were provided, tilted the opposite way, it would be possible to modify the motion of the wagon in such a way that the ball going down this board would stop motionless when it hit the ground, and even sometimes run in the opposite direction to the wagon.

But we have been off the subject too long, and if Simplicio is satisfied with the solution of this first argument against the mobility of the earth, derived from vertically falling bodies, the rest of them may be taken up.

SALV. The digressions made up to this point are not so foreign to the matter in hand as to be called entirely apart from it. Moreover, from such things there result trains of reasoning awakened in the minds of not one of us alone, but all three. Besides, we are arguing for our own amusement, and are not obligated to any such strictness as one would be who was methodically treating a subject for professional reasons, with the intention of publishing it. I do not want this epic of ours to adhere so closely to poetic unity as to leave no room for episodes, for the introduction of which the slightest relevance ought to suffice. It should be almost as if we had met to tell stories, so that it is permitted for me to relate anything which hearing yours may call to my mind.

SAGR. That suits me perfectly. And since we are being so expansive it may be all right for me to ask you, Salviati, before going further, whether you have ever thought about what one may believe with regard to the line which is described by a heavy body falling naturally from the top of a tower to its base. If you have reflected on this, please tell me your thoughts.

SALV. I have thought about it at times, and I have not the slightest doubt that if one were certain about the nature of the motion with which a heavy body descends in order to get to the center of the terrestrial globe, then by combining this with the common circular motion of the diurnal rotation, one would discover exactly what sort of a line it is that the center of gravity of the body describes as a composite of those two movements.

SAGR. As to the simple movement toward the center, depending on gravity, I think that one may believe absolutely without error that it is a straight line, exactly as it would be if the earth were immovable.

SALV. As to this part one may not only believe it, but experience renders it certain.

SAGR. But how does experience assure us of this if we never do see any motion except that which is composed of the two, circular and downward?

SALV. Rather, Sagredo, we never see anything but the simple downward one, since this other circular one, common to the earth, the tower, and ourselves, remains imperceptible and as if nonexistent. Only that of the stone, not shared by us, remains perceptible; and of this our senses show that it is along a straight line always parallel to a tower which is built upright and perpendicular on the surface of the earth.

SAGR. You are right, and indeed I have shown myself to be a dunce, such a simple matter not having occurred to me. But now that this is evident, what else do you say you want to have understood about the nature of this downward movement?

SALV. It is not enough to understand that it is straight. It is required to know whether it is uniform or variable; that is, whether the same velocity is always maintained or whether there is a slowing down or an acceleration.

SAGR. Surely it is clear that there is continual acceleration.

SALV. Nor is this enough; it would be needful to know the ratio according to which such acceleration takes place. This problem, I believe, has not been known up to now by any philosopher or mathematician whatever; although by philosophers, especially Peripatetics, entire volumes — and large ones — have been written on the subject of motion.

SIMP. Philosophers occupy themselves principally about universals. They find definitions and criteria, leaving to the mathematicians certain fragments and subtleties, which are then rather curiosities. Aristotle contented himself with defining excellently what motion in general is, and showing the main attributes of local motion; that is, that sometimes it is natural, sometimes forcible, sometimes it is simple, other times composite, on some occasions uniform and on others accelerated; and for the accelerated motions he was content to supply the causes of acceleration, leaving to mechanics or other low artisans the investigation of the ratios of such accelerations and other more detailed features.

SAGR. All right, Simplicio. But you, Salviati, descending sometimes from the throne of His Peripatetic Majesty, have you ever toyed with the investigation of these ratios of acceleration in the motion of falling bodies?

SALV. I have not needed to think them out, because our common friend the Academician showed me a treatise of his on motion[†] in which this is worked out along with many other questions. But it would be too great a digression for us to interrupt with this our present discussion, which for that matter is a digression itself; it would make, so to speak, a play within a play.

SAGR. I am content to excuse you from this recital for the time being, on condition that this shall be one of the propositions saved, among others, for examination in some special session, since such information is highly desirable to me. In the meanwhile let us get back to the line described by the body falling from the top of a tower to its base.

SALV. If the straight movement toward the center of the earth were uniform, and the circular motion toward the east were also uniform, the two could be compounded into a spiral line; one of those defined by Archimedes in his book about the spirals bearing his name, which are those generated when a point moves uniformly along a straight line which is being uniformly rotated about a fixed point at one of its extremities. But since the motion of the falling weight is continually accelerated, the line compounded of the two movements must have an ever-increasing ratio of successive distances from the circumference of that circle which would have been marked out by the center of gravity of the stone had it always remained on the tower. It is also required that this departure be small at the beginning — or rather minimal, even the least possible. For leaving from rest (that is, from the privation of downward motion) and entering into motion straight down, the falling weight must pass through every degree of slowness that exists between rest and any speed of motion. These degrees are infinite, as was discussed at length and decided already.

Supposing, then, that such is the progress of acceleration; it being further true that the descending weight tends to end at the center of the earth, then the line of its compound motion must be such as to travel away from the top of the tower at an ever-increasing rate. To put it better, this line leaves from the circle described by the top of the tower because of the revolution of the earth, its departure from that circle being less *ad infinitum* according as the moving body is found to be less and less removed from the point where it was first placed. Moreover, this line of compound motion must tend to terminate at the center of the earth. Now, making these two assumptions, I draw the circle BI with A as a center and radius AB, which represents the terrestrial globe. Next, prolonging AB to C, the height of the tower BC is drawn; this, carried by the earth along the circumference BI, marks out with its top the arc CD.

The line described by a natural falling body, assuming the earth's motion about its own center, would probably be the circumference of a circle.

Now dividing line CA at its midpoint E, and taking E as a center and EC as radius, the semicircle CIA is described, along which I think it very probable that a stone dropped from the top of the tower C will move, with a motion composed of the general circular movement and its own straight one.[†]

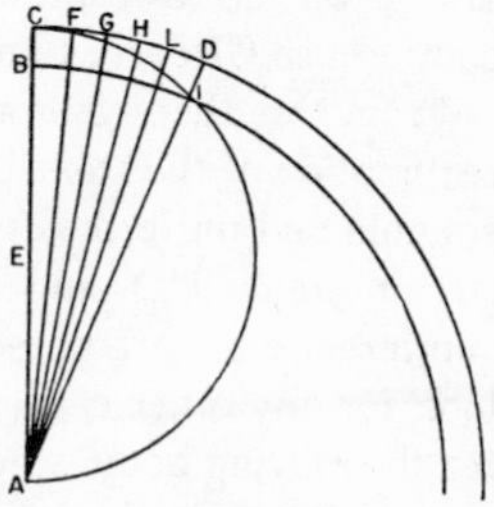

FIG. 8

For if equal sections CF, FG, GH, HL are marked on the circumference CD, and straight lines are drawn to the center A from the points F, G, H, and L, the parts of these intercepted between the two circles CD and BI represent always the same tower CB, carried by the earth's globe toward DI. And the points where these lines are cut by the arc of the semicircle CIA are the places at which the falling stone will be found at the various times. Now these points become more distant from the top of the tower in an ever-increasing proportion, and that is what makes its straight motion along the side of the tower show itself to be always more and more rapid. You may also see how, thanks to the infinite acuteness of the angle of contact between the two circles DC and CI, the departure of the stone from the circumference CFD (that is, from the top of the tower) is very, very small at the beginning, which is the same as saying that the downward motion is extremely slow; in fact, slower and slower *ad infinitum* according to its closeness to the point C, the state of rest. Finally, one may understand how such motion tends eventually to terminate at the center of the earth.

SAGR. I understand the whole thing perfectly, and I cannot think that the center of gravity of the falling body follows any other line but one such as this.

SALV. Hold on, Sagredo; I have also in store for you three little reflections of mine which may not displease you. The first is that if we consider the matter carefully, the body really moves in nothing other than a simple circular motion, just as when it rested on the tower it moved with a simple circular motion.

The second is even prettier;[†] it moves not one whit more nor less than if it had continued resting on the tower; for the arcs CF, FG, GH, etc., which it would have passed through staying always on the tower, are precisely equal to the arcs of the circumference CI corresponding to the same CF, FG, GH, etc.

From this there follows a third marvel — that the true and real motion of the stone is never accelerated at all, but is always equable and uniform. For all these arcs marked equally on the circumference CD, and corresponding arcs marked on the circumference CI, are passed over in equal times. So we need not look for any other causes of acceleration or any other motions, for the moving body, whether remaining on the tower or falling, moves always in the same manner; that is, circularly, with the same rapidity, and with the same uniformity.

Now tell me what you think of these curiosities of mine.

SAGR. I tell you that I cannot find words to express the admiration they cause in me; and so far as my mind can make out at present, I do not believe that there is any other way in which these things can happen. I sincerely wish that all proofs by philosophers had half the probability of this one. Just to complete my satisfaction, I should like to hear the proof that those arcs are equal.

SALV. The demonstration is very easy. Suppose a line to be drawn from I to E; now the radius of the circle CD, that is the line CA, being double the radius CE of the circle CI, the circumference of the former will be double that of the latter, and every arc of the larger circle will be double the similar arc of the smaller. Thus half the arc of the larger circle is equal to the arc of the lesser. And since the angle CEI, made at the center E of the lesser circle and subtending the arc CI, is double the angle CAD, made at the center A of the larger circle and subtending the arc CD, then the arc CD is one-half of the arc in the larger circle similar to the arc CI. Hence the two arcs CD and CI are equal; and the same may be demonstrated in the same way for all the other parts. But that the descent of heavy bodies does take place in exactly this way, I will not at present declare; I shall only say that if the line described by a falling body is not exactly this, it is very near to it.

SAGR. Well, Salviati, there is another remarkable thing which I have just been reflecting about. It is that, according to these considerations, straight motion goes entirely out the window and nature never makes any use of it at all. Even that use which you granted to it at the beginning, of restoring to their places such integral, natural bodies as were separated from the whole and badly disorganized, is now taken away and assigned to circular motion.

Straight motion seems entirely excluded from nature.

SALV. This would necessarily follow if the terrestrial globe were proved to move circularly, which I do not claim has been done. Up to this point I have only been considering, and shall go on considering, the cogency of the reasons that have been assigned by philosophers as proofs of the immobility of the earth. The first of these, taken from the fall of perpendicular bodies, has suffered under all the difficulties that you have been hearing, but I don't know how much importance Simplicio attaches to these. So before going on to the testing of the other arguments, it would be good for him to set forth anything he has to say against these.

SIMP. As to this first argument, I really must admit I have been listening to various subtleties that I have not thought about, and since they are new to me I cannot answer them right now. But I have never taken this argument based upon vertically falling bodies to be one of the strongest arguments in favor of the immobility of the earth. I am wondering what is going to happen to the argument from cannon shots, especially those opposite to the diurnal motion.

SAGR. If only the flying of the birds didn't give me as much trouble as the difficulties raised by cannons and all the other experiments mentioned put together! These birds, which fly back and forth at will, turn about every which way, and (what is more important) remain suspended in the air for hours at a time — these, I say, stagger my imagination. Nor can I understand why with all their turning they do not lose their way on account of the motion of the earth, or how they can keep up with so great a velocity, which after all much exceeds that of their flight.

SALV. As a matter of fact, your point is well taken. Perhaps Copernicus himself was unable to find a solution which entirely satisfied him, and for that reason he remained silent on it. Though indeed he was very brief in his examination of the other adverse arguments; by reason of the profundity of his mind, I suppose, and his preoccupation with the most abstruse reflections, just as a lion is but little impressed by the insistent baying of small dogs. Therefore let us save the objection of the birds for the last, and meanwhile try to satisfy Simplicio as to the others by showing him that, as usual, he has the solutions at his fingertips though he does not notice them.

First, let us take the flight of shots made with the same cannon, powder, and ball, now toward the east and now to the west. Tell me what it is that moves you to believe that, if the diurnal revolution were the earth's, the westward shot would have to carry much farther than the eastward one?

SIMP. I am inclined to believe this because on the eastward shot the ball is followed by the cannon while it is outside the cannon. The latter, carried by the earth, travels rapidly in the same direction; hence the fall of the ball to earth takes place but a short way from the cannon. In the westward shot, on the other hand, before the ball hits the earth the gun is removed far to the east, wherefore the space between the ball and the cannon — that is, the length of this shot — will appear greater than the other by the length of the cannon's path (that is, the earth's) during the time the two balls are in the air.

Why it appears that the cannon ball shot toward the west should go farther than that toward the east.

SALV. I should like to find some way of setting up an experiment which corresponds to the motion of these projectiles as that of the ship corresponded to the motion of falling bodies. I am trying to think how to do so.

SAGR. I believe it would turn out very satisfactorily to take a little open carriage, place a crossbow in it with the bolt at half-elevation (since in that way the shot goes farthest of all), and then, while the horses are running, to shoot once in the direction of their motion and again the opposite way. Taking careful note where the carriage is at the moment the arrow strikes the ground in each case, it could be seen exactly how much farther the one carried than the other.

SIMP. It seems to me that such an experiment would be very suitable, and I have no doubt that the shot (that is, the space between the arrow and the place where the carriage was when the arrow struck the ground) would be much less when it went in the direction of the carriage than when it went the other way. Let the shot in itself be 300 yards, for example, and the travel of the carriage while the arrow is in the air, 100 yards. Then, when the shooting is with its course, the carriage will pass 100 of the 300 yards of the shot, so that at the time the arrow strikes the ground the space between it and the carriage will be only 200 yards. But on the other hand in the shot with the carriage running opposite to the arrow, when the arrow shall have passed over its 300 yards and the carriage its 100 additional the other way, the distance between them will be found to be 400 yards.

SALV. Would there be any way to make these two shots travel equally?

SIMP. I don't know of any other way than to make the carriage stand still.

SALV. That, of course; but I mean with the carriage going full speed.

SIMP. Only by bending the bow harder with the course and more weakly against the course.

SALV. Then there is another way, and this is it. But how much would you need to strengthen your bow, and later to weaken it?

SIMP. In our example, in which we have assumed that the bow would shoot 300 yards, it would be required for the shot along the course to strengthen the bow so as to shoot 400 yards, and the other way to weaken it so as to shoot no more than 200. Thus each shot would go out 300 yards with respect to the carriage, which, with its travel of 100 yards which is to be subtracted from the shot of 400 and added to that of 200, would reduce both to 300.

SALV. But what effect does the greater or lesser strength of the bow have upon the arrow?

SIMP. The strong bow shoots it with greater speed; the weaker with less. The same arrow goes as much farther one time than the other as the speed with which its nock goes forth is greater at one time than the other.

SALV. So that to shoot the arrow in one direction as well as the other and have it depart equally from the moving carriage, it is necessary that if on the first shot of the given example it leaves with, say, four degrees of speed, then on the other shot it must leave with only two. But if the same bowing is employed, three degrees will always be received from that.

SIMP. That is it. That is why the shots cannot go forth equally if shot with the same bowing while the carriage is running.

SALV. I forgot to ask at what speed it is assumed that the carriage is going in this particular experiment.

SIMP. The speed of the carriage must be assumed as one degree in comparison with the three of the bow.

SALV. Yes, this makes the accounts balance. But, tell me, when the carriage is running, don't all the things in the carriage move with that same speed?

SIMP. No doubt about it.

SALV. Also the bolt, and the bow, and the string with which this is strung?

SIMP. That is right.

SALV. Then when the bolt is discharged in the direction of the carriage, the bow impresses its three degrees of speed upon a bolt which already possesses one degree, thanks to the carriage which carries it at that speed in that direction. Thus when the nock leaves the string it does so with four degrees of speed. And on the other hand, shooting the other way, the same bow confers its three degrees upon a bolt moving with one degree in the opposite direction, so that at its separation from the string only two degrees of speed remain with it. But you yourself have already declared that in order to make the shots equal it is required that the bolt leave with four degrees in one case, and with two in the other. Hence, without changing the bow, the course of the carriage itself regulates the flights, and this experiment clinches the matter for those who would not or could not be convinced of it by reason.

Now apply this argument to the cannon, and you will find that whether the earth moves or whether it stands still, shots made with the same force must always carry equally no matter in what direction they are sent. Aristotle's error, and Ptolemy's, and Tycho's, and yours, and that of all the rest, is rooted in a fixed and inveterate impression that the earth stands still; this you cannot or do not know how to cast off, even when you wish to philosophize about what would follow from assuming that the earth moved. Thus in the other argument, without reflecting that when the stone is on the tower it does whatever the terrestrial globe does about moving or not moving, and having it fixed in your mind that the earth stands still, you always argue about the fall of the rock as if it were leaving a state of rest, whereas you ought to say: "If the earth is fixed, the rock leaves from rest and descends vertically; but if the earth moves, the stone, being likewise moved with equal velocity, leaves not from rest but from a state of motion equal to that of the earth. With this it mixes its supervening downward motion, and compounds out of them a slanting movement."

Solution of the argument taken from the cannon shots to east and west.

SIMP. But, good heavens, if it moves slantingly, why do I see it move straight and perpendicular? This is a bald denial of manifest sense; and if the senses ought not to be believed, by what other portal shall we enter into philosophizing?

SALV. With respect to the earth, the tower, and ourselves, all of which all keep moving with the diurnal motion along with the stone, the diurnal movement is as if it did not exist; it remains insensible, imperceptible, and without any effect whatever. All that remains observable is the motion which we lack, and that is the grazing drop to the base of the tower. You are not the first to feel a great repugnance toward recognizing this nonoperative quality of motion among the things which share it in common.

SAGR. There has just occurred to me a certain fantasy which passed through my imagination one day while I was sailing to Aleppo, where I was going as consul for our country. Perhaps it may be of some help in explaining how this motion in common is nonoperative and remains as if nonexistent to everything that participates in it. If it is agreeable with Simplicio, I should like to discuss with him what I fancied to myself at that time.

Sagredo's striking example of the ineffectiveness of motion in common.

SIMP. The novelty of the things I am hearing makes me not merely tolerant of listening, but curious; please go on.

SAGR. If the point of a pen had been on the ship during my whole voyage from Venice to Alexandretta and had had the property of leaving visible marks of its whole trip, what trace — what mark — what line would it have left?

SIMP. It would have left a line extending from Venice to there; not perfectly straight — or rather, not lying in the perfect arc of a circle — but more or less fluctuating according as the vessel would now and again have rocked. But this bending in some places a yard or two to the right or left, up or down, in a length of many hundreds of miles, would have made little alteration in the whole extent of the line. These would scarcely be sensible, and without an error of any moment it could be called part of a perfect arc.

SAGR. So that if the fluctuation of the waves were taken away and the motion of the vessel were calm and tranquil, the true and precise motion of that pen point would have been an arc of a perfect circle. Now if I had had that same pen continually in my hand, and had moved it only a little sometimes this way or that, what alteration should I have brought into the main extent of this line?

SIMP. Less than that which would be given to a straight line a thousand yards long which deviated from absolute straightness here and there by a flea's eye.

SAGR. Then if an artist had begun drawing with that pen on a sheet of paper when we left the port and had continued doing so all the way to Alexandretta, he would have been able to derive from the pen's motion a whole narrative of many figures, completely traced and sketched in thousands of directions, with landscapes, buildings, animals, and other things. Yet the actual, real, essential movement marked by the pen point would have been only a line; long, indeed, but very simple. But as to the artist's own actions, these would have been conducted exactly the same as if the ship had been standing still. The reason that of the pen's long motion no trace would remain except the marks drawn upon the paper is that the gross motion from Venice to Alexandretta was common to the paper, the pen, and everything else in the ship. But the small motions back and forth, to right and left, communicated by the artist's fingers to the pen but not to the paper, and belonging to the former alone, could thereby leave a trace on the paper which remained stationary to those motions.

Thus it is likewise true that the earth being moved, the motion

of the stone in descending is actually a long stretch of many
hundred yards, or even many thousand; and had it been able
to mark its course in motionless air or upon some other surface,
it would have left a very long slanting line. But that part of all
this motion which is common to the rock, the tower, and ourselves remains insensible and as if it did not exist. There remains
observable only that part in which neither the tower nor we are
participants; in a word, that with which the stone in falling
measures the tower.

SALV. A very subtle idea for explaining this point, which for
many people is rather difficult to understand.

Now, unless Simplicio has something to say in reply, we may
pass on to the other experiments, the unraveling of which will
be not a little assisted by the things explained up to now.

SIMP. I have nothing special to say. I was half bemused by this
sketching, and by thinking how these lines, drawn in so many
directions here and there, up and down, back and forth, and
complicated by many turnings, are essentially and in reality only
parts of one single line drawn in a single direction, with no variation except an occasional bending of the straight mark a tiny
bit to the right or left and the moving of the pen point faster
or slower, but with a minimum of unevenness. Now I am thinking how a letter might be written in the same way, and how those
most elegant writers who, to show the dexterity of their hands,
draw a beautiful knot with thousands of turnings in a single
stroke without taking pen from paper, would convert into one
flourish all the motion of the pen (which is essentially a single
line all drawn in the same direction and little bent or sloped from
perfect straightness) while they were in a swiftly sailing boat.
I am very glad that Sagredo has awakened this thought in me.
But let us proceed, for the hope of hearing the rest of this will
keep me most attentive.

SAGR. If you are curious to hear similar ingenuities, which do not
occur thus to everyone, there is no lack of them for us, especially
in this matter of navigation. Will it not seem to you a great idea
that struck me on this same trip, when it occurred to me that the
topgallant of the ship, without the mast breaking or bending,
had made a longer voyage than the foot of the mast? For the
top, being farther from the center of the earth than the foot, had
to describe an arc of a greater circle than that passed by the
latter.

Ironic recitation
of very puerile
conclusions
taken from a
certain
encyclopedia.

SIMP. And thus when a man goes walking, his head travels farther than his feet?

SAGR. You have seen right through it for yourself, by your own ingenuity. But let us not interrupt Salviati.

SALV. I am pleased to see Simplicio exercising himself — if indeed the idea is his, and he has not borrowed it from a certain handbook of conclusions in which there are others no less elegant and ingenious.†

Objection to the earth's diurnal motion, taken from perpendicular cannon shots.

Let us, then, proceed with the discussion of the vertical cannon shot toward the zenith, and the return of the ball by the same line to the same gun, despite the fact that during its long separation from the cannon the earth has carried the latter many miles to the east. It seems that the ball ought to fall an equal distance to the west of the gun, or, since this does not occur, that the cannon must have awaited it without being moved.

The objection answered, showing the fallacy.

The solution is the same as that of the stone falling from the tower, and the whole fallacy and equivocation consists in constantly assuming as true that which is in question. For the adversary has it always fixed in his mind that the ball starts from rest on being shot from the piece; but it cannot leave from a state of rest unless rest is assumed for the terrestrial globe, which is the very conclusion in question.

In replying to this, those who make the earth movable answer that the cannon and the ball which are on the earth share its motion, or rather that all of them together have the same motion naturally. Therefore the ball does not start from rest at all, but to its motion about the center joins one of projection upward which neither removes nor impedes the former. In such a way, following the general eastward motion of the earth, it keeps itself continually over the same gun during both its rise and its return. You will see the same thing happen by making the experiment on a ship with a ball thrown perpendicularly upward from a catapult. It will return to the same place whether the ship is moving or standing still.

Another solution to the same objection.

SAGR. This satisfies me entirely; but as I have noticed that Simplicio takes delight in certain ingenuities that serve to catch the unwary, so to speak, I shall ask him whether, supposing the earth to stand still for the moment, and upon it to be a cannon pointed at the zenith, he has any trouble in understanding that it is truly shot perpendicularly, and that the ball on leaving and on returning goes by the same straight line — always assuming all external and accidental impediments to be removed.

SIMP. I understand that this is exactly what must happen.

SAGR. Now if the cannon is not placed perpendicularly, but tilted in some direction, what must be the motion of the ball? Would it perhaps go like the other shot, along a perpendicular line, and return then by the same line?

SIMP. Not so; leaving the cannon, its motion would follow a straight line continuing the alignment of the cannon, except in so far as its own weight would make it incline from that direction toward the earth.

SAGR. Then the alignment of the cannon is what regulates the ball's motion; and the ball does not move, or would not move out of that line, if its own weight did not make it incline downward. Therefore if the cannon were placed vertically and the ball were shot upward, it would return by the same straight line downward, for the ball's motion due to its weight is downward along the same perpendicular. Hence the travel of the ball outside the gun continues the alignment of that portion of the trip which is made inside the gun. Is that not so?

SIMP. That is the way it looks to me.

SAGR. Now picture to yourself the cannon erect and perpendicular, and the earth turning upon itself with the diurnal motion, carrying the piece with it; tell me what the motion of the ball will be inside the cannon, supposing this to be fired.

SIMP. It will be a straight and perpendicular motion, the cannon being aimed at the zenith.

SAGR. Think it over carefully, because I think that it will not be perpendicular at all. It would indeed be perpendicular if the earth stood still, because then the ball would not have any motion except that given to it by the charge. But if the earth is turning, the ball inside the cannon has also the diurnal motion, so that the impulse of firing being superimposed on this, it travels with two motions from the breech to the mouth of the piece,† the compounding of which results in the motion made by the center of gravity of the ball being a slanting line.

For a clearer comprehension of this, let the cannon AC be erect and the ball B be within it. It is obvious that if the gun stands still and is fired, the ball will go out by the mouth A, its center traveling along the piece describing the perpendicular

Projectiles continue their motion along the straight line which follows the direction of the motion that they had together with the thing projecting them while they were connected with it.

Assuming the earth to revolve, a cannon ball shot vertically does not move along a perpendicular line, but along a slanted one.

line BA, and it will go on following this alignment outside the gun, moving toward the zenith. But if the earth goes round and consequently carries the cannon with it, then during the time in which the ball impelled by the charge is moving through the gun, the cannon carried by the earth will pass to the place DE, and the ball B upon emerging will be at the muzzle D. The motion of the center of the ball will be according to the line BD — no longer perpendicular, but inclined toward the east. And, as was decided before, being obliged to continue its motion in the air according to the direction of the motion made within the piece, the ball's movement will remain in agreement with the slope of the line BD. This will not be perpendicular, but inclined toward the east, in which direction the cannon also is traveling; hence the ball will be able to follow the motion of the earth and of the gun. Now this, Simplicio, shows you how the shot which seems to be vertical is not so at all.

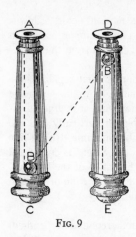

FIG. 9

SIMP. I am not quite convinced about this. And you, Salviati?

SALV. Partly so, but I feel some uneasiness which I wish to heaven I knew how to put in words. It seems to me that in accordance with what has been said, if the gun is vertical and the earth moves, then the ball will fall neither to the west of the piece as Aristotle and Tycho would have it, nor yet upon it as I should like, but rather somewhat to the east. For according to your explanation it would have two motions which would agree in casting it in that direction; that is, the general motion of the earth which carries the cannon and the ball from CA to ED, and that of the charge which hurls it along the slanted line BD, both motions being toward the east and therefore greater than the motion of the earth.

SAGR. No, sir; the motion which carries the ball toward the east comes entirely from the earth, and the firing has no part in this. The motion which impels the ball upward is entirely from the charge, and with this the earth has nothing to do. For surely if you do not fire the charge, the ball will never get out of the gun, nor will it rise a hairsbreadth; and likewise if you hold the earth still and fire the charge, the ball will go vertically without the slightest deviation. Thus, though the ball does have two motions, one up and the other around, from which are compounded the diagonal BD, the upward impulse comes entirely from the firing, while the circular impulse comes wholly from the earth and is equal to that of the earth. And since it is equal, the ball will always maintain itself vertically over the mouth of the gun and will ultimately return into it. And keeping itself always over the alignment of the cannon, it would also continually appear to be overhead to anyone near the gun, and would therefore seem to him to leave it exactly at right angles toward our zenith.

SIMP. Another difficulty remains with me. This is that the motion of the ball within the gun is so extremely fast as to make it seem impossible that in the moment of time during which the cannon goes from CA to ED it would confer such an inclination upon the diagonal line CD that the ball, thanks to this alone, could keep up with the course of the earth while in the air.

SAGR. You are mistaken on several counts. First, I believe that the inclination of the diagonal CD would be much greater than you imagine, for I consider it unquestionable that the velocity of the earth's motion, not only at the equator but even in our latitude, is greater than that of the ball when that is moving within the cannon; hence the interval CE would be absolutely greater than the whole length of the piece, and the inclination of the diagonal would consequently be more than half a right angle. But it is immaterial whether the velocity of the earth is greater or less than that of the shot, since if the velocity of the earth is small and consequently the slope of the diagonal is small, then little inclination is needed to make the ball continue to keep itself over the cannon in its flight. In brief, if you think it over carefully you will understand that the motion of the earth, by transferring the cannon along with it from CA to ED, confers upon the diagonal CD whatever great or little inclination is required to adjust the shot to its demands.

But you err in the second place by wanting to consider the ball's property of keeping up with the earth as coming from the impetus of the firing. You are falling back into the error which Salviati appeared to commit a short time ago. Keeping up with

How hunters aim
at birds in the
air.

the earth is the primordial and eternal motion ineradicably and inseparably participated in by this ball as a terrestrial object, which it has by its nature and will possess forever.

SALV. Let us give in, Simplicio, for the matter stands just as he says. And now from this argument I begin to understand the hunter's problem† — that of those marksmen who kill birds in the air with their guns. I once thought that because of the birds' flight, aim must be taken some distance from the bird, anticipating it by a certain interval, more or less according to the speed of flight and the distance of the bird, to the end that the ball when fired would go along the direct line of sight and arrive at the same time and the same point as the bird would in its flight, and they would meet. Therefore I asked one of these men whether that was their practice, and he told me no, that the device used was much easier and surer. They work in exactly the same way as if shooting at a stationary bird; that is, they fix their sights on a flying bird and follow it by moving the fowling piece, keeping the sights always on it until firing; and thus they hit it just as they would a motionless one. So the turning motion made by the fowling piece in following the flight of the bird with the sights, though slow, must be communicated to the ball also; and this is combined with the other motion, from the firing. Thus the ball would have from the firing a motion straight upward, and from the barrel a slant according to the motion of the bird, exactly as has already been said about the cannon shot. There the ball was impelled upward toward the zenith by the charge, and inclined toward the east by the earth's motion; from the two motions compounded, it followed the course of the earth and appeared to onlookers merely to go straight up, thereafter returning by the same line downward. Therefore, to hold the sights continually directed at the mark makes the shot carry properly. In order to hold the sights on the target if the mark is standing still, the barrel must be held still; and if the target is moving, the barrel will be held on the mark with that motion.

Solution of the
objection taken
from cannon
shots toward
north and south.

Upon this depends the proper answer to that other argument, about shooting with the cannon at a southerly or northerly mark. It was objected there that if the earth moved, the shots would all fall wide to the west, because during the time the ball was going through the air toward the target after leaving the cannon, the target, being carried toward the east, would leave the cannon ball to the west. I reply, then, by asking whether it is not true that once the cannon was aimed at a mark and left so, it would continue to point at that same mark whether the earth moved or stood still. It must be answered that the sighting changes in no way; for if the mark is fixed, the cannon is likewise fixed; and if it moves, being carried by the earth, the cannon also moves in the same way. And if the sights are so maintained, the shot always travels true, as is obvious from what has been said previously.

SAGR. Just a minute please, Salviati, while I bring up something which occurs to me about these hunters and the flying birds. I believe that their way of operating is as you said, and I likewise think that it results in hitting the birds, but it does not seem to me that these actions exactly agree with those of shooting a cannon, which must hit just as accurately when gun and target are moving as when both are at rest. The disparity seems to me to be that in shooting the cannon, it and the target are moving with equal speed, both being carried by the motion of the terrestrial globe. Although the cannon will sometimes be placed closer to the pole than the target and its motion will consequently be somewhat the slower, being made along a smaller circle, this difference is insensible because of the small distance from the cannon to the mark. But in the marksman's shooting, the motion of the fowling piece with which he is following the bird is very slow in comparison with the bird's flight. It seems to me to follow from this that the small motion conferred upon the shot by the turning of the barrel cannot multiply itself in the air up to the speed of the bird's flight, once the ball has left, in such a way that it always stays aimed at the bird. Rather, it seems to me that the bullet would necessarily be anticipated and left behind. It may be added that in this action the air through which the ball passes is not assumed to have the bird's motion, whereas the cannon, the target, and the intervening air have equally the diurnal motion. So I believe that among the reasons that the marksman hits the bird, besides that of his following its flight with the gun barrel, there is that of anticipating it somewhat by keeping the sights ahead. Moreover, I believe the shooting is done not with a single ball but with a large number of pellets which, spreading out in the air, occupy a very large space. And on top of this there is the very great speed with which they go toward the bird upon leaving the gun.

SALV. See how far the flight of Sagredo's wit anticipates and gets ahead of the crawling of mine, which might perhaps have noticed these distinctions, but not without long mental application.

Reply to the
argument taken
from shots
point-blank to
east and west.

Now returning to the subject, it remains to consider the point-blank shots toward east and west. The former, if the earth moves, ought to travel always high over the mark, and the latter beneath it, since the eastern parts of the earth (because of the diurnal motion) are always dropping below the tangent parallel to the horizon, for which reason the stars in the east appear to be rising; the western parts are rising, so that the stars in the west seem to go down. Hence the shots which are aimed along this tangent toward an eastern target (which is going down while the ball is traveling along that tangent) ought to arrive high; and those to the west, low, because of the rising of the target while the ball

Solution of the
objection taken
from shots to
east and west.

goes along the tangent. The explanation is similar to the others; just as the eastern target is continually setting because of the motion of the earth under a motionless tangent, so also the cannon for the same reason continually declines and keeps on pointing always at the same mark so that the shots carry true.†

The Copernicans
too liberally
admit as true
some very
questionable
propositions.

And this seems to me an appropriate time to take notice of a certain generosity on the part of the Copernicans toward their adversaries when, with perhaps too much liberality, they concede as true and correct a number of experiments which their opponents have never really made. Such, for example, is that of the body falling from the mast of a ship while it is in motion, and there are many others, among which I am positive is this one of cannon shots to the east carrying high and those to the west, low. And because I believe it has never been done, I should like to have them tell me just what difference they think ought to be perceived between the same shots, taking the earth first as motionless and then as moving. Simplicio, reply to this for them.

SIMP. I cannot pretend to answer as soundly as perhaps someone might who was better informed than I, but I shall say what seems to me at the moment would be their reply. It is in fact just what has already been shown — that if the earth were moving, eastward shots would always carry high (and so forth) provided that the ball was compelled to move along the tangent, as seems probable.

SALV. And if I should say that that is what actually happens, how would you go about refuting my statement?

SIMP. An experiment would be required to clear it up.

SALV. But do you think that there is a cannoneer who is so skillful that he could hit the mark every time at, say, 500 yards?

SIMP. Goodness, no; I doubt if there is one, no matter how expert, who could promise to err proportionately no more than a yard.

SALV. Then how could we settle our question, with such inconclusive shooting?

SIMP. We could resolve it in two ways; one, by firing many shots, and the other, from the fact that in view of the tremendous velocity of the earth, the deviation from the mark would, I think, be enormous.

SALV. Enormous — that is, much greater than a yard; for so much variation, or even more, is granted to occur ordinarily even if the earth is at rest.

SIMP. I am sure that the variation would be very much greater.

SALV. Now, if you are willing, let us make for our own satisfaction a rough calculation; if it comes out as I expect, it will serve us also as a warning in the future not to be taken in by other people's shouting, so to speak, and yield to whatever happens first to strike our fancy. Moreover, to give every advantage to the Peripatetics and Tychonians, let us imagine ourselves to be at the equator, shooting a cannon point-blank toward the west at a target 500 yards distant. First let us see approximately how much time can elapse while the ball, having left the gun, is going toward the mark. We know this to be brief, certainly no more than that in which a pedestrian takes two steps, and this in turn is short of one second. For suppose the pedestrian to walk three miles an hour; that is, nine thousand yards; since an hour contains 3,600 seconds, he will take two and a half steps a second. So one second is longer than the time the ball is in motion. And since the diurnal revolution takes twenty-four hours, the western horizon rises fifteen degrees in an hour, or fifteen minutes of arc in a minute of time, or fifteen seconds of arc in a second of time. Now since one second is the time required for the shot, the western horizon rises in this time fifteen seconds of arc, and the target an equal amount. Hence it rises fifteen seconds of the arc of that circle whose radius is 500 yards,[†] this being supposed to be the distance of the target from the cannon. Now let us see, in a table of arcs and chords[†] (here it is, right in Copernicus's book) what the chord of fifteen seconds is for a radius of 500

Calculation of how much cannon shots would vary from the mark, assuming the earth's motion.

yards. Here, you see, the chord of one minute is less than thirty parts where the radius is 100,000. Then for the same radius, the chord of one second would be less than one-half of one such part; that is, less than one part where the radius is 200,000; therefore the chord of fifteen seconds would be less than fifteen parts in 200,000. But that which is less than fifteen parts in 200,000 is nevertheless greater than four one-hundredths of one part in 500. Hence the rising of the target while the ball is in motion is less than four one-hundredths — that is, one twenty-fifth — of a yard, or about an inch. Therefore just one inch would be the entire variation of a westward shot if the earth made the diurnal motion.

Now if I say to you that this variation actually occurs in all the shots (I mean going one inch below where they would go if the earth did not move), how would you go about convincing me otherwise, Simplicio, and showing by experiment that this did not happen? Don't you see that it is impossible to refute me without first finding a method of shooting with such precision at a mark that you never miss by a hairsbreadth? For when the shots vary by a yard, as they do in fact, I shall always tell you that each one of these variations contains one of one inch caused by the motion of the earth.

SAGR. Excuse me, Salviati, but you are too generous. I can tell the Peripatetics that if every shot hit square in the center of the target, it would not contradict the motion of the earth one bit; for cannoneers are always so experienced in adjusting the sights to the target and so expert at pointing the gun at the mark that the shot would hit it despite the motion of the earth. And I say that if the earth should stop, their shots would *not* hit the mark, but those to the west would carry high and those to the east, low.† Now let Simplicio persuade me of the contrary.

SALV. A paradox worthy of Sagredo. But it must be seen that this variation due to the rest or motion of the earth, since it can only be very small, cannot but be submerged in the large ones which continually occur on account of accidents. And this is all said and granted for good measure to Simplicio merely as a warning of how carefully we must tread in conceding the truth of many experiments to those who have never performed them, but who boldly would produce such as are needed to serve their purposes. I say that this is thrown into the bargain for Simplicio, because the plain truth is that with regard to the effects of these shots the same thing exactly must happen with or without the motion of the earth. And such will be the fate of all other experiments put forth or capable of being put forth, though they have at first glance an appearance of truth, inasmuch as the ancient idea of the earth's immobility keeps us in the midst of equivocations.

SAGR. For my part I am fully satisfied, and I understand perfectly that anyone who will impress upon his mind this general communication to all terrestrial things of the diurnal motion (which suits them all naturally, just as in the ancient idea it was considered that rest with respect to the center suited them) will discern without any trouble the fallacy and the equivocation that make the arguments appear conclusive.

There remains for me only that doubt which I hinted at before, about the flight of birds. Since these have the lively faculty of moving at will in a great many ways, and of keeping themselves for a long time in the air, separated from the earth and wandering about with the most irregular turnings, I am not entirely able to see how among such a great mixture of movements they can avoid becoming confused and losing the original common motion. Once having been deprived of it, how could they make up for this or compensate for it by flying, and keep up with all the towers and trees which run with such a precipitous course toward the east? I say "precipitous," because for the great circle of the globe it is little less than a thousand miles an hour, while I believe that the swallow in flight makes no more than fifty.

SALV. If birds had to keep up with the course of the trees by means of their wings, they would soon fall behind; and if they were deprived of the universal rotation, they would remain so much behind and their westward course would be so furious that, to anyone who could see it, it would surpass that of an arrow by a great deal. But I think we should not be able to perceive it, just as cannon balls are not seen when they race through the air, driven by the energy of the charge. Now the fact is that the birds' own motion — I mean that of flight — has nothing to do with the universal motion, from which it receives neither aid nor hindrance. What keeps that motion unaltered in the birds is the air itself through which they wander. This, following naturally the whirling of the earth, takes along the birds and everything else that is suspended in it, just as it carries the clouds. So the birds do not have to worry about following the earth, and so far as that is concerned they could remain forever asleep.

SAGR. I am easily convinced that the air can take the clouds along with it, they being of material which is very tractable by reason of its lightness and its lack of any contrary tendency; indeed, they are of a material which shares in the qualities and properties of the earth. But birds, being animate, can also move contrary to the diurnal motion; and that the air can restore this to them once they have interrupted it seems problematical to me, especially since they are solid and heavy bodies. As was said before, we see rocks and other heavy bodies remain defiant to the impetus of the wind, and when they do give in to it they are never moved with any such speed as that of the wind which pushes them.

SALV. Let us not grant to the moving air so little force, Sagredo; it is able to drive heavily laden ships and to uproot trees and to overthrow towers when it moves swiftly. Yet in such violent actions as these, its motion cannot be said by a long way to be as fast as the diurnal rotation.

SIMP. You see, then; moving air will be able to keep up the motion of projectiles also, in accordance with Aristotle's teaching. It did seem strange to me that he should have erred in this particular.

SALV. It certainly would be able to do so if it could keep up its own motion. But just as ships stop and trees cease to bend when the wind slackens, so the motion of the air does not keep on after the stone has left the hand and the arm is stopped. Hence it remains true that something besides the air makes the projectile move.

SIMP. What do you mean, the ship stops when the wind slackens? It is often seen that the wind has stopped, and the sails have even been furled, and yet the vessel continues to travel for miles on end.

SALV. This argues against you, Simplicio, if the air, which by carrying the sails propels the ship, is stopped, and without help of any kind from the medium the ship continues its course.

SIMP. It might be said that the water was the medium which propelled the ship and maintained its motion.

SALV. Well, that certainly might be said, but it would be the exact opposite of the truth. For the truth is that the water has such a strong resistance to being separated by the ship's hull that it works against this with much foaming and does not let the ship receive a large part of that velocity which the wind would confer upon it if the hindrance of the water were not there. You must never have considered, Simplicio, the fury with which the water strikes against a boat when, rapidly driven by oars or by the wind, the boat runs through still water; if you had paid attention to this effect you would not have thought up such a silly idea now. I see that you have hitherto been one of that herd who, in order to learn how matters such as this take place, and in order to acquire a knowledge of natural effects, do not betake themselves to ships or crossbows or cannons, but retire into their studies and glance through an index and a table of contents to see whether Aristotle has said anything about them; and, being assured of the true sense of his text, consider that nothing else can be known.

SAGR. Happy are they, and much to be envied for this. For if a knowledge of everything is naturally desired, and if being informed is the same thing as taking credit for being informed, then they enjoy a very great knowledge. They can persuade themselves that they know and understand everything, in complete defiance of those who recognize their own ignorance of what they do not know. These latter, perceiving that they know only the tiniest portion of what is knowable, exhaust themselves in waking and studying, and mortify themselves with experiments and observations.

A great joy, much to be envied, is that of people who think they know everything.

But please let us return to our birds, with regard to which you have said that the air, moving very speedily, can restore that part of the diurnal movement which they may have lost in the sportings of their flight. To this I reply that the moving air does not seem able to confer upon a solid and heavy body so much as its own velocity, and since that of the air is that of the earth, it does not appear that the air would be sufficient to supply the deficit of that lost by the birds in flight.

SALV. Your argument puts up an appearance of much probability, and your doubt is not one that is raised by ordinary intelligences; yet outside of its appearance, I do not believe that essentially it has a bit more force than those already considered and disposed of.

SAGR. There is not the slightest doubt that unless it is rigorously conclusive, it is absolutely ineffective; for it is only when a conclusion is inescapable that no worthwhile argument can be produced against it.

SALV. Your having more trouble with this objection than with the others seems to me to depend upon birds being animate, and thereby being able to use force at will against the original inherent motion of terrestrial objects. In just the same way, we see them fly upward when they are alive; a motion impossible to them as heavy bodies, so that when dead they can only fall downward. From this you assume that the causes which hold for all other sorts of projectiles previously discussed cannot hold for birds. Well, this is true enough, Sagredo; and because it is true we do not see other projectiles do what birds do; for if you drop a dead bird and a live one from the top of a tower, the dead one will do the same as a stone; that is, it will follow first the general diurnal motion, and then the motion downward, being heavy. But as to the live bird, the diurnal motion always remaining in it, what is to prevent it from sending itself by the beating of its wings to whatever point of the compass it pleases? And such a new motion being its own, and not being shared by us, it must make itself noticeable. If the bird moves off toward the west in its flight, what is there to prevent it from returning once more to the tower by means of a similar beating of its wings? For after all, its leaving toward the west in flight was nothing but the subtraction of a single degree from, say, ten degrees of diurnal motion, so that nine degrees remain to it while it is flying. And if it alighted on the earth, the common ten would return to it; to this it could add one by flying toward the east, and with the eleven it could return to the tower. In sum, when we consider well and reflect more closely upon the effects of flight in birds, these do not differ in any way from those of projectiles directed toward any part of the earth, except that the latter are moved by an external source and the former by an internal principle.

Argument taken from the flight of birds against the earth's motion is resolved.

For a final indication of the nullity of the experiments brought forth, this seems to me the place to show you a way to test them all very easily. Shut yourself up with some friend in the main cabin below decks on some large ship, and have with you there some flies, butterflies, and other small flying animals. Have a large bowl of water with some fish in it; hang up a bottle that empties drop by drop into a wide vessel beneath it. With the ship standing still, observe carefully how the little animals fly with equal speed to all sides of the cabin. The fish swim indifferently in all directions; the drops fall into the vessel beneath;

Experiment which alone shows the nullity of all those adduced against the motion of the earth.

and, in throwing something to your friend, you need throw it no more strongly in one direction than another, the distances being equal; jumping with your feet together, you pass equal spaces in every direction. When you have observed all these things carefully (though there is no doubt that when the ship is standing still everything must happen in this way), have the ship proceed with any speed you like, so long as the motion is uniform and not fluctuating this way and that. You will discover not the least change in all the effects named, nor could you tell from any of them whether the ship was moving or standing still. In jumping, you will pass on the floor the same spaces as before, nor will you make larger jumps toward the stern than toward the prow even though the ship is moving quite rapidly, despite the fact that during the time that you are in the air the floor under you will be going in a direction opposite to your jump. In throwing something to your companion, you will need no more force to get it to him whether he is in the direction of the bow or the stern, with yourself situated opposite. The droplets will fall as before into the vessel beneath without dropping toward the stern, although while the drops are in the air the ship runs many spans. The fish in their water will swim toward the front of their bowl with no more effort than toward the back, and will go with equal ease to bait placed anywhere around the edges of the bowl. Finally the butterflies and flies will continue their flights indifferently toward every side, nor will it ever happen that they are concentrated toward the stern, as if tired out from keeping up with the course of the ship, from which they will have been separated during long intervals by keeping themselves in the air. And if smoke is made by burning some incense, it will be seen going up in the form of a little cloud, remaining still and moving no more toward one side than the other. The cause of all these correspondences of effects is the fact that the ship's motion is common to all the things contained in it, and to the air also. That is why I said you should be below decks; for if this took place above in the open air, which would not follow the course of the ship, more or less noticeable differences would be seen in some of the effects noted. No doubt the smoke would fall as much behind as the air itself. The flies likewise, and the butterflies, held back by the air, would be unable to follow the ship's motion if they were separated from it by a perceptible distance. But keeping themselves near it, they would follow it without effort or hindrance; for the ship, being an unbroken structure, carries with it a part of the nearby air. For a similar reason we sometimes, when riding horseback, see persistent flies and horseflies following our horses, flying now to one part of their bodies and now to another. But the difference would be small as regards the falling drops, and as to the jumping and the throwing it would be quite imperceptible.

SAGR. Although it did not occur to me to put these observations to the test when I was voyaging, I am sure that they would take place in the way you describe. In confirmation of this I remember having often found myself in my cabin wondering whether the ship was moving or standing still; and sometimes at a whim I have supposed it going one way when its motion was the opposite. Still, I am satisfied so far, and convinced of the worthlessness of all experiments brought forth to prove the negative rather than the affirmative side as to the rotation of the earth.

Now there remains the objection based upon the experience of seeing that the speed of whirling has a property of extruding and discarding material adhering to the revolving frame. For that reason it has appeared to many, including Ptolemy,† that if the earth turned upon itself with great speed, rocks and animals would necessarily be thrown toward the stars, and buildings could not be attached to their foundations with cement so strong that they too would not suffer similar ruin.

SALV. Before coming to the solution of this objection, I cannot help mentioning something I have noticed many times, and not without amusement. It occurs in nearly everyone who hears for the first time of the earth's motion. Such people so firmly believe the earth to be motionless that not only do they have no doubt of its being at rest, but they really believe that everyone else has always agreed with them in thinking it to have been created immovable and kept so in all past ages. Rooted in this idea, they are stupefied to hear that someone grants it to have motion, as if such a person, after having held it to be motionless, foolishly imagined it to have been set in motion when Pythagoras (or whoever it was) first said that it moved, and not before. Now that a silly idea like this, of supposing that those who admit the earth's motion believe it first to have been stable, from its creation up to the time of Pythagoras, and then made movable only after

Stupidity of
some who think
the earth to have
begun to move
when Pythagoras
commenced say-
ing that it
moved.

Pythagoras deemed it to be so, should find a place in the giddy minds of common people is no marvel to me; but that the Aristotles and the Ptolemies should also have fallen into this puerility truly seems to me strange and inexcusable simple-mindedness.

SAGR. Then you believe, Salviati, that Ptolemy thought he needed to support the stability of the earth only by arguments directed against people who concede it to have been immovable up to the time of Pythagoras, and who affirm it to have been made movable only when Pythagoras attributed motion to it?

SALV. I cannot help believing so, when we consider well the attitude he takes in refuting their proposition. His refutation is to be found in the demolition of buildings and the flinging of stones, animals, and men themselves toward the sky. Now such ruin and havoc could not be visited upon edifices and animals unless these existed on the earth in the first place, and men could not be located or edifices built upon the earth unless it was standing still. So it is obvious that Ptolemy is arguing against those who, having granted quiescence to the earth for some time — that is, while animals and stones and masons could remain on it and build palaces and cities — suddenly make it movable afterward, to the ruin and destruction of the buildings, animals, etc. For if he had undertaken to dispute with those who attributed a whirling to the earth ever since its original creation, he would have refuted them by saying that if the earth had always moved, there never could have been beasts or men or stones upon it; much less buildings erected, cities founded, etc.

SIMP. I am not convinced of any Aristotelian or Ptolemaic impropriety here.

SALV. Ptolemy argues either against those who considered the earth always movable or against those who thought it to be stable for a time and then to be set in motion. If against the former, he ought to have said: "The earth has not always moved, for there would never have been men nor animals nor edifices on earth, the terrestrial whirling having not permitted them to stay." But since his reasoning is, "The earth does not move, because beasts and men and buildings placed on the earth would be precipitated from it," he assumes the earth to have been once in that state which would have allowed beasts and men to stay and build them. From this the conclusion is drawn that the earth has been fixed at some time; that is, adapted to the stay of animals and the building of edifices. Now do you understand what I mean?

Aristotle and Ptolemy appear to have refuted the earth's mobility against those who would believe that, having stood still a long time, it began to move in Pythagoras's time.

SIMP. Yes and no; but this has little to do with the merit of the case, nor can a slight error of Ptolemy's, committed by inadvertence, suffice to move the earth if it is immovable. But all joking aside, let us get to the heart of this argument, which to me appears unanswerable.

SALV. And I, Simplicio, wish to tie it even tighter and make it more binding by showing more sensibly how true it is that heavy bodies, whirled quickly around a fixed center, acquire an impetus to move away from that center even when they have a natural tendency to go toward it. Tie one end of a cord to a bottle containing water and, holding the other end firmly in your hand (making your arm and the cord the radius, and your shoulder knot the center), cause the vessel to go around swiftly so that it describes the circumference of a circle. Whether this is parallel to the horizon, or vertical, or slanted in any other way, the water will not spill out of the bottle in any event; rather, he who swings it will always feel the cord pull forcibly to get farther away from his shoulder. And if a hole is made in the bottom of the bottle, the water will be seen to spurt forth no less toward the sky than laterally or toward the ground. And if in place of water you put pebbles in the bottle, upon your turning it in the same manner it will be felt to exert the same force against the cord. Finally, small boys may be seen throwing rocks a great distance by whirling a slotted stick with a stone in the end. All these are arguments of the truth of the conclusion that whirling confers an impetus upon the moving body toward the circumference, if the motion is swift. And since, if the earth revolved upon itself, the motion of its surface (especially near the equator) would be incomparably faster than the objects mentioned, it would necessarily throw everything into the sky.

SIMP. The objection does indeed seem to be much better established and tied down, and to my mind it will be a difficult thing to remove it or unravel it.

SALV. The unraveling depends upon some data well known and believed by you just as much as me, but because they do not strike you, you do not see the solution. Without teaching them to you then, since you already know them, I shall cause you to resolve the objection by merely recalling them.

SIMP. I have frequently studied your manner of arguing, which gives me the impression that you lean toward Plato's opinion that

Fast whirling has a property of extruding and dissipating things.

nostrum scire sit quoddam reminisci. So please remove all question for me by telling me your idea of this.

SALV. How I feel about Plato's opinion I can indicate to you by means of words and also by deeds. In my previous arguments I have more than once explained myself with deeds. I shall pursue the same method in the matter at hand, which may then serve as an example, making it easier for you to comprehend my ideas about the acquisition of knowledge if there is time for them some other day, and if Sagredo will not be annoyed by our making such a digression.

SAGR. Rather, I shall be much obliged. For I remember that when I was studying logic, I never was able to convince myself that Aristotle's method of demonstration, so much preached, was very powerful.

SALV. Then let us proceed. Simplicio, tell me what motion is made by that little rock, tight in the notch of the stick, when the boy moves it so as to cast it a long way?

SIMP. The motion of the stone while it is in the notch is circular; that is, it travels along the arc of a circle whose fixed center is the shoulder knot and whose radius is the stick and the arm.

SALV. And when the stone escapes from the stick, what is its motion? Does it continue to follow its previous circle, or does it go along some other line?

SIMP. It certainly does not go on moving around, for then it would not fly away from the thrower's shoulder, and we should not see it go extremely far.

SALV. Well, then, what is its motion?

SIMP. Let me think a moment here, for I have not formed a picture of it in my mind.

SALV. Listen to that, Sagredo; here is the *quoddam reminisci* in action, sure enough.

Well, Simplicio, you are thinking a long time.

SIMP. So far as I can see, the motion received on leaving the notch can only be along a straight line. Or rather, it is necessarily along a straight line, so far as the adventitious impetus is concerned. Seeing that it described an arc caused me some little trouble, but since that arc bends always downward, and not in any other direction, I recognized that this inclination comes from the weight of the stone which naturally pulls it down. The impressed impetus, I say, is undoubtedly in a straight line.

Our knowledge is a kind of reminiscence, according to Plato.

Motion impressed by the thrower is along a straight line only.

SALV. But what straight line? An infinity of them can be drawn, in every direction from the notch and point of separation between the stone and the stick.

SIMP. It moves along the one which is in alignment with the motion which the stone made together with the stick.

SALV. You have just finished telling us that the motion of the stone in the notch was circular. Now circularity and alignment exclude each other, no part of a circular line being straight.

SIMP. I do not mean that the projectile's motion is in alignment with the whole circular motion, but with that of the last point, where the circular motion ended. I understand it completely in my own mind, but I do not know how to express it.

SALV. I also see that you understand the thing itself, but lack the proper terms for expressing it. Now these I can indeed teach you; that is, I can teach you the words, but not the truths, which are things. And so that you may plainly feel that you know the thing and merely lack terms to express it, tell me: When you shoot a bullet with a gun, in what direction does it receive an impetus to go?

SIMP. It acquires an impetus to go along that straight line which continues the alignment of the barrel, slanting neither to right nor to left, up nor down.

SALV. Which is as much as to say that it makes no angle whatever with the straight line of its motion through the barrel.

SIMP. That is what I meant.

SALV. Then if the line of motion of the projectile must extend so as to make no angle with the circular line it was describing while it was with the thrower; if that circular motion must pass into straight motion; what must this straight line be?

SIMP. It can be no other than that line which touches the circle at the point of separation. For all others would, it seems to me, intersect the circumference if produced, and would therefore make some angle with it.

SALV. You have reasoned well, and have shown yourself half a geometer. Keep it in mind, then, that your real concept is revealed in these words; that is, that the projectile acquires an impetus to move along the tangent to the arc described by the motion of the projectile at the point of its separation from the thing projecting it.

SIMP. I understand perfectly, and this is just what I meant.

SALV. Which point of a straight line touching a circle is closest of all to the center of that circle?

SIMP. The point of contact, without doubt; for that is on the circumference of the circle and the others are outside it. And points on the circumference are all equally distant from the center.

SALV. Then a moving body leaving from the point of contact and moving along the straight tangent will go continually farther from the contact and also from the center of the circle.

SIMP. This is surely so.

SALV. Now, if you have kept in mind the propositions which you have told me, collect them all together, and tell me what you gather from them.

SIMP. I do not think I am so forgetful as to be unable to recall them. From what has been said, I gather that a projectile, rapidly rotated by someone who throws it, upon being separated from him retains an impetus to continue its motion along the straight line touching the circle described by the motion of the projectile at the point of separation. By this motion the projectile goes always farther from the center of the circle described by the motion which projects it.

Projectile moves along the tangent to its previous circle of motion at the point of separation.

SALV. Then up to this point you know the reason for heavy bodies located on the surface of a rapidly moved wheel being cast off and thrown out from its circumference, always farther from the center.

SIMP. I believe I can say I am certain of that. But this new knowledge only increases my incredulity that the earth could revolve with such great speed and not throw to the skies all stones, animals, etc.

SALV. In the same way that you knew what went before, you will know — or rather, do know — the rest too. And by thinking it over for yourself you would likewise recall it by yourself. But to save time, I shall help you to remember it.

Up to this point you knew all by yourself that the circular motion of the projector impresses an impetus upon the projectile to move, when they separate, along the straight line tangent to the circle of motion at the point of separation, and that continuing with this motion, it travels ever farther from the thrower. And you have said that the projectile would continue to move along that line if it were not inclined downward by its own weight, from which fact the line of motion derives its curvature. It seems to me that you also knew by yourself that this bending always tends toward the center of the earth, for all heavy bodies tend that way.

Now I shall pass on a little further and ask you whether the moving body in continuing its straight motion after the separation goes uniformly farther from the center (or from the circumference, if you like) of that circle of which its previous motion was a part. That is to say, do you believe a body which leaves from the point of tangency and moves along the tangent goes uniformly away from the point of contact and from the circumference of the circle?

SIMP. No, indeed; because the tangent when close to the point of contact is very little distant from the circumference, with which it makes an extremely small angle. But as it goes farther and farther away, the distance increases in an increasing ratio. Thus in a circle that might have, for example, a diameter of ten yards, a point on the tangent two or three feet away from the contact will be three or four times as far from the circumference as a point one foot away, and a point only half a foot away I believe likewise would be hardly a quarter of the distance of the latter. As close to the contact as an inch or two, the tangent could scarcely be distinguished from the circumference.

SALV. Then the departure of the projectile from the circumference of its previous circular motion is extremely small at first?

SIMP. Almost imperceptible.

SALV. Now tell me something else. How far away after the separation would the projectile commence to sink downward, having received from the thrower's motion an impetus to move straight along the tangent, as indeed it would move if its own weight did not draw it down?

SIMP. I think it would begin at once, for having nothing to sustain it, its own weight could not help acting.

SALV. So that if the rock thrown from a rapidly moving wheel had any such natural tendency to move toward the center of the wheel as it has to go toward the center of the earth, it might very well return to the wheel, or rather never leave it.† For the distance traveled being so extremely small at the beginning of its separation (because of the infinite acuteness of the angle of contact), any tendency that would draw it back toward the center

of the wheel, however small, would suffice to hold it on the circumference.

SIMP. I have no doubt at all that by assuming something which
is not and cannot be so (that is, that the tendency of the heavy
body is to go toward the center of the wheel), it would not be
extruded or flung away.

SALV. I do not assume, and have no need to assume, that which
is not; for I do not wish to deny that rocks are flung out. I am
speaking thus only by way of hypothesis, so that you may tell me
the rest. Now picture to yourself the earth as a huge wheel which,
moving with great speed, must cast off the stones. You have
already been able to tell me that the motion of the projectile must
be along that straight line which touches the earth at the point
of separation. And how noticeably does this tangent recede from
the surface of the terrestrial globe?

SIMP. I doubt if it gets an inch away in a thousand yards.

SALV. And didn't you say that the projectile, drawn by its own
weight, sinks from the tangent toward the center of the earth?

SIMP. That is what I said, and now I shall say the rest, too; I
understand completely that the stone would not be separated
from the earth, because its motion away in the beginning would
be so very minute that its inclination toward the center of the
earth would be a thousand times stronger. The center in this case
would be that of the earth as well as that of the wheel, so it must
truly be conceded that stones, animals, and other heavy bodies
could not be thrown off.

But now a new difficulty is created for me by things which
are very light. These have a very weak tendency to descend
toward the center, and since they lack the property of drawing
back to the surface, I do not see why they do not have to be
extruded; and as you know, *ad destruendum sufficit unum.*

SALV. You shall have satisfaction as to this, too. But first tell me
what you mean by "light things." Do you mean material so light
that it actually goes up, or merely that which, while not absolutely light, weighs so little that although it goes down, it does
so but slowly? For if you mean absolutely light things, I shall
grant them to be extruded as readily as you do.

SIMP. I mean the other, such as feathers, wool, cotton, and the
like, which the slightest force is sufficient to lift up, and yet
which are seen to remain quietly on the earth.

SALV. Since these feathers do have some natural tendency to descend toward the center of the earth, however small it is, I tell you that this is enough to prevent them being lifted up. Nor is this unknown to you. Tell me: If a feather were thrown off by the whirling of the earth, what direction would it take?

SIMP. That of the tangent at the point of separation.

SALV. And if it were forced to return and rejoin the earth, along what line would it move?

SIMP. Along that line going through it to the center of the earth.

SALV. So that two motions come under our consideration: a motion of projection, commencing at the point of contact and following the tangent,† and another of downward tendency, commencing at the projectile and going along the secant toward the center. To have projection occur, it is required that the impetus along the tangent prevail over the tendency along the secant. Is that not so?

SIMP. It seems so to me.

SALV. But what do you think would have to exist in the projecting motion in order for it to prevail over the downward tendency, so that the detachment of the feather and its departure from the earth would follow?

SIMP. I don't know.

SALV. How can you help knowing? Here the moving body is one and the same — that is, the feather. Now how can the same moving body exceed its own motion and prevail over itself?

SIMP. I don't see how it can prevail over or yield to itself in motion except by moving faster or slower.

SALV. There; you see, you did know how. Now if there is to be a projection of the feather, and if its motion along the tangent is to prevail over its motion along the secant, then what must their velocities be?

SIMP. The motion along the tangent must be greater than that along the secant. But how stupid of me! Isn't the former many thousands of times greater, not merely than the downward motion of the feather, but than that of the stone too? And I allowed myself, simple-mindedly enough, to be convinced that stones would not be extruded by the whirling of the earth! I take it back, then, and declare that if the earth did move, then stones, elephants, towers, and cities would necessarily fly toward the heavens; and since that does not happen, I say that the earth does not move.

SALV. Oh, Simplicio, you yourself rise up so fast that I begin to fear more for you than for the feather. Relax a little, and listen.

If, in order for the stone or feather resting on the surface of the earth to be retained, it were necessary that its descent should be greater than or equal to its motion made along the tangent, then you would be right in saying that it would have to move as fast or faster along the secant downward than along the tangent eastward. But didn't you tell me a little while ago that a thousand yards along the tangent from the point of contact, it would be scarcely an inch away from the circumference? So it is not enough for the tangential motion (which is that of the diurnal rotation) to be simply faster than the motion along the secant (which is that of the feather downward). The former must be so much faster that the time required to carry the feather a thousand yards along the tangent shall be less than that of its moving a single inch downward along the secant; which I tell you it will never be, though you make the latter motion as fast and the former as slow as you please.

SIMP. And why couldn't the motion along the tangent be so fast that it would not give the feather time to arrive at the surface of the earth?

SALV. Try to state your case quantitatively (*in termini*), and I shall answer you. Say, then, how much faster you think the latter motion should be made than the former in order to suffice.

SIMP. I shall say that if, for example, the latter were a million times faster than the former, the feather (and the stone likewise) would be extruded.

SALV. Saying this, you say what is false; not from any deficiency in logic or physics or metaphysics, but merely in geometry. For if you were aware of only its first principles, you would know that a straight line may be drawn from the center of a circle to a tangent, cutting this in such a way that the portion of the tangent lying between the contact and the secant will be a million, or two, or three million times greater than that portion of the secant which remains between the tangent and the circumference; and by degrees as the secant approaches the contact, this proportion becomes greater *ad infinitum*. So there is no danger, however fast the whirling and however slow the downward motion, that the feather (or even something lighter) will begin to rise up. For the tendency downward always exceeds the speed of projection.

A geometrical
demonstration to
prove the impos-
sibility of extru-
sion by terres-
trial whirling.

SAGR. I am not quite convinced on this matter.

SALV. I shall give you a perfectly general and yet a very easy demonstration of it.

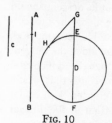

FIG. 10

Given the ratio of BA to C, BA being as much greater than C as you please, and let there be a circle with center D, from which it is required to draw a secant so that the tangent shall have the same ratio to this secant as BA has to C. With respect to BA and C take the third proportional AI; as BI is to IA, make the diameter FE to EG. From the point G draw the tangent GH. I say that this is what was required, and that as BA is to C, so HG is to GE. For FE being to EG as BI is to IA, by composition FG is to GE as BA is to AI; and since C is the mean proportional between BA and AI, GH is the mean between FG and GE. Therefore as BA is to C, so FG is to GH; that is, HG is to GE; which is what was required to be done.

SAGR. I am satisfied with the demonstration, but it still does not entirely remove my doubts. Rather, I find a certain confusion turning over in my mind, like so many dense and dark clouds, and it prevents my seeing clearly the necessity of the conclusion with that lucidity which belongs to mathematical reasoning alone. What confuses me is this: It is true that the space between the tangent and the circumference decreases infinitely in the direction of the point of contact. But on the other hand it is also true that the tendency in the moving body to descend always diminishes as the body approaches the limiting boundary (*primo termine*) of its descent; that is, the state of rest. This is obvious from what you have said about it when showing that descending bodies departing from rest must pass through all degrees of slowness between rest and any assigned degree of speed, these being less and less *ad infinitum*.

It may be added that this speed and this tendency to motion may diminish infinitely for yet another reason, arising from the weight of the moving body being capable of infinite diminution.†‎ Hence the causes which reduce its tendency to descend (and consequently favor its being thrown off) are two — the lightness of the moving body, and its closeness to the point of rest; and both are infinitely susceptible of increase. But opposing these (which favor projection) there is but a single cause, and I do not understand how this, although it likewise is infinitely augmentable, can hold out alone against the conjunction and combination of the others, which are still two in number, both being infinitely augmentable.

SALV. The objection does you credit, Sagredo, and in order to shed light on it so that we can more clearly comprehend it (for you also mentioned holding it confusedly), let us define it by reducing it to a diagram, which will perhaps also bring it more easily to a solution. So let us mark thus a perpendicular line toward the center, AC, and let the horizontal line AB be at right angles to this, along which the motion of projection is made, and which the projectile would continue to follow with uniform motion if its weight did not bend it downward.

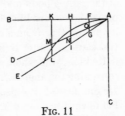

FIG. 11

Now suppose a straight line AE to be drawn from A, making any desired angle with AB, and let us mark off on AB some equal spaces AF, FH, and HK, drawing from these the perpendiculars FG, HI, and KL, down as far as AE. And since as we have remarked on other occasions the falling body starting from rest acquires always a greater degree of speed as time goes on, according to the time elapsed, we can picture the spaces AF, FH, and HK as representing equal times, and the perpendiculars FG, HI, and KL as representing the degrees of speed acquired in the said times.† Thus the degree of speed acquired in the whole time AK will, by the line KL, be represented relatively to the degree HI acquired in the time AH, and to the degree FG acquired in the time AF; which degrees KL, HI, and FG obviously have the same ratios as the times KA, HA, and FA. And if other perpendiculars are drawn from arbitrary points marked on the line FA, smaller and smaller degrees will be found *ad infinitum*, always proceeding toward the point A, which represents the first instant of time and the original state of rest. This withdrawal toward A represents the infinite diminution of the original tendency toward downward motion with the approach of the moving body to the original state of rest, which approach is infinitely augmentable.

Now let us find that other diminution of speed which can be made *ad infinitum* by decreasing the weight of the body. This will be represented by drawing another line from the point A, making a smaller angle than BAE; let this be AD. This, cutting the parallels KL, HI, and FG in the points M, N, and O, shows us the degrees FO, HN, and KM acquired in the times AF, AH, and AK to be less than the other degrees FG, HI, and KL acquired in the same times but by a heavier body, this being a lighter one. And it is obvious that by withdrawal of the line EA toward AB, restricting the angle EAB (which can be done *ad infinitum*, just as weight can be infinitely decreased), the speed of the falling body and consequently the cause that impeded its projection comes likewise to be diminished *ad infinitum*. Hence it appears that from a combined diminution *ad infinitum* of the two causes counter to it, projection cannot be impeded.

Reducing the whole argument to a few words, let us say: By restricting the angle EAB, the degrees of speed LK, IH, and GF are diminished. By also withdrawing the parallels KL, HI, and FG toward the angle A, these same degrees are diminished, and both diminutions may proceed *ad infinitum*. Therefore the downward speed of motion can indeed be diminished so much (admitting of a twofold diminution *ad infinitum*) that it no longer suffices to restore the moving body to the surface of the wheel, and consequently to impede its projection or prevent it.

On the other hand, then, in order to prevent projection taking place it is necessary that those spaces through which the projectile has to descend in order to get back to the wheel must be made so short and close that however slow the descent of the moving body may be, even if infinitely diminished, it still suffices to take it back there. Hence it would be necessary to find a diminution of these spaces which was not merely infinite, but of an infinity such as to overcome the double infinity accomplished in decreasing the downward speed of the body. But how is a magnitude to be diminished still more than one which is doubly diminished *ad infinitum?* Take note, Simplicio, just how far one may go without geometry and philosophize well about nature!

The degrees of speed, infinitely diminished by the decrease of the weight of the moving body and by the approach to the first point of motion (the state of rest), are always determinate. They correspond proportionately to the parallels included be-

tween the two straight lines meeting in an angle such as the
angle BAE, or BAD, or some other angle infinitely acute but
still rectilinear. But the diminution of the spaces through which
the moving body must go to return to the surface of the wheel is
proportional to another sort of diminution included between lines
which contain an angle infinitely narrower and more acute than
any rectilinear angle whatever,† which is as follows: Take some
point C on the perpendicular AC, and with it as center describe
the arc AM of radius CA. This will cut the parallels which determine the degrees of speed, no matter how compressed they
may be within the most acute rectilinear angle. Of those parallels,
the parts which lie between the arc and the tangent AB are the
amounts of the spaces of return to the wheel. They grow always
less than these parallels of which they are parts, and diminish
in an increasing ratio as they approach the point of contact.

Now, the parallels included between the straight lines, as they
retreat toward the angle, always diminish in the same ratio; that
is, AH being divided in the middle by the point F, the parallel HI
will be double FG, and dividing FA in the middle, the parallel
drawn from the point of division will be one-half FG. Continuing
this division *ad infinitum,* each subsequent parallel will be half
of the next preceding one. But it is not thus with the line intercepted between the tangent and the circumference of the circle;
for making the same division on FA and assuming, for example,
that the parallel through H to the arc is double that through F,
this latter will then be more than double the next one. And
continually as we come closer to the contact A, the preceding line
will contain the following line three, four, ten, a hundred, a
thousand, a hundred thousand, a hundred million times, and
more *ad infinitum.* Thus the shortness of such lines is reduced
until it far surpasses what is needed to make the projectile, however light, return to (or rather be kept on) the circumference.

SAGR. I am well satisfied with the entire argument and with its
binding force. Yet it seems to me that if anyone wanted to pursue
it further, he could raise some difficulties. He might say that, of
the two causes which make the descent of the moving body slower
and slower *ad infinitum,* it is obvious that the one which depends
upon proximity to the first point of descent increases in a constant ratio, just as the parallels always maintain the same ratio
to one another, and so on, but that it is not so obvious that the

diminution of speed dependent upon the decrease of weight in the body — which is the second cause — would also be made in this same ratio. And who guarantees that this would not be made according to the ratios of the lines intercepted between the tangent† and the circumference, or in some even greater proportion?

SALV. I have been taking it as true that the speeds of naturally falling bodies follow the proportions of their weights,† out of regard to Simplicio and Aristotle, who declares this in many places as an evident proposition. You question this in favor of my opponents, saying that it might be that the speed increases in a greater ratio than that of the weights, even infinitely greater. With this, the whole preceding argument falls to the ground. It remains for me to sustain it by telling you that the proportion of the speeds is much less than that of the weights, and in this way not only to support but to strengthen what has been said.

Of this I adduce experiment as the proof, which will show us that a weight thirty or forty times heavier than another (for example a ball of lead and another of cork) will scarcely move more than twice as fast. Now if no projection would occur when the speed of the falling body was diminished in the proportion of the weights, still less will it do so when the speed is but little diminished by much reducing the weight.

But even assuming that the speed would decrease in a much greater ratio than that with which the weight was reduced, and even if this ratio were that with which the parallels between the tangent and the circumference were diminished, I am not necessarily convinced that even the lightest materials you can think of would necessarily be projected. Indeed, I declare that they would not be; understanding, of course, not intrinsically light materials (that is, devoid of all weight and going upward by nature), but those which descend very slowly and have very little weight. What makes me believe this is that a diminution of weight made according to the ratio of the parallels between the tangent and the circumference has as its ultimate and highest term the absence of weight, just as those parallels have for their ultimate term of reduction precisely that contact which is an indivisible point. Now weight never does diminish clear to its last term, for then the moving body would be weightless; but the space of return for the projectile to the circumference does reduce to its ultimate smallness, which happens when the moving body rests upon the circumference at that very point of contact, so that no space whatever is required for its return. Therefore let the tendency to downward motion be as small as you please, yet it will always be more than enough to get the moving body back to the circumference from which it is distant by the minimum distance, which is none at all.

SAGR. The argument is truly very subtle, but nonetheless convincing, and it must be admitted that trying to deal with physical problems without geometry is attempting the impossible.

SALV. Simplicio will not say so, though I do not believe he is one of those Peripatetics who discourage their disciples from the study of mathematics as a thing that disturbs the reason and renders it less fit for contemplation.

SIMP. I would not do Plato such an injustice, although I should agree with Aristotle that he plunged into geometry too deeply and became too fascinated by it. After all, Salviati, these mathematical subtleties do very well in the abstract, but they do not work out when applied to sensible and physical matters. For instance, mathematicians may prove well enough in theory that *sphaera tangit planum in puncto,* a proposition similar to the one at hand; but when it comes to matter, things happen otherwise. What I mean about these angles of contact and ratios is that they all go by the board for material and sensible things.

SALV. Then you do not believe that the tangent touches the surface of the terrestrial globe in one point?

SIMP. Not just in one point; I believe that a straight line would go for tens and hundreds of yards touching even the surface of water, let alone the ground, before separating from it.

SALV. But don't you see that if I grant you this, it will be so much the worse for your case? For if even assuming that the tangent lies removed from the earth except at one point, it has been proven that the projectile would not be separated, because of the extreme acuteness of the angle of contact (if it can indeed be called an angle), how much less cause will it have for becoming separated if that angle is completely closed and the surface united with the tangent? Do you not see that in this way the projection would take place along the very surface of the earth, which is as much as to say that it would not be made at all? So you see that the power of truth is such that when you try to attack it, your very assaults reinforce and validate it.

Truth sometimes gains strength from being contradicted.

But since I have removed this one error for you, I should not like to leave you in that other error of considering a material sphere not to touch a plane in a single point alone. I certainly hope that a conversation of only a few hours with persons who have some knowledge of geometry will make you appear a little more knowing among those completely ignorant of it. Now to show you how great the error is of those who say, for example, that a sphere of bronze does not touch a steel plate in one point, let me ask you what you would think of anyone who might say — and stubbornly insist — that the sphere was not truly a sphere?

SIMP. I should consider him quite bereft of reason.

Even a material
sphere touches a
material plane in
but one point.

SALV. That is the state of anyone who says that the material sphere does not touch a material plane in one point, for saying so is the same as saying that the sphere is not a sphere. And to see that this is the case, tell me what the essence of a sphere consists in; that is, what is it that makes a sphere different from all other solid bodies?

Definition of a
sphere.

SIMP. I believe that the essence of a sphere consists in its having all the straight lines drawn from its center to its circumference equal.

SALV. So that if such lines were not equal, the solid would not be a sphere at all.

SIMP. No.

SALV. Next, tell me whether you believe that of many lines which may be drawn between two points, more than a single one can be straight.

SIMP. Certainly not.

SALV. But still you understand that this one straight line will necessarily be shorter than all the others.

SIMP. I understand that, and I have a clear proof of it, offered by a great Peripatetic philosopher. It seems to me, if I remember correctly, that he set it forth as a reproach to Archimedes, who assumed this to be known when he might have proved it.

SALV. This must have been a great mathematician, being able to prove what Archimedes did not know how to prove and could not prove.† If you happen to remember the demonstration, I should like to hear it; I recall quite well that Archimedes, in his books on the sphere and the cylinder, places the proposition among the postulates, so I am certain that he took it to be incapable of demonstration.

SIMP. I think I remember it, for it is very short and simple.

SALV. So much the greater the shame of Archimedes and the glory of this philosopher.

SIMP. I shall draw the figure for it.

Between the points A and B draw the straight line AB and the curve ACB, of which it is to be proved that the straight line is the shorter; the proof is this. Take a point

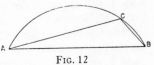

FIG. 12

A Peripatetic's proof that the straight line is the shortest of all.

C on the curve, and draw two more straight lines AC and CB, the two of which are longer than the single line AB; for Euclid proves this. But the curve ACB is greater than the two straight lines AC and CB; therefore, *a fortiori*, the curve ACB will be much greater than the straight line AB, which was to be proved.

SALV. If you were to look through all the paralogisms in the world, I do not believe that a better example than this could be found to illustrate the most majestic fallacy of all fallacies; that of proving *ignotum per ignotius*.

Paralogism of this Peripatetic who proves *ignotum per ignotius*.

SIMP. In what way?

SALV. What do you mean, "in what way?" Isn't this the unknown conclusion which you wish to prove: That the curve ACB is longer than the straight line AB? And isn't this the middle term, which you take as known: That the curve ACB is greater than the two lines AC and CB, which are known to be greater than AB? And if it is unknown that the curve is greater than the single straight line AB, why won't it be even more unknown that it is greater than the two straight lines AC and CB, which are known to be greater than just AB? Yet you take this as known.

SIMP. I still do not see what the fallacy consists in.

SALV. Since the two straight lines are greater than AB, just as Euclid knew, then whenever the curve is greater than the two straight lines ACB, will it not be greater than the single line AB?

SIMP. Certainly.

SALV. That the curve ACB is greater than the straight line AB is the conclusion; this is better known than the middle term, which is that the same curve is greater than the two straight lines AC and CB. Now when the middle term is less well known than the conclusion, one must be proving *ignotum per ignotius*.

Now back to our purpose. It is sufficient that you understand that the straight line is the shortest of all lines that can be drawn between two points. And as to the main conclusion, you say that a material sphere does not touch a plane in a single point. Then what contact does it have?

SIMP. It will be part of the surface of the sphere.

SALV. And likewise the contact of one sphere with another equal one will still be a similar portion of its surface?

SIMP. There is no reason that it should not be.

SALV. Then also the two spheres will touch each other with the same two portions of their surfaces, since each of these being adapted to the same plane, they must be adapted to each other.

Proof that the
sphere touches
the plane in a
single point.

Now imagine two spheres touching whose centers are A and B, and let their centers be connected by the straight line AB passing through their contact. Let it pass through the point C,

and take another point D in this contact, connecting the two straight lines AD and DB so that they form the triangle ADB. Then the two sides AD and DB will be equal to the other single side ACB, each of them containing two radii, which are all equal

FIG. 13

by definition of the sphere. And thus the straight line AB drawn between the two centers A and B will not be the shortest of all, the two lines AD and DB being equal to it; which you will admit is absurd.

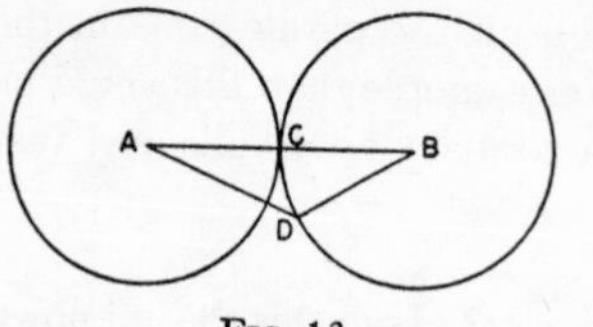

SIMP. This proves it for abstract spheres, but not material ones.

SALV. Show me then where the fallacy of my argument lies, so that it is not conclusive for material spheres although it is for immaterial and abstract ones.

Why in the ab-
stract a sphere
touches a plane
in one point
but a material
one does not in
reality.

SIMP. Material spheres are subject to many accidents to which immaterial spheres are not subjected. Why might it not be that a metallic sphere being placed upon a plane, its own weight would press down so that the plane would yield somewhat, or indeed that the sphere would be mashed at the contact? Besides it is hard to find such a perfect plane, since matter is porous, or a sphere so perfect that all its radii are exactly equal.

SALV. Oh, I readily grant you all these things, but they are beside the point. For when you want to show me that a material sphere does not touch a material plane in one point, you make use of a sphere that is not a sphere and of a plane that is no plane. By your own statement, spheres and planes are either not to be found in the world, or if found they are spoiled upon being used for this effect. It would therefore have been less bad for you to have granted the conclusion conditionally; that is, for you to have said that if there were given a material sphere and plane which were perfect and remained so, they would touch one another in a single point, but then to have denied that such were to be had.

SIMP. I think that the philosopher's proposition is to be taken in that sense, because doubtless it is the imperfection of matter which prevents things taken concretely from corresponding to those considered in the abstract.

SALV. What do you mean, they do not correspond? Why, what you are saying right now proves that they exactly correspond.

SIMP. How is that?

SALV. Are you not saying that because of the imperfection of matter, a body which ought to be perfectly spherical and a plane which ought to be perfectly flat do not achieve concretely what one imagines of them in the abstract?

SIMP. That is what I say.

SALV. Then whenever you apply a material sphere to a material plane in the concrete, you apply a sphere which is not perfect to a plane which is not perfect, and you say that these do not touch each other in one point. But I tell you that even in the abstract, an immaterial sphere which is not a perfect sphere can touch an immaterial plane which is not perfectly flat in not one point, but over a part of its surface, so that what happens in the concrete up to this point happens the same way in the abstract. It would be novel indeed if computations and ratios made in abstract numbers should not thereafter correspond to concrete gold and silver coins and merchandise. Do you know what does happen, Simplicio? Just as the computer who wants his calculations to deal with sugar, silk, and wool must discount the boxes, bales, and other packings, so the mathematical scientist (*filosofo geometra*), when he wants to recognize in the concrete the effects which he has proved in the abstract, must deduct the material hindrances, and if he is able to do so, I assure you that things are in no less agreement than arithmetical computations. The errors, then, lie not in the abstractness or concreteness, not in geometry or physics, but in a calculator who does not know how

Things in the abstract have precisely the same requirements as in the concrete.

to make a true accounting. Hence if you had a perfect sphere and a perfect plane, even though they were material, you would have no doubt that they touched in one point; and if it is impossible to have these, then it was quite beside the purpose to say *sphaera aenea non tangit in puncto.*

But I have something else to add, Simplicio: Granted that a perfect material sphere cannot be given, nor a perfect plane, do you believe it would be possible to have two material bodies with their surfaces curved in some places as irregularly as you pleased?

SIMP. I believe there is no shortage of such ones.

SALV. If there are such, then they also touch in one point; for meeting in a single point is not at all a special privilege of the perfect sphere and a perfect plane. Rather, anyone who got to the bottom of this matter would find that it is a great deal harder to discover two bodies which touch with parts of their surfaces than with a point alone. For to have two surfaces fit together well, either both must be exactly flat, or if one is convex, the other must be concave with a curvature which exactly corresponds to the convexity of the other. Such conditions are much more difficult to find, because of their too strict determinacy, than those others in which their random shapes are infinite in number.

SIMP. Then you think that two stones or bits of iron taken at random and brought together will touch each other in a single point most of the time?

SALV. In casual encounters I think not, as there will usually be some little yielding dirt on them, and because they are not brought together carefully without any striking, a very little of which suffices to make one surface yield to the other a bit so that they mutually take on each other's imprint, at least in some small portion. But if the surfaces were well scoured and both were placed upon a table so that one could not bear down upon the other, and then if one were gently pushed toward the other, I have no doubt that they could be brought into simple contact at a single point.

SAGR. With your permission, I must bring up a certain difficulty of mine, inspired in me by hearing Simplicio adduce the impossibility of finding a material and solid body which would be perfectly spherical in shape, and by seeing Salviati lend assent

To touch in one point is a property not of the perfect sphere alone, but of all curved figures.

It is more difficult to find figures which touch with parts of their surfaces than with one point alone.

to this by not contradicting it. Now I should like to know whether there would be the same difficulty about forming a solid of some other shape; or, to express myself better, whether the greater trouble would be encountered in forming from a block of marble a perfect sphere or pyramid, or a perfect horse or grasshopper.

SALV. I shall give you an answer to your first question, but first I apologize for the apparent assent I gave to Simplicio. I did that merely for the time being, because before I went into this matter I had it in mind to say what is perhaps the same idea as yours, or one very much like it. Replying to your first question, I say that if any shape can be given to a solid, the spherical is the easiest of all, as it is the simplest, and holds that place among all solid figures which the circle holds among surfaces — the description of the circle, being easiest of all, having been considered by mathematicians as alone worthy of being placed among the postulates underlying the description of all other shapes. The formation of a sphere is so easy that if a circular hole is bored in a flat metal plate and any very roughly rounded solid is rotated at random within it, it will without any other artifice reduce itself to as perfect a spherical figure as possible, so long as the solid is not smaller than a sphere which would pass through the hole. And what is even more worthy of consideration is that spheres of various sizes may be formed within the same hole. But when it comes to forming a horse or, as you say, a grasshopper, I leave it to you to judge, for you know that few sculptors in the world are equipped to do that. I believe that Simplicio will not disagree with me as to this particular.

SIMP. I do not know that I disagree with you at all. My opinion is that none of the shapes named can be perfectly obtained, but to approximate one as nearly as possible to the most perfect degree, I believe that it would be incomparably easier to reduce a solid to a spherical shape than to the form of a horse or a grasshopper.

SAGR. And upon what do you think that this higher degree of difficulty would depend?

SIMP. Just as the great ease of forming a sphere stems from its absolute simplicity and uniformity, so an extreme irregularity makes the production of the other figures difficult.

SAGR. Then since the irregularity is the cause of the difficulty, even the shape of a rock broken at random with a hammer would

Spherical shape more easily formed than any other.

Circular shape alone is placed among the postulates.

Spherical figures of different sizes may be formed with a single instrument.

Irregular forms difficult to produce.

be among the shapes hard to produce, this being perhaps even more irregular than a horse?

SIMP. It should be as you say.

SAGR. But tell me: Whatever form this rock has, does it have this perfectly, or not?

SIMP. That which it has, it has so perfectly that nothing else corresponds to it so exactly.

SAGR. Well, if of the shapes which are irregular, and hence hard to obtain, there is an infinity which are nevertheless perfectly obtained, how can it be right to say that the simplest and therefore the easiest of all is impossible to obtain?

SALV. Please, gentlemen, it seems to me that we have gone off woolgathering. Since our arguments should continue to be about serious and important things, let us waste no more time on frivolous and quite trivial altercations. Please let us remember that to investigate the constitution of the universe is one of the greatest and noblest problems in nature, and it becomes still grander when directed toward another discovery; I refer to that of the cause of the flow and ebb of the sea, which has been sought by the greatest men who ever lived and has perhaps been revealed by none. Therefore if nothing remains to be brought up for the complete explanation of the objection derived from the whirling of the earth, which was the last thing adduced as an argument for its being motionless with respect to its own center, let us get on to the scrutiny of the evidence for and against its annual motion.

SAGR. Salviati, I should not like you to measure the minds of us others with the yardstick of your own. You, having always occupied yours with the highest meditations, consider low and frivolous those which we take to be food for thought. But sometimes, just to please us, do not disdain to unbend and grant something to our curiosity. Thus, as to the explanation of the last objection, taken from the casting off of things by the diurnal whirling, much less would have satisfied me than what you produced; yet even the extra materials were so fascinating to me that not only did they not weary my mind, but by their novelty they have drawn me along with as much delight as I could wish for. So if any other reflections remain to be added by you, bring them forth, and for my part I shall be very glad to hear them.

SALV. I have always taken great joy in the things I have found

Constitution of
the universe
among the
noblest of
problems.

out, and next to this greatest pleasure I rank that of discussing them with a few friends who understand them and show a liking for them. Now, since you are one of these, I shall loosen the reins a little on my ambition (which much enjoys itself when I am showing myself to be more penetrating than some other person noted for his acuity) and I shall for good measure add to the last discussion one more fallacy on the part of the followers of Ptolemy and Aristotle, selected from an argument already set forth.

SAGR. You may see how eagerly I await to hear it.

SALV. Up to this point we have made no issue about granting it to Ptolemy as an unquestionable fact that since the casting off of the stone is caused by the speed of the moving wheel about its center, the cause of this casting off is augmented as the speed of whirling is increased. From this it was inferred that on account of the rapidity of the terrestrial whirling being very much greater than that of any machine which we can rotate artificially, the consequent extrusion of stones, animals, etc. should be very violent.

I now take note that there is a very gross fallacy in this argument when we indiscriminately compare such speeds with each other absolutely. It is true that if I make a comparison between speeds of the same wheel, or of two equal wheels, then that which is turned the more rapidly will hurl stones with the greater impetus, and when the speed increases the cause of projection will increase also in the same ratio. But now suppose the speed to be made greater not by increasing the speed of a given wheel (which would be done by making it have a larger number of revolutions in the same time), but by increasing the diameter and enlarging the wheel, preserving the same time for each revolution of the large wheel as of the small one. The velocity would now be greater in the large wheel merely by reason of its greater circumference. No one would suppose the cause for extrusion to increase in the ratio of the speed of its rim to that of the smaller wheel; that would be quite false, as may be shown at once by a ready experiment, roughly as follows. We can throw a stone better with a stick a yard long than with one six yards long, even if the motion of that end of the long stick where the stone is stuck is more than twice as fast as the motion of the end of the shorter stick — as it would be if the speeds were such that during one

Cause for projection does not grow in proportion to speed when this is increased by enlarging the wheel.

complete revolution of the larger stick, the smaller one made three turns.

SAGR. I completely understand that what you are telling me must necessarily take place as you say, Salviati. But I do not readily see why equal speeds should operate unequally in the extrusion of projectiles, being much more active in casting off from smaller than from larger wheels. Therefore I beg you to disclose to me how this takes place.

SIMP. Well, Sagredo, this time you do not seem to be quite up to your own standard. Usually you see through everything in an instant, yet now you are overlooking a fallacy that has crept into the stick experiment which I have been able to detect. This is the different manner of operation in making a cast with a short stick and with a long one. For in order to have the stone fly out of the notch, you must not continue the motion uniformly, but just when it is fastest you must check your arm and restrain the speed of the stick. By this means the swiftly moving stone will fly off impetuously. Now, you could not thus check the longer stick, which, because of its length and flexibility, would not completely obey the restraint of your arm but would continue to accompany the rock through some distance, keeping in contact with a gentle restraint and not letting it escape as it would if the stick had struck against some solid obstacle. For if both sticks struck against some restraint which checked them, I believe that the stone would fly from one just as from the other, even though their motions were of equal speed.

SAGR. With your permission, Salviati, I shall make some reply to Simplicio since he has challenged me. I say that in his argument there is both good and bad; good, in that most of it is true, and bad because it is entirely beside the point. It is quite true that the stones will travel forward impetuously if that which is swiftly carrying them strikes against an immovable obstacle. This agrees with the effect which is seen every day in a boat traveling briskly which runs aground or strikes some obstacle; everyone aboard, being caught unawares, tumbles and falls suddenly toward the front of the boat. If the terrestrial globe should encounter an obstacle such as to resist completely all its whirling and stop it, I believe that at such a time not only beasts, buildings, and cities would be upset, but mountains, lakes, and seas, if indeed the globe itself did not fall apart. But all this has nothing to do with our purpose. We are speaking of what may

Given that the diurnal whirling is the earth's and that it were suddenly stopped by some obstacle or hindrance, then buildings, mountains, and perhaps the whole globe would dissolve.

follow from the earth's motion of turning uniformly and placidly upon itself, however great its speed may be.

Likewise what you say about the sticks is partly true, but Salviati did not bring this up as something exactly corresponding to the matters we are dealing with. It is merely a rough example which is able to arouse the mind to investigate more accurately whether the speed, in whatever manner it is increased, increases the cause of projection in the same ratio. For instance, if a wheel ten yards in diameter moved in such a manner that a point on its circumference traveled one hundred yards per minute, and thereby had the power (*impeto*) with which to hurl a stone, would that power be increased a hundred thousand times in a wheel one million yards in diameter? Salviati denies that it would, and I am inclined to agree with him; but not knowing the reason for this, I have asked him for it and am awaiting it with interest.

SALV. What I am here for is to give you as much satisfaction as my abilities permit, and although it may have seemed to you at first that I was investigating things foreign to our purpose, still I believe that as the argument progresses we shall find that to be not so at all. But let Sagredo tell me those things in which he has observed the resistance of any moving body to consist.

SAGR. At present the only internal resistance to being moved which I see in a movable body is the natural inclination and tendency it has to an opposite motion. Thus in heavy bodies, which have a tendency toward downward motion, the resistance is to upward motion.

I said "internal resistance" because I believe that this is what you meant, and not external resistances, which are many and accidental.

SALV. It is what I meant, and your perspicacity has defeated my cunning. But if I have held back something in asking the question, I wonder whether Sagredo has been completely adequate in satisfying it with his answer, or whether there is not in the movable body, besides a natural tendency in the opposite direction, another intrinsic and natural property which makes it resist motion. So tell me once more: Do you not believe that the tendency of heavy bodies to move downward, for example, is equal to their resistance to being driven upward?†

SAGR. I believe it to be exactly so, and it is for this reason that two equal weights in a balance are seen to remain steady and in

The tendency of heavy bodies to downward motion equals their resistance to upward motion.

equilibrium, the heaviness of one weight resisting being raised by the heaviness with which the other, pressing down, seeks to raise it.

SALV. Very well. So to have one raise the other, it would be necessary to add weight to the one pressing down, or subtract weight from the other. But if the resistance to upward motion consists only in heaviness, how does it happen that in a balance with unequal arms (that is, in a steelyard), a weight of one hundred pounds with its downward pressure (*gravare*) may be insufficient to raise one of four pounds which resists it, and that this latter one of four pounds, by sinking, may raise up one hundred? For such is the effect of the steelyard's counterweight upon the heavy object that we wish to weigh. If resistance to being moved resided in heaviness alone, how could the steelyard counterweight of only four pounds resist the weight of a bale of wool or silk which will be eight hundred or a thousand, or even be able to overcome the bale with its moment (*momento*) and raise it up? So one must admit, Sagredo, that another resistance and another power (*forza*) than that of simple heaviness are being dealt with here.

SAGR. There is no escaping it, but tell me what this second force (*virtù*) is.

SALV. It is that which did not exist in the equal-armed balance. Consider what there is that is new in the steelyard, and therein lies necessarily the cause of the new effect.†

SAGR. I believe that your probing has caused something to stir vaguely in my mind. In both instruments, weight and motion are involved; in the balance, the movements are equal and therefore one weight must exceed the other in heaviness in order to move. In the steelyard, the lesser weight moves the greater only when the latter moves very little, being weighed at the lesser distance, and the former moves quite a way, hanging at the greater distance. One must say, then, that the smaller weight overcomes the resistance of the greater by moving much when the other moves little.

SALV. Which is to say that the speed of the less heavy body offsets the heaviness of the weightier and slower body.

Greater speed
exactly compen-
sates for in-
creased weight. SAGR. But do you believe that this speed exactly compensates that heaviness? That is, that the moment and the power of a moving body of say four pounds weight are as much as those of

a body weighing one hundred, whenever the former has one
hundred units of speed and the latter only four units?

SALV. Certainly, as I can show you by many experiments. But
for the present let this single confirmation by the steelyard be
enough for you. In this you see that the light steelyard counterweight is enough to sustain and balance the very heavy bale,
when its distance from the center on which the steelyard is suspended, and about which it turns, is as many times the lesser
distance from there to where the bale hangs, as is the absolute
weight of the bale when compared with that of the steelyard
counterweight. And for this inability of the huge bale with its
weight to lift up the counterweight, so much lighter, one can see
no other cause than the disparity in the movements which must
be made by each of them when the bale, by descending a single
inch, makes the counterweight go up a hundred inches. It is here
assumed that the bale weighs one hundred counterweights, and
that the distance of the counterweight from the center of the
steelyard is one hundred times the distance between that same
center and the suspension point of the bale. For it is the same
to say that the counterweight is to be moved a space of one hundred inches while the bale is moved a single inch, as to say that
the speed of motion of the counterweight is one hundred times
the speed of motion of the bale.

Now fix it well in mind as a true and well-known principle that
the resistance coming from the speed of motion compensates
that which depends upon the weight of another moving body,
and consequently that a body weighing one pound and moving
with a speed of one hundred units resists restraint as much as
another of one hundred pounds whose speed is but a single unit.†
And two equal movable bodies will equally resist being moved if
they are to be made to move with equal speed. But if one is to
move faster than the other, it will make the greater resistance
according as the greater speed is to be conferred upon it.

These things asserted, let us get to the explanation of our
problem, and for easier comprehension let us make a little diagram for it.

Let there be two unequal wheels around this center A, BG
being on the circumference of the smaller, and CEH on that of
the larger, the radius ABC being vertical to the horizon. Through
the points B and C we shall draw the tangent lines BF and CD,

and in the arcs BG and CE we take two arcs of equal length, BG and CE. The two wheels are to be understood as rotating about their center with equal speed in such a way that two moving bodies will be carried along the circumferences BG and CE with equal speeds. Let the bodies be, for instance, two stones placed at B and C, so that in the same time during which stone B travels over the arc BG, stone C will pass the arc CE.

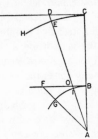

Fig. 14

Now I say that the whirling of the smaller wheel is much more powerful at projecting the stone B than is the whirling of the larger wheel at projecting the stone C. And since, as already explained, the projection would be along the tangent, if the stones B and C should be separated from their wheels and commence motions of projection from the points B and C, they would be flung along the tangents BF and CD by the impetus received from whirling. The two stones therefore have equal impetuses for traveling along the tangents BF and CD, and if no other power were to deviate them, it is along these that they would travel. Isn't that so, Sagredo?

SAGR. That is the way it seems to me the thing takes place.

SALV. But what power do you think could deviate the stones from moving along the tangents, where the impetus of whirling actually casts them?

SAGR. Either their own weight, or some glue which may hold them in place attached to the wheels.

SALV. But to deviate a moving body from a motion for which it has the impetus, is not a greater or a lesser power needed, according as the deviation must be greater or less?† That is, according as they must in this deviation pass through a greater or a lesser space in a given time?

SAGR. Yes. For it was already concluded above that in order to make a body move, the faster it is to be moved the greater must be the moving force.

SALV. Well, consider how in order to deviate the stone on the smaller wheel from the motion of projection that would be made along the tangent BF, and to keep it attached to the wheel, its weight would have to be pulled back as far as the secant FG, or rather the perpendicular drawn from the point G to the line BF, whereas on the larger wheel the withdrawal would need to be no more than the secant DE, or rather the perpendicular drawn from the point E to the tangent DC. This is much less than FG, and always less and less, the larger the wheel is made. And since these withdrawals have to be made in equal times (i.e., while the two equal arcs BG and CE are being traversed), that of stone B (viz., the retraction FG) will have to be much faster than the other, DE. Therefore much more force is needed to hold the stone B joined to its small wheel than the stone C to its large one, which is the same as to say that a smaller thing will hinder projection from the large wheel than will prevent it on the small one. And thus it is obvious that the larger the wheel becomes, the more the cause for projection is diminished.†

SAGR. From what I now understand, thanks to your lengthy analysis, I think I can satisfy my own mind with a very brief argument. Thus an equal impetus along the tangents being impressed upon both stones by the equal speed of the wheels, the large circumference is seen by reason of its small separation from the tangent to favor, in a way, and to cloy with dainty bites the appetite (so to speak) that the stone has for leaving the circumference; hence any small retention, whether from the stone's own tendency or from some glue, is enough to keep it joined. This remains useless for accomplishing the same on the small wheel, which, little favoring the direction of the tangent, tries too greedily to retain the stone, and (the glue being no stronger than that which holds the other stone united to the larger wheel) this stone loses hold and runs along the tangent.

Meanwhile I am not only convinced that I was wrong about all this, having believed that the cause for projection grew according as the speed of whirling increased, but now I have begun to consider the following. Since the casting off diminishes with the enlargement of the wheel, it might be true that to have the large wheel extrude things as does the small one, its speed would have to be increased as much as its diameter, which would be the case when their entire revolutions were finished in equal times. And thus it might be supposed that the whirling of the earth would no more suffice to throw off stones than would any other wheel, as small as you please, which rotated so slowly as to make but one revolution every twenty-four hours.

SALV. We shall not look further into this right now; it suffices that we have abundantly shown (unless I am much mistaken) the ineffectiveness of the argument which at first glance appeared to be very conclusive and has been deemed so by many great men. I shall consider the time and the words well spent if I have made some headway toward convincing even Simplicio, I shall not say of the motion of the earth, but at least that the conviction of those who do believe in it is not as ridiculous and foolish as the rank and file of philosophers hold it to be.

SIMP. The solutions produced so far to the objections raised against the diurnal revolution of the earth (i.e., those taken from the fall of heavy bodies from the top of a tower, from projections perpendicularly upward or with any lateral inclination toward the east, west, south, north, etc.) have to some extent diminished in me the ancient disbelief leveled against such an opinion. But now there are other great difficulties turning over in my mind, from which I shall certainly never be able to escape. I believe that perhaps you yourselves would not be able to solve them, and it may be that they have never come to your ears, being quite recent. These are the refutations by two authors who write *ex professo* against Copernicus; the first are to be read in a booklet of scientific theses,† and the others have been inserted by a great philosopher and mathematician in a treatise of his in favor of Aristotle and his opinion about the inalterability of the heavens. In this is proved that not only the comets but the new stars (that is, the one of 1572 in Cassiopeia and that of 1604 in Sagittarius†) were not above the spheres of the planets at all, but were actually beneath the moon's orbit in the elemental sphere. And he proves this against Tycho, Kepler, and many other astronomical observers, beating them with their own weapons; that is, by means of parallaxes. If you like, I can produce arguments from both authors, because I have attentively read them more than once, and you can examine their force and say how they look to you.

SALV. Our main goal being to bring forth and consider everything that has been adopted for and against the two systems, Ptolemaic and Copernican, it would not be good to pass by anything written on this subject.

SIMP. Then I shall begin with the objections contained in the booklet of theses and later take up the others. First, the author cleverly calculates how many miles per hour a point on the

Further objections by two modern authors against Copernicus.

earth's surface travels at the equator, and how many at other points, in other latitudes. Not content with investigating such movements in hourly times, he finds them also in minutes, and still unsatisfied with minutes, he pursues them down to a single second. Moreover, he goes on to show precisely how many miles would be traveled in such a time by a cannon ball placed in the moon's orbit, assuming this orbit to be as large as figured by Copernicus himself, so as to take away every subterfuge from his adversary. These very ingenious and elegant reckonings made, he shows that a heavy body falling from there would consume rather more than six days to get to the center of the earth, toward which heavy bodies tend naturally.

Now if by Divine power, or by means of some angel, a very large cannon ball were miraculously transported there and placed vertically over us and released, it is indeed a most incredible thing (in his view and mine) that during its descent it should keep itself always in our vertical line, continuing to turn with the earth about its center for so many days, describing at the equator† a spiral line in the plane of the great circle, and at all other latitudes spiral lines about cones, and falling at the poles in a simple straight line.

The great improbability of this he then establishes and confirms by advancing, through his method of interrogation, many difficulties which it is impossible for the followers of Copernicus to remove; these are, if I remember correctly . . .

SALV. Just a moment, please, Simplicio. You do not want to lose me with so many new things at one stretch; I have a poor memory, so I have to go step by step. And since I remember having already calculated how long it would take such a heavy body falling from the moon's orbit to arrive at the center of the earth, and seem to recall that it would not take this long, it would be good for you to explain what rule this author employed in his computation.

SIMP. In order to prove his point *a fortiori,* he has made matters very advantageous for the opposing side by assuming that the speed of the body falling in a vertical line to the center of the earth would equal that of its circular motion in the great circle of the moon's orbit, which would be equivalent to going 12,600 German miles† per hour — a thing which really smacks of the impossible. Still, in order to excel in caution and to give every

First objection from the modern author of the booklet of theses.

A cannon ball would take more than six days in falling from the moon's orbit to the center of the earth, according to the opinion of this modern author.

advantage to the other side, he supposes this to be true, and he concludes that the time of fall in any case would be more than six days.

SALV. And is that all there is to his method? Does he prove in this way that the time of fall must be more than six days?

SAGR. I think he conducted himself too discreetly, since having it within his arbitrary power to give any speed he wished to such a falling body, and consequently to make it get to the earth in six months, or even six years, he contented himself with six days. But please, Salviati, restore my good humor somewhat by telling me in what manner your calculation proceeded, since you say you have once made it, for I am satisfied that if the question had not required some brilliant work you would not have put your mind to it.

SALV. Sagredo, it is not enough for a conclusion to be noble and great; the point is in treating it nobly. Who does not know that in the dissection of some organ of an animal, there may be discovered infinite marvels of provident and most wise nature? Yet for every animal that the anatomist cuts up, a thousand are quartered by the butcher. Now in trying to satisfy your request, I do not know in which of the two costumes I shall make my appearance on the stage; yet taking heart from the spectacle put on by this author of Simplicio's, I shall not hold back from telling you — if I can remember it — the method which I used.

But before setting to work, I cannot help saying that I very much doubt whether Simplicio has faithfully related the method by which this author of his found that the cannon ball would consume more than six days in coming from the moon's orbit clear to the center of the earth. For if he assumed that its speed of descent was equal to its speed in the orbit, as Simplicio says he assumed, he would stand exposed as quite ignorant of even the most elementary and simplest knowledge of geometry. It is indeed remarkable to me that Simplicio himself, in granting this assumption he tells us of, does not see the enormous absurdity contained in it.

SIMP. It may be that I have erred in relating it, but it is certain that I perceive no fallacy in it.

SALV. Maybe I did not quite understand what you recited. Didn't you say that this author makes the speed of the ball in descent equal to that which it would have going around in the moon's orbit, and that falling with such a velocity it would get to the center in six days?

SIMP. I think that that is what he wrote.

SALV. And you do not see so gross an absurdity? But of course you are pretending, for you cannot be ignorant that the radius of a circle is less than one-sixth of its circumference, and that consequently the time in which the moving body would pass over the radius would be less than one-sixth the time in which, moving with the same speed, it would travel around the circumference. Therefore the ball, descending with the speed with which it moved in the curve, would arrive at the center in less than four hours; that is, assuming that in the curve it would complete one revolution in twenty-four hours, as would have to be supposed in order for it to remain always in the same vertical line.

SIMP. Now I understand the error quite well, but I do not wish to attribute it to him undeservedly. It must be that I made a mistake in reciting this argument of his, and, in order to avoid taking responsibility for the others, I should like to have his book. If there were anyone to go and fetch it, I should appreciate it very much.

SAGR. A servant can be sent in haste, and there need be no time wasted at all, for meanwhile Salviati will favor us with his computation.

SIMP. Let him go, for he will find it open on my desk, together with that other one which argues against Copernicus.

SAGR. Let us have that brought too, just to make sure.

And now Salviati will make his calculation; I have dispatched a servant.

SALV. First of all, it is necessary to reflect that the movement of descending bodies is not uniform,† but that starting from rest they are continually accelerated. This fact is known and observed by all, except the modern author mentioned, who, saying nothing about acceleration, makes the motion uniform. But this general knowledge is of no value unless one knows the ratio according to which the increase in speed takes place, something which has been unknown to all philosophers down to our time. It was first discovered by our friend the Academician, who, in some of his yet unpublished writings,† shown in confidence to me and to some other friends of his, proves the following.

The acceleration of straight motion in heavy bodies proceeds

Gross absurdity in the argument based upon the ball falling from the moon's orbit.

Precise computation of the time of fall of the cannon ball from the moon's orbit to the earth's center.

Acceleration of
natural motion
of heavy bodies
is in proportion
to the odd num-
bers commencing
from unity.

Spaces passed
over by the fall-
ing heavy body
are as the squares
of the times.

Whole new
science of the
Academician
regarding local
motion.

according to the odd numbers beginning from one. That is, mark-
ing off whatever equal times you wish, and as many of them, then
if the moving body leaving a state of rest shall have passed
during the first time such a space as, say, an ell, then in the second time it will go three ells; in the third, five; in the fourth,
seven, and it will continue thus according to the successive odd
numbers. In sum, this is the same as to say that the spaces passed
over by the body starting from rest have to each other the ratios
of the squares of the times in which such spaces were traversed.
Or we may say that the spaces passed over are to each other as
the squares of the times.

SAGR. This is a remarkable thing that I hear you saying. Is there
a mathematical proof of this statement?

SALV. Most purely mathematical, and not only of this, but of
many other beautiful properties belonging to natural motions
and to projectiles also, all of which have been discovered and
proved by our friend. I have seen and studied them all, to my
very great delight and amazement, seeing a whole new science
arise around a subject on which hundreds of volumes have been
written; yet not a single one of the infinite admirable conclusions
within this science had been observed and understood by anyone
before our friend.

SAGR. You are taking away from me my desire to proceed with
the discussions we have commenced, in order just to hear some
of the demonstrations you hint of. So tell them to me at once,
or at least give me your word that you will hold a special session
with me, Simplicio being present if he should wish to learn the
properties and attributes of the most basic effect in nature.

SIMP. Indeed I should; though as to what belongs to physical
science, I do not believe it necessary to get down to minute details. A general knowledge of the definition of motion and of the
distinction between natural and constrained motion, uniform
and accelerated motion, and the like, is sufficient. For if these
were not enough, I do not believe that Aristotle would have neglected to teach us everything that was lacking.

SALV. That might be. But let us waste no more time on this, for
I promise to spend half a day on it separately for your satisfaction. Indeed, I now remember having once before promised you
this same satisfaction. Getting back to our calculation, already
begun, of the time in which a heavy body would fall from the

moon's orbit all the way to the center of the earth, and in order not to proceed arbitrarily or at random, but with a rigorous method, let us first seek to make sure by experiments repeated many times how much time is taken by a ball of iron, say, to fall to earth from a height of one hundred yards.

SAGR. And taking for this purpose a ball of determinate weight, the same as that for which we shall make the computation of the time of descent from the moon.

SALV. That makes no difference at all, for a ball of one, ten, a hundred, or a thousand pounds will all cover the same hundred yards in the same time.

SIMP. Oh, that I do not believe, nor does Aristotle believe it either; for he writes that the speeds of falling heavy bodies have among themselves the same proportions as their weights.†

SALV. Since you want to admit this, Simplicio, you must also believe that a hundred-pound ball and a one-pound ball of the same material being dropped at the same moment from a height of one hundred yards, the larger will reach the ground before the smaller has fallen a single yard. Now try, if you can, to picture in your mind the large ball striking the ground while the small one is less than a yard from the top of the tower.

Aristotle's error in affirming that heavy falling bodies move in proportion to their weights.

SAGR. I have no doubt in the world that this proposition is utterly false, but I am not quite convinced that yours is completely true; nevertheless I believe it because you affirm it so positively, which I am sure you would not do unless you had definite experiments or rigid proofs.

SALV. I have both, and when we deal separately with the subject of motion I shall communicate them to you. Meanwhile, in order not to break the thread again, let us suppose we want to make the computations for an iron ball of one hundred pounds which in repeated experiments falls from a height of one hundred yards in five seconds.† Since, as I have told you, the distances measured by the falling body increase according to the squares of the times, and one minute being twelve times five seconds, if we multiply 100 yards by the square of 12, which is 144, we shall get 14,400 as the number of yards which the same moving body will travel in one minute. And following the same rule, since an hour is 60 minutes, multiplying 14,400 (the number of yards passed in one minute) by the square of 60, that is, by 3,600, the number of yards passed in one hour becomes 51,840,000, which is 17,280 miles. And if we wish to know the space covered in four hours, we may multiply 17,280 by 16, which is the square of 4, and this becomes 276,480 miles, which is much greater than the distance from the lunar orbit to the center of the earth. The latter is 196,000 miles, taking the distance of that orbit to be 56 times the radius of the earth (as this modern author does) and the radius of the earth to be 3,500 miles of 3,000 yards to the mile, these being our Italian miles.

Therefore, Simplicio, you see that that space from the orbit of the moon to the center of the earth, which your computer said could not be passed over in six days, would be passed in much less than four hours, when the calculation is made from experiment and not by rule of thumb. Making the computation exactly, it is covered in 3 hours, 22 minutes, and 4 seconds.

SAGR. My dear sir, please do not cheat me out of this exact calculation, for it must be a very elegant affair.

SALV. So it is, really. Therefore having, as I said, by careful experiment observed that such a moving body falling from a height of 100 yards covers this in 5 seconds, let us say: If 100 yards are passed in 5 seconds, 588,000,000 (for that is 56 radii of the earth) would be covered in how many seconds? The rule for this operation is to multiply the third number by the square of the second; this comes out to 14,700,000,000, which must be divided by the first number, that is by 100, and the square root of the quotient — which is 12,124 — is the number sought. This is 12,124 seconds, which is 3 hours, 22 minutes, and 4 seconds.

SAGR. Now I have seen the operations, but I understand nothing of the reasons for working thus, nor does this seem to be the time to ask about them.

SALV. Indeed I wish to tell you, even though you do not ask, for it is very easy. Let us designate these three numbers by the letters A for the first, B for the second, and C for the third; A and C are the numbers for the spaces, and B is the number for the time; the fourth number is sought, which is also a time.

100	5	588,000,000	
A	B	C	25
1		14,700,000,000	
22		359 56	
241		10	60) 12124
2422			202
24244			3

We know that whatever proportion the space A has to the
space C, the square of the time B must have to the square of the
time sought. Therefore, by the Rule of Three, the number C is
multiplied by the square of the number B, this product is divided
by the number A, and the quotient will be the square of the number sought, its square root being that same required number.
Now you see how easy it is to understand.

SAGR. So are all truths, once they are discovered; the point is
in being able to discover them. I am quite convinced, and much
obliged to you. If any more curiosities remain in this matter, I
beg you to tell me them. For if I may speak frankly I may say,
saving Simplicio's presence, that from your discussions I always
learn something new and beautiful, whereas from those of his
philosophers I don't know that I have ever learned anything of
importance.

SALV. Plenty remains to be said about these local motions, but
according to our agreement we should reserve them for a separate
session. Right now I shall say something pertaining to this author produced by Simplicio, to whom it appears that he has given
a great advantage to his opponents by conceding that this cannon
ball, in falling from the moon's orbit, would go with the same
speed with which it would move around if it remained there and
partook of the diurnal rotation. Now I tell him that this ball
falling from the orbit to the center would acquire a degree of
speed far more than double that of the diurnal rotation in the
lunar orbit, and I shall demonstrate this with assumptions that
are quite correct and not arbitrary.

You must therefore know that the falling body, ever acquiring
new speed according to the ratios already mentioned, wherever
it may be in the line of its motion it will have such a degree of
velocity that were it to continue to move uniformly with this,
then in a second time equal to that of its previous descent it
would traverse twice the distance already passed over. Thus, for
example, if this ball in falling from the lunar orbit to its center
has consumed 3 hours, 22 minutes, and 4 seconds, I say that at
the center it will be found to have such a degree of speed that
without increasing this further it could continue to move uniformly and pass over in another 3 hours, 22 minutes, 4 seconds
the double of that space, which is as much as the entire diameter
of the lunar orbit.

The falling body,
moving uniform-
ly for an equal
time with the
degree of veloc-
ity acquired,
would pass over
double the space
passed during its
accelerated
motion.

Since from the moon's orbit to its center is some 196,000 miles, which the ball covers in 3 hours, 22 minutes, 4 seconds, then according to what has been said, if the ball continued to move with the speed which it has on arriving at the center, it would travel in another 3 hours, 22 minutes, 4 seconds a space of twice that, or 392,000 miles. But the same ball staying in the moon's orbit, which is 1,232,000 miles around, and moving along it in the diurnal motion, would make in the same 3 hours, 22 minutes, 4 seconds 172,880 miles, which is much less than half of 392,000 miles. So you see that the motion of the orbit is not what this modern author says; that is, a velocity impossible for the falling ball to participate in.

SAGR. All would be well with this argument, and it would satisfy me, if I could be sure of that part about the body moving through double the space already fallen, in another time equal to that of its descent, when it continued to move uniformly with the maximum speed acquired in descending. This proposition was once before assumed by you to be true, but was not proved.

SALV. This is one of our friend's proofs, and you will see it in good time. Meanwhile I wish to set forth some conjectures, not to teach you anything new, but to take away from you a certain contrary belief and to show you how matters may stand. Have you not observed that a ball of lead suspended from the ceiling by a long, thin thread, when we remove it from the perpendicular and release it, will spontaneously pass beyond the perpendicular almost the same amount?

SAGR. I have indeed observed that, and I have seen (especially when the ball is very heavy) that it rises so little less than it descends that I have sometimes thought the ascending arc would be equal to the descending one, and wondered whether the oscillations could perpetuate themselves. And I believe that they would, if the impediment of the air could be removed, which, with its resistance to being parted, holds back a little and would impede the motion of the pendulum. But the hindrance is small indeed, as argued by the large number of vibrations made before the moving ball is completely stopped.

SALV. The motion would not perpetuate itself, Sagredo, even if the impediment of the air were completely removed, for there is another one which is much more recondite.

SAGR. And what is that? None other occurs to me.

SALV. It will please you very much to learn of it, but I shall tell

Motion of
pendant heavy
bodies would be
perpetual if
impediments
were removed.

it to you later; meanwhile, let us continue. I have put forth the observation of the pendulum so that you would understand that the impetus acquired in the descending arc, in which the motion is natural, is able by itself to drive the same ball upward by a forced (*violento*) motion through as much space in the ascending arc; by itself, that is, if all external impediments are removed. I believe also that you understand without any trouble that just as in the descending arc the velocity goes on increasing to the lowest point of the perpendicular, so in the ascending arc it keeps diminishing all the way to the highest point. The latter speed diminishes in the same ratio in which the former is augmented, so that the degrees of speed at points equally distant from the lowest point are equal to each other. From this it seems possible to me (arguing with a certain latitude) to believe that if the terrestrial globe were perforated through the center, a cannon ball descending through the hole would have acquired at the center such an impetus from its speed that it would pass beyond the center and be driven upward through as much space as it had fallen, its velocity beyond the center always diminishing with losses equal to the increments acquired in the descent; and I believe that the time consumed in this second ascending motion would be equal to its time of descent. Now if, in progressively diminishing until totally extinguished, the highest speed which the ball has at the center conducts it in as much time through as much space as it had passed through in acquiring speed — from none at all up to the highest degree — it certainly seems reasonable that if it were always to move with this highest degree of speed, it would pass through both these distances in an equal amount of time. For if we mentally divide these speeds into increasing and decreasing degrees — as for example in the numbers to the right — so that the first increase up to 10, and the rest decrease down to 1; and then if the former (of the descending time) and the others (of the ascending time) are added together, it is seen that they make the same sum as if one of the two parts had been made up of the highest degree throughout. Therefore all the space passed through with all the degrees of speed, increasing and decreasing (which in this case is the entire diameter), must be equal to the space passed in as many of the maximum speeds as number one-half the total of the increasing and the decreasing ones. I

If the terrestrial globe were tunneled through, a heavy body would pass beyond the center and ascend by as much space as that of its descent.

1
2
3
4
5
6
7
8
9
10
10
9
8
7
6
5
4
3
2
1

know I have expressed this obscurely, and only hope that it is understood.

SAGR. I think I understood it well enough; indeed, I can show in a few words that I did so. You meant that commencing from rest and progressively increasing the velocity by equal ad- 0
ditions, which are those of the successive integers beginning 1
with 1, or rather with 0 (which represents the state of rest), 2
and arranging these thus and taking consecutively as many 3
as you please, so that the minimum degree is 0 and the maxi- 4
mum is 5, for example, then all these degrees of speed with 5
which the body moves make a sum of 15. And if the body were moved at this maximum degree for the same number as there are of these, the total of all these speeds would be double the above; that is, 30. Hence if the body moved for the same time with a uniform speed of this maximum degree of 5, it would have to pass through double the space which it passed during the time in which it was accelerated and started from the state of rest.

SALV. In accordance with your very swift and subtle compre-hension, you have expressed the whole thing much more clearly than I did, and you also made me think of something else to add. For the increases in the accelerated motion being continuous, one cannot divide the ever-increasing degrees of speed into any determinate number; changing from moment to moment, they are always infinite. Hence we may better exemplify our meaning by imagining a triangle, which shall be this one, ABC. Taking in the side AC any number of equal parts AD, DE, EF, and FG,

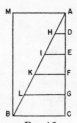

FIG. 15

and drawing through the points D, E, F, and G straight lines parallel to the base BC, I want you to imagine the sections marked along the side AC to be equal times. Then the parallels drawn through the points D, E, F, and G are to represent the degrees of speed, accelerated and increasing equally in equal times. Now A repre-sents the state of rest from which the moving body, departing, has acquired in the time AD the velocity DH, and in the next period the speed will have increased from the degree DH to the degree EI, and will progressively become greater in the succeeding times, according to the growth of the lines FK, GL, etc. But since the acceleration is made continuously† from moment to moment, and not discretely (*intercisamente*) from one time to another, and the point A is assumed as the instant of minimum speed (that is, the state of rest and the first instant of the subsequent time AD), it is obvious that before the degree of speed DH was acquired in the time AD, infinite others of lesser and lesser degree have been passed through. These were achieved during the infinite instants that there are in the time DA corresponding to the infinite points on the line DA. Therefore to represent the infinite degrees of speed which come before the degree DH, there must be understood to be infinite lines, always shorter and shorter, drawn through the infinity of points of the line DA, parallel to DH. This infinity of lines is ultimately represented here by the surface of the triangle AHD. Thus we may understand that whatever space is traversed by the moving body with a motion which begins from rest and continues uniformly accelerating, it has consumed and made use of infinite degrees of increasing speed corresponding to the infinite lines which, starting from the point A, are understood as drawn parallel to the line HD and to IE, KF, LG, and BC, the motion being continued as long as you please.

Now let us complete the parallelogram AMBC and extend to its side BM not only the parallels marked in the triangle, but the infinity of those which are assumed to be produced from all the points on the side AC. And just as BC was the maximum of all the infinitude in the triangle, representing the highest degree of speed acquired by the moving body in its accelerated motion, while the whole surface of the triangle was the sum total of all the speeds with which such a distance was traversed in the time AC, so the parallelogram becomes the total and aggregate of just as many degrees of speed but with each one of them equal to the maximum BC. This total of speeds is double that of the total of the increasing speeds in the triangle, just as the parallelogram is double the triangle. And therefore if the falling body makes use of the accelerated degrees of speed conforming to the triangle ABC and has passed over a certain space in a certain time, it is indeed reasonable and probable that by making use of the uniform velocities corresponding to the parallelogram it would pass with uniform motion during the same time through double the space which it passed with the accelerated motion.

SAGR. I am entirely persuaded. But if you call this a probable

In natural science one need not seek mathematical evidence.

A pendulum hanging from a longer cord makes its vibrations less frequently than one from a short cord.

Vibrations of the same pendulum are made with the same frequency, whether the vibrations are large or small.

Cause that would impede the pendulum and reduce it to rest.

argument, what sort of thing would rigorous proofs be? I wish to Heaven that in the whole of ordinary philosophy there could be found even one proof this conclusive!

SIMP. In physical science there is no occasion to look for mathematical precision of evidence.

SAGR. Well, isn't this question of motion a physical one? Yet I do not notice that Aristotle proves to me even the most trivial property of it. But let's not get farther afield. Salviati, please do not neglect to tell me what you hinted to me about that other cause for the pendulum stopping in addition to the resistance of the medium against being separated.

SALV. Tell me: of two pendulums of unequal length, doesn't the one which is hanging by the longer cord perform its oscillations the more infrequently?

SAGR. Yes, if they are swinging an equal distance from the perpendicular.

SALV. Oh, that makes no difference, for the same pendulum makes its oscillations in equal times,† whether they are long or short (that is, whether the pendulum is removed a long way or very little from the perpendicular). Or, if they are not exactly equal, the difference is insensible, as experiment will show you. But even if they were quite unequal, that would help rather than hinder my case. For let us denote the perpendicular AB, and hang from the point A on the cord AC the weight C, and still another, higher up on the same, which shall be E. Drawing the cord AC aside from the perpendicular and letting it loose, the weights C and E will move through the arcs CBD and EGF, and the weight E, hanging at the lesser distance and also being moved aside less, as you said, would try to go back sooner and to make its vibrations more frequently than the weight C. Therefore it would impede the latter from going back as far toward the point D as it would do if it were free, and, being thus an impediment to it in every oscillation, would finally bring it to rest.

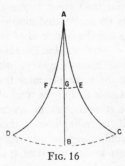

FIG. 16

Now this cord, with the middle weight removed, is itself a compound of many weighted pendulums; that is, each of its parts is just such a pendulum, attached closer and closer to the point A, and therefore arranged so as to make its vibrations more and more frequent, and consequently each is able to place a continual hindrance on the weight C. An indication of this is that as we observe the cord AC, we see it stretch not tightly, but in an arc; and if in place of the cord we put a chain, we see this effect much more evidently; most of all when the weight C is quite far from the perpendicular AB. For the chain is composed of many linked parts, each of which is heavy, and the arcs AEC and AFD will be seen to be noticeably curved. Therefore since the parts of the chain try to make their vibrations the more frequent according to their closeness to the point A, the lowest part cannot travel as much as it would naturally. And with the continual lessening of the vibrations of the weight C, they would finally stop even if the impediment of the air were taken away.

Cord or chain to which the pendulum is attached is bent in an arc during its vibrations, and not stretched straight.

SAGR. Well, here come the books right now. Take them, Simplicio, and find the place that was in question.

SIMP. Here it is — where he begins to argue against the diurnal motion of the earth, having first refuted its annual motion. *Motus Terrae annuus asserere Copernicanos cogit conversionem eiusdem quotidianam; alias idem Terrae hemispherium continenter ad Solem esset conversum, obumbrato semper averso.* ("The annual motion of the earth asserted by the Copernicans compels them to assert its diurnal rotation; otherwise the same hemisphere of the earth would be continually turned toward the sun, the opposite side being always in shade.") Thus half the earth would never see the sun.

SALV. It seems to me from his very opening that this man has no very clear idea of the Copernican position; for if he had noticed that therein the axis of the terrestrial globe is made always parallel to itself, he would have said not that half the earth would never see the sun, but that the year would be one single natural day. That is, in all parts of the earth it would be day for six months and night for six months, as happens with the inhabitants near the poles. But let us excuse him for this and go on with the rest.

SIMP. He continues: *Hanc autem gyrationem Terrae impossibilem esse, sic demonstramus.* ("That such gyration of the earth is impossible, we prove thus.") This next is the explanation of the figure which follows, in which we see depicted many descending heavy bodies and ascending light ones, and birds which are keeping themselves in the air, etc.

SAGR. Show me, please. My, what pretty pictures; what birds, what balls! And what are these other beautiful things?

SIMP. Those are balls which are coming from the moon's orbit.

SAGR. And what is this, here?

SIMP. It is a snail which they call *buovoli* here in Venice; it also is coming from the moon.

SAGR. Oh, indeed. So that is why the moon has such a great influence over these shellfish, which we call armored fish.

SIMP. Next comes that calculation which I told you about, of the travel in one natural day, one hour, one minute, and one second, which would be made by a point on the earth placed at the equator, as well as at a latitude of 48 degrees. And then follows this, of which I was wondering whether I had erred in the recital, so let us read it: *His positis, necesse est, Terra circulariter mota, omnia ex aëre eidem etc. Quod si hasce pilas aequales ponemus pondere, magnitudine, gravitate, et in concavo spherae lunaris positas libero descensui permittamus, si motum deorsum aequemus celeritate motui circum (quod tamen secus est, cum pila A etc.), elabentur minimum (ut multum cedamus adversariis) dies sex: quo tempore sexies circa Terram, etc.* ("These things being supposed, it is necessary if the earth moves circularly, that all things from the air do the same, etc. So that if we suppose these balls to be equal in size and weight and placed in the hollow of the moon's orbit and permit them a free descent, and if we make the motion downward equal to the motion around (which however is otherwise, since the ball A, etc.) they will fall at least (that we may grant a good deal to our adversaries) six days, in which time they will be turned about the earth six times, etc.")

SALV. You have recited this fellow's objection only too faithfully. From which you may see, Simplicio, how carefully those should tread who wish to make others believe things which perhaps they themselves do not credit. For it seems impossible to me that this author did not perceive that he was imagining a circle whose diameter, which among mathematicians is less than one-third of the circumference, was more than 12 times as great as that; an error which puts as more than 36 that which is less than one.[†]

SAGR. Maybe these mathematical ratios which are true in the abstract do not exactly correspond when applied in the concrete to physical and elemental circles. Though it does seem to me that a cooper, in determining the radius of the bottom to be made for a barrel, makes use of the abstract rules of the mathematicians despite such bottoms being very material and concrete things. But let Simplicio make this author's excuses for him, and tell us whether he thinks that physics differs as much from mathematics as all that.

SIMP. This refuge appears insufficient to me because the variation is too great; in this case I can only say *quandoque bonus etc.* But supposing Salviati's calculation to be more correct, and that the time of the ball's descent is no more than three hours, it seems to me a remarkable thing in any case that in coming from the moon's orbit, distant by such a huge interval, the ball should have a natural tendency to keep itself always over the same point of the earth which it stood over at its departure, rather than to fall behind in such a very long way.

SALV. The effect might be remarkable or it might be not at all remarkable, but natural and ordinary, depending upon what had gone on before. If, in agreement with the supposition made by the author, the ball had possessed the twenty-four-hour circular motion while it remained in the moon's orbit, together with the earth and everything else contained within that orbit, then that same force which made it go around before it descended would continue to make it do so during its descent too. And far from failing to follow the motion of the earth and necessarily falling behind, it would even go ahead of it, seeing that in its approach toward the earth the rotational motion would have to be made in ever smaller circles, so that if the same speed were conserved in it which it had within the orbit, it ought to run ahead of the whirling of the earth, as I said.

But if the ball had no rotation in the orbit, it would not in descending be obliged to remain perpendicularly over that point of the earth which was beneath it when the descent began. Nor does Copernicus or any of his adherents say it would.

SIMP. But the author will object, as you see, asking upon what principle this circular motion of heavy and light bodies depends — whether upon an internal or an external principle.

SALV. Keeping to the problem in hand, I say that the principle which would make the ball revolve while in the lunar orbit is the same one which would maintain this revolving also during the descent. I shall leave it to the author to make this be internal or external, at his pleasure.

SIMP. The author will prove that it cannot be either internal or external.

SALV. And I shall reply that the ball was not moving in the orbit, and thus be freed from any responsibility of explaining why, in descending, it remains vertically over the same point, since it will not remain so.

SIMP. Very well, but as heavy and light bodies can have neither an internal nor an external principle of moving circularly, then neither does the earth move circularly. And thus we have his meaning.

SALV. I did not say that the earth has neither an external nor an internal principle of moving circularly; I say that I do not know which of the two it has. My not knowing this does not have the power to remove it.

But if this author knows by which principle other world bodies are moved in rotation, as they certainly are moved, then I say that that which makes the earth move is a thing similar to whatever moves Mars and Jupiter, and which he believes also moves the stellar sphere. If he will advise me as to the motive power of one of these movable bodies, I promise I shall be able to tell him what makes the earth move. Moreover, I shall do the same if he can teach me what it is that moves earthly things downward.†

SIMP. The cause of this effect is well known; everybody is aware that it is gravity.†

SALV. You are wrong, Simplicio; what you ought to say is that everyone knows that it is called "gravity." What I am asking you for is not the name of the thing, but its essence, of which essence you know not a bit more than you know about the essence of whatever moves the stars around. I except the name which has been attached to it and which has been made a familiar household word by the continual experience that we have of it daily. But we do not really understand what principle or what force it is that moves stones downward, any more than we understand what moves them upward after they leave the thrower's hand, or what moves the moon around. We have merely, as I said, assigned to the first the more specific and definite name "gravity," whereas to the second we assign the more general term "impressed force" (*virtù impressa*), and to the last-named we give "spirits" (*intel-*

No more is known of what moves heavy bodies downward than of what moves the stars around; we know no more about these causes than the names we have given them.

ligenza), either "assisting"† (*assistente*) or "abiding" (*informante*); and as the cause of infinite other motions we give "Nature."

SIMP. It appears to me that this author is asking much less than what you are refusing to answer. He does not ask you in name and in detail for the principle which moves light and heavy bodies around; letting that be what it may, he asks only whether you consider it to be intrinsic or extrinsic. Thus, for example, although I do not know what entity gravity is, by which earth descends, I do know that it is an internal principle, since earth, if unimpeded, moves spontaneously. And on the contrary I know that the principle which moves it upward is external, though I do not know what that force is which is impressed upon it by the thrower.

SALV. How many questions we should have to be diverted into, if we wished to settle all the difficulties that are linked together, one in consequence of another! You call that principle external, preternatural, and constrained which moves heavy projectiles upward, but perhaps it is no less internal and natural than that which moves them downward. It may perhaps be called external and constrained while the movable body is joined to its mover; but once separated, what external thing remains as the mover of an arrow or a ball? It must be admitted that the force which takes this on high is no less internal than that which moves it down. Thus I consider the upward motion of heavy bodies due to received impetus to be just as natural as their downward motion dependent upon gravity.

The force which conducts heavy bodies thrown upward is no less natural than the heaviness which moves them downward.

SIMP. This I shall never admit, because the latter has a natural and perpetual internal principle while the former has a finite and constrained external one.

SALV. If you flinch from conceding to me that the principles of motion of heavy bodies downward and upward are equally internal and natural, what would you do if I were to tell you that they may also be one and the same (*medesimo in numero*)?

SIMP. I leave it to you to judge.

SALV. Rather, I want you to be the judge. Tell me, do you believe that contradictory internal principles can reside in the same natural body?

Contrary principles cannot reside naturally in the same subject.

SIMP. Absolutely not.

SALV. What would you consider to be the natural intrinsic tendencies of earth, lead, and gold, and in brief of all very heavy materials? That is, toward what motion do you believe that their internal principle draws them?

SIMP. Motion toward the center of heavy things; that is, to the center of the universe and of the earth, whither they would be conducted if not impeded.

SALV. So that if the terrestrial globe were pierced by a hole which passed through its center, a cannon ball dropped through this and moved by its natural and intrinsic principle would be taken to the center, and all this motion would be spontaneously made and by an intrinsic principle. Is that right?

SIMP. I take that to be certain.

SALV. But having arrived at the center is it your belief that it would pass on beyond, or that it would immediately stop its motion there?

SIMP. I think it would keep on going a long way.

SALV. Now wouldn't this motion beyond the center be upward, and according to what you have said preternatural and con-strained? But upon what other principle will you make it depend, other than the very one which has brought the ball to the center and which you have already called intrinsic and natural? Let me see you find an external thrower who shall overtake it once more to throw it upward.

And what is said thus about motion through the center is also to be seen up here by us. For the internal impetus of a heavy body falling along an inclined plane which is bent at the bottom and deflected upward will carry the body upward also, without interrupting its motion at all. A ball of lead hanging from a thread and moved from the perpendicular descends spontane-ously, drawn by its internal tendency; without pausing to rest it goes past the lowest point and without any supervening mover it moves upward. I know you will not deny that the principle which moves heavy bodies downward is as natural and internal to these as the principle which moves light ones upward is to those. Hence I ask you to consider a ball of wood which, descend-ing through the air from a great height and therefore moved by a natural principle, meets with deep water and continues its descent; without any other external mover it submerges for a long stretch, and yet the downward motion through water is preternatural to it. Still, it depends upon a principle which is

Natural motion converts itself into motion that is called preter-natural and forcible.

internal and not external to the ball. Thus you see how a movable body may be moved with contrary motions by the same internal principle.

SIMP. I believe there are answers for all these objections, though for the moment I do not remember them. However that may be, the author goes on to ask upon what principle this circular motion of heavy and light bodies may depend; that is, whether upon an internal or an external principle; and following this, he proves that it can be neither the one nor the other, saying: *Si ab externo, Deusne illum excitat per continuum miraculum? an vero angelus? an aër? Et hunc quidem multi assignant. Sed contra. . . .* ("If upon an external principle, is it God who excites them, in a continual miracle? Or rather an angel? Or the air? And indeed many thus assign the cause. But against this. . . .")

SALV. Do not bother to read the objection, for I am not one of those who assign such a principle to the surrounding air. As to the miracle or the angel, I rather lean that way, because whatever begins with a Divine miracle or an angelic operation, such as the transportation of a cannon ball to the moon's orbit, is not unlikely to do everything else by means of the same principle. But so far as the air is concerned, it suffices for me that this shall not impede the circular motion of bodies which are supposed to move through it. And for this it is enough (and more is not to be looked for) that the air be moved with the same motion, and its revolutions proceed with the same speed, as the terrestrial globe.

SIMP. And he rebels likewise against this, asking what takes the air around; nature, or constraint? He then refutes nature as a cause, by saying that this is contrary to truth, experience, and Copernicus himself.

SALV. It certainly is not contrary to Copernicus, who wrote no such thing, and this author attributes it to him only out of an excess of courtesy. Rather what Copernicus said (and it seems to me he did well to say it) was that the part of the air close to the earth, absorbing terrestrial vapors more readily, might have the same nature as the earth and follow its motion naturally. Or, being contiguous to the earth, the air might follow it in the way in which the Peripatetics say that the upper parts of the element of fire follow the motion of the moon's orbit. So it is up to them to explain whether such a motion would be natural or constrained.

Tendency of ele-
mental bodies to
follow the earth
has a limited
sphere.

SIMP. The author would reply that if Copernicus makes only the lower part of the air move, the upper part lacking this motion, then there can be no reason for quiet air being able to take heavy bodies along with it and make them follow the earth's motion.

SALV. Copernicus will say that this natural tendency of elemental bodies to follow the terrestrial motion has a limited sphere, outside of which such a natural tendency ceases. Besides, as I have said, it is not the air that carries along with it those moving things which follow the earth's motion when separated from it. So all the objections which this author adduces to prove that the air cannot be the cause of such effects are worthless.

SIMP. Then if this alternative cannot be the case, one will have to admit that such effects depend upon an internal principle, against which position *oboriuntur difficillimae, immo inextricabiles, quaestiones secundae* ("there arise most difficult, even insoluble secondary questions"), which are the following: *Principium illud internum vel est accidens, vel substantia. Si primum, qualenam illud? nam qualitas loco motiva circum hactenus nulli videtur esse agnita.* ("This internal principle is either an accidental property or a substance. If the former, what can it be? Up to the present, no quality of changing place about a center has been acknowledged to be seen by anyone.")

SALV. What does he mean, not noticed by anyone; not by us? All these elemental materials which move around together with the earth? Look at the way this author assumes as true that which is in question!

SIMP. He says this is not seen, and it seems to me he is right about that.

SALV. Not seen by us, because we are going around together with them.

SIMP. Listen to this other objection: *Quae etiam si esset, quomodo tamen inveniretur in rebus tam contrariis? in igne ut in aqua? in aëre ut in terra? in viventibus ut in anima carentibus?* ("Which even if it were, how could it be found in such contrary things? In fire as in water? In the air as in the earth? In living creatures as in nonliving?")

SALV. Assuming for the moment that water and fire are contraries, as well as air and earth (yet much could be said on this subject), the most that might follow from this would be that they could not have motions in common which are contrary to one another. Thus for example the motion upward, which belongs naturally to fire, could not belong to water, and just as water is by nature contrary to fire, that motion is proper to it which is naturally contrary to the motion of fire; this will be motion *deorsum*. But circular motion is contrary neither to motion *sursum* nor to that *deorsum*; indeed, it may mix with either, as Aristotle himself affirms. So why may it not belong equally to heavy and to light bodies?

Next, the most that cannot be common to the living and the nonliving are those things which depend upon the soul. Must not bodily things, so far as they are elemental and consequently share in the elemental qualities, be common to the corpse and the living body? And therefore if circular motion belongs to the elements, it must be common to their compounds also.

SAGR. This author must believe that if a dead cat falls out of a window, a live one cannot possibly fall too, since it is not a proper thing for a corpse to share in qualities that are suitable for the living.

SALV. Thus this author's argument is not conclusive against those who say that some internal event is the principle of the circular motion of heavy and light bodies.

I do not know to what extent he has proved that this cannot be a substance.

SIMP. This he combats with many arguments, of which the first is this one: *Si secundum (nempe si dicas tale principium esse substantiam), illud est aut materia, aut forma, aut compositum; sed repugnant iterum tot diversae rerum naturae, quales sunt aves, limaces, saxa, sagittae, nives, fumi, grandines, pisces, etc.; quae tamen omnia, specie et genere differentia, moverentur a natura sua circulariter, ipsa naturis diversissima, etc.* ("If the latter (that is, should you say that this principle is a substance), it is either matter, form, or a compound of both. But such diverse natures of things are again repugnant; such are birds, snails, stones, arrows, snows, smokes, hails, fishes, etc.; all of which, notwithtanding differences in species and kind, are moved of their own nature circularly, their natures being most diverse, etc.")

SALV. If these things named are naturally diverse, and things naturally diverse cannot have a common motion, then in order to accommodate all of them it will be necessary to think of more motions than just the two, upward and downward. And if one motion must be found for arrows, one for snails, another for rocks, and still another for fish, then you will also have to consider worms, topazes, and mushrooms, which are no less different in their natures than hail and snow.

SIMP. You seem to take this argument as a joke.

SALV. Not at all, Simplicio; but it has already been answered before. That is, if a motion either up or down can suit the things named, then so can a circular motion suit them. Being of the Peripatetic persuasion, do you not pose a greater difference between an elemental comet and a heavenly star than between a fish and a bird? Yet both the former move circularly.

Now go on with the second argument.

SIMP. *Si Terra staret per voluntatem Dei, rotarenturne caetera annon? Si hoc, falsum est a natura gyrari; si illud, redeunt priores quaestiones. Et sane mirum esset, quod gavia pisciculo, alauda nidulo suo, et corvus limaci petraeque etiam volens imminere non posset.* ("If the earth should stop by the will of God, would the rest of these things rotate or not? If not, then it is false that they rotate naturally. If so, then the earlier questions arise once more; and it would be truly remarkable if the seagull could not hover over the small fish, the skylark over her nest, and the crow over the snail and the rock, though wishing to do so.")

SALV. For my part I should give a general answer: That if by God's will the earth should stop its diurnal whirling, the birds would do whatever that same will of God desired. But if this author should wish a more detailed answer, then I should say that all these things would do the opposite from what would be done if, when they were keeping themselves in the air separated from the earth, the terrestrial globe were set unexpectedly in very precipitous motion by God's will. Now it is up to this author to advise you what would happen in that case.

SAGR. Salviati, please concede to this author at my request that if the earth were stopped by God's will, these other things separated from it would go right on around in their natural movement, and let us hear what impossibilities or inconveniences would follow from that. For I cannot see myself any greater disorders than these which the author himself produces; namely, that skylarks, though wishing to, would be unable to stay over their nests, nor crows over snails or rocks, so that then crows

would have to contain their appetite for snails and the young of
skylarks would perish of hunger and cold, their mothers being
no longer able to feed them or to brood them. This is all the ruina-
tion that I can deduce would happen according to this author's
statement. Look and see whether any greater troubles would
have to follow, Simplicio.

SIMP. I cannot discover any greater ones, but it may be assumed
that the author found other disorders in Nature which perhaps he
did not wish to adduce on account of his deep respect for her.

I shall continue with the third objection: *Insuper quî fit, ut
istae res tam variae tantum moveantur ab occasu in ortum paral-
lelae ad aequatorem? ut semper moveantur, numquam quiescant?*
("Besides, how is it that these things, so diverse, are moved only
from west to east, parallel to the equator? And that they are
always moving and never still?")

SALV. They are moved from west to east, parallel to the equator,
without stopping, in just the same way in which you believe the
fixed stars are moved from east to west, parallel to the equator,
without stopping.

SIMP. *Quare quo sunt altiores celerius, quo humiliores tardius?*
("Why are the higher the swifter, and the lower the slower?")

SALV. Because in a sphere or circle which turns about its center,
the more distant parts describe larger circles and the nearer
describe smaller ones, in the same time.

SIMP. *Quare quae aequinoctiali propriores in maiori, quae re-
motiores in minori, circulo feruntur?* ("Why are those near the
equinoctial plane carried about in larger circles and those more
remote in smaller ones?")

SALV. In order to imitate the stellar sphere, in which those things
closest to the equinoctial plane move in larger circles than those
more distant from it.

SIMP. *Quare pila eadem sub aequinoctiali tota circa centrum
Terrae, ambitu maximo, celeritate incredibili; sub polo vero,
circa centrum proprium gyro nullo, tarditate suprema volvetur?*
("Why does the same ball circulate clear around the earth's
center in the great circle with incredible speed at the equinoctial
plane, but at the pole turn around its own center without any
circulation and slow to the last degree?")

SALV. By copying the stars of the firmament, which would do the
same if the diurnal motion were theirs.

SIMP. *Quare eadem res, pila, verbi caussa, plumbea, si semel
Terram circuivit descripto circulo maximo, eandem ubique non
circummigret secundum circulum maximum, sed translata extra
aequinoctialem in circulis minoribus agetur?* ("Why does not the
same thing, for example a lead ball, go around everywhere in the
same great circle, if once describing the great circle it has en-
compassed the earth, but moves instead in lesser circles when
removed from the equinoctial plane?")

SALV. Because thus would do — or indeed have done, in Ptolemy's doctrine — some fixed stars which were once very close
to the equinoctial plane and described very large circles, while
now that they are farther from it, they describe smaller ones.

SAGR. Oh, if only I could keep all these beautiful things in my
mind, I should consider that I had made a great achievement!
Simplicio, you must lend me this little book, for within it there
must be an ocean of rare and exquisite things.

SIMP. I shall make you a present of it.

SAGR. Oh, no; not that; I would never deprive you of it. But are
the interrogations finished yet?

SIMP. No, indeed. Listen to this one: *Si latio circularis gravibus
et levibus est naturalis, qualis est ea quae fit secundum lineam
rectam? Nam si naturalis, quomodo et is motus qui circum est,
naturalis est, cum specie differat a recto? Si violentus, quî fit ut
missile ignitum sursum evolans, scintillosum caput sursum a
Terra, non autem circum, volvat, etc.?* ("If a circular bearing
is natural to heavy and light bodies, what about that which is
made along a straight line? For if natural, how then is motion
about the center natural, seeing it differs in kind from straight
motion? If constrained, how is it that a fiery arrow flying upward,
sparkling over our heads at a distance from the earth, does not
turn about, etc.?")

SALV. It has been said already many times that circular motion
is natural for the whole and for the parts when they are in the
optimum arrangement; straight motion is to restore disorderly
parts to order. Though it would be better to say that they never
move in a straight motion, whether ordered or disordered, but
in a mixed motion, which might even be a plain circle. But only
a part of this mixed motion is visible and observable to us, which
is the straight part; the circular remainder stays imperceptible
because we also share in it. This applies to rockets, which do
move up and around, but we cannot distinguish the circular mo-

Of mixed motion,
we do not see the
circular part be-
cause of our par-
ticipation in it.

tion because we also are moving with it. But I do not believe that this author understood this compounding, for you see how positively he says that rockets go straight up and do not revolve at all.

SIMP. *Quare centrum sphaerae delapsae sub aequatore, spiram describit in eius plano, sub aliis parallelis spiram describit in cono? sub polo descendit in axe, lineam gyralem decurrens in superficie cylindrica consignatam?* ("Why does the center of a falling sphere describe a spiral over† the equator in the plane of the latter, and in other latitudes a conical spiral? Why does it descend along the axis at the poles, running a gyrating line about a cylindrical surface?")

SALV. Because among the lines along which heavy bodies descend, drawn from the center to the circumference of the sphere, the one which passes through the equator describes a circle, and those which pass through other parallels describe conical surfaces, and the axis describes nothing else, but remains itself. And if I must tell you what I really think, I shall say that I cannot put any construction upon all these interrogations which would take away the motion of the earth. For if I should ask this author (granting him that the earth does not move) what would happen in all these instances if it did move as Copernicus would have it, then I am quite sure that he would say that all these effects would go right on which he is busy raising as obstacles to counter its mobility. Hence necessary consequences are accounted absurdities in this fellow's mind.

But please, if there is anything more, let us hurry on through this tedious stuff.

SIMP. This which comes next opposes Copernicus and his followers for holding that the motion of the parts when separated from their whole serves merely to reunite them with their whole, whereas circular movement is absolutely natural to their diurnal whirling. Against this he urges that in the opinion of such men, *Si tota Terra, una cum aqua, in nihilum redigeretur, nulla grando aut pluvia e nube decideret, sed naturaliter tantum circumferretur; neque ignis ullus aut igneum ascenderet, cum in illorum non improbabili sententia, ignis nullus sit supra.* ("If the whole earth, together with the water, were reduced to nothing, no hail or rain would fall from the clouds, but would only be carried naturally around; nor would any fire or flaming thing ascend, since in their not improbable view† there is no fire above.")

SALV. The providence of this philosopher is admirable and

worthy of great praise, he not contenting himself with thinking of things that might happen in the course of nature, but trying to provide himself against occasions on which things happened which are absolutely known never to happen. Well, in order to hear a few fine subtleties, I wish to concede to him that if the earth and the water were annihilated, no hails or rains would fall any more, nor would flaming matter go up, but would just keep on going around. Now, what of it? What does the philosopher say in reply to me?

SIMP. The objection is in the words which immediately follow. Here they are: *Quibus tamen experientia et ratio adversatur.* ("Which, however, experience and reason refute.")

SALV. Now I ought to give up, since he has such a big advantage over me; namely, experience which I lack. For as yet I have never happened to see the terrestrial globe and all the element of water annihilated, so as to have been able to see what the hail and water did in this little cataclysm. But at least he tells us, for our information, what they did do?

SIMP. He does not say another word about it.

SALV. I should gladly pay to have a chat with this fellow, in order to ask him whether when this globe vanished it took away also the common center of gravity, as I suppose it would. In that case, I think that the hail and water would remain senseless and stupid among the clouds, not knowing what to do with themselves. Or it might be that, attracted by such a large empty space left by the departure of the terrestrial globe, all the surrounding things would be rarefied — especially the air, which is extremely distractable — and would rush with great speed to refill it. And perhaps the more solid and material bodies, such as birds (for it is reasonable that many of these would be in the air), would draw back more toward the center of this huge empty space, as it seems likely that the more confined spaces would be assigned to substances which contained much material in less bulk; and that there, dead at last from hunger and reduced to earth, they would form a new little globe, with such little water as was in the clouds at that time.

Or it might be that the same materials, being insensitive to light, would not discover the earth's departure, and would blindly descend as usual, expecting to encounter it; and that one step at a time they would betake themselves to the center, whither they would go at present if the globe itself did not hinder them.

And finally, to give this philosopher a less indefinite reply, I say to him that I know just as much about what would happen after the earth was annihilated as he would have known about what was going to take place on it and around it before it was created. Since I am sure that he would say that he would not even have been able to imagine anything of what was to follow, since only experience has given him knowledge of this, he should not refuse to pardon me, but should forgive me for not knowing as much as he does about the things which would happen after the annihilation of this globe, seeing that I lack this experience which he possesses.

Now tell me whether there is anything else.

SIMP. There is this figure, which represents the terrestrial globe with a huge cavity in its center, full of air. And to show that heavy bodies do not move downward in order to unite with the terrestrial globe, as Copernicus says, he puts this stone here in the center, and asks what it would do if released. And he places another here on the concave surface of this great cavern, and makes the same interrogation, saying as to the first: *Lapis in centro constitutus aut ascendet ad Terram in punctum aliquod, aut non. Si secundum, falsum est partes ob solam seiunctionem a toto ad illud moveri. Si primum, omnis ratio et experientia renititur; neque gravia in suae gravitatis centro conquiescent. Item, si suspensus lapis liberatus decidat in centrum, separabit se a toto, contra Copernicum; si pendeat, refragatur omnis experientia, cum videamus integros fornices corruere.* ("The stone placed in the center either ascends to the earth at some point, or not. If not, it is false that the parts separated from the whole move toward that. If the former, it contradicts all reason and experience, nor does the heavy body rest at its center of gravity. And if the hanging stone is set free and descends to the center, it will separate from its whole, contrary to Copernicus; if it goes on hanging, this contradicts all experience, since we see entire arches collapse.")

SALV. I shall reply, although I am at a great disadvantage, being in the hands of someone who has learned from experience what these stones in this huge cavern would do; a thing which I have never seen. And I shall say that I believe that heavy things exist prior to the common center of gravity; hence it is not a center (which is nothing but an indivisible point and therefore in-

Heavy things exist prior to the center of gravity.

A large aggrega-
tion of heavy
bodies being
transported, its
separate particles
would follow it.

capable of acting) that attracts heavy materials to itself, but simply that these materials, coöperating naturally toward a juncture, would give rise to a common center, this being that around which parts of equal moments are arranged. From this I suppose that the large aggregation of heavy bodies being transferred to any place, the particles which were separated from the whole would follow; and if not impeded, they would penetrate wherever they might find parts less heavy than themselves. But upon arriving where they met with heavier material, they would descend no farther. Therefore I think that in this cavern full of air, the entire vault would press inward, and would sustain itself upon that air by force (*violentemente*) only when the hardness of this vault could not be overcome and broken under its weight. But I believe that detached stones would get down to the center and would not float above upon the air. Nor may it be said from this that they do not move toward their whole, all the parts being moved toward the place where the whole would go if it were not hindered.

SIMP. What now remains is a certain mistake which he has noticed in one of the followers of Copernicus who, making the earth move with the annual and diurnal motions in the same way in which a cartwheel moves upon the circle of the earth and about itself, would be making the terrestrial globe too large or the orbit too small, since 365 revolutions of the equator are much less than the circumference of the earth's orbit.

SALV. Pay attention lest you equivocate, and say the opposite of what must be written in the booklet. It must say that this author was making the terrestrial globe too small or the orbit too great; not the terrestrial globe too large and the annual orbit too small.

SIMP. There is no mistake on my part; look here at the words in the book: *Non videt quod vel circulum annuum aequo minorem, vel orbem terreum iusto multo fabricet maiorem.* ("He does not see that either he is making the annual circle less than is proper, or the earth's globe much larger than is suitable.")

SALV. I have no way of knowing whether the original author erred, since the author of this booklet does not identify him; but the error of the booklet is indeed obvious and inexcusable, whether or not the original follower of Copernicus made a mistake; for the author of the booklet passes by so material an error without noticing it or amending it.* But let him be pardoned this as an error of inadvertence rather than anything else. Besides, if I were not already sick and tired of occupying myself with all these petty quibbles and spending time on them to so little purpose, I could show that it is not impossible for a circle no bigger even than a cartwheel, making not 365 but less than 20 revolutions,† to describe or measure not merely the circumference of the earth's orbit, but one a thousand times as large. I say this in order to show that there is no lack of subtleties which are much greater than this one, with which this author notes the error of Copernicus. But I beg of you, let us pause for breath a moment, and then go on to the other philosophical opponent of this same Copernicus.

Nothing prevents the measuring and describing of a line greater than any given circle by means of a small circle many times revolved.

SAGR. Really, I need breath too, though as for me it is only my ears that have been wearied. If I thought I were going to hear nothing cleverer from that other author, I don't know but what I should decide to go and take the air in a gondola.

SIMP. I believe you are about to hear stronger arguments, for this author is a consummate scientist and also a great mathematician. He has refuted Tycho in the matters of comets and novas.

SALV. Is he perhaps that same author of the *Anti-Tycho?*

SIMP. The very same. But the refutation concerning the new stars is not in the *Anti-Tycho*, in which he merely demonstrated that they were not prejudicial to the inalterability and ingenerability of the heavens, as I told you before. Then after the *Anti-Tycho*, having by means of parallaxes found a way of proving the new stars also to be elemental things and contained within the moon's orbit, he wrote another book, *De tribus novis stellis*, *etc.*, and in this he inserted the arguments against Copernicus as well. I previously gave you what he wrote about the new stars in the *Anti-Tycho*, where he did not deny that they were in the heavens, but he demonstrated that their production did not affect the inalterability of the heavens. This he did by means of a purely philosophical argument in the manner I described to you; it did not occur to me to tell you how afterward he found a method of removing them from the heavens. Since he proceeded in this refutation by means of calculations and parallaxes, subjects of which I understand but little or nothing, I had not

---

*[Here the error is attributed to the author of the booklet, but really the error is not his.]†

read them, and had studied only these objections to the earth's motion, which are purely physical.

SALV. I understand quite well; it will be proper, after we have heard his opposition to Copernicus, that we judge, or at least see, the manner in which these new stars are proved by means of parallaxes to have been elemental. So many astronomers of such great reputation all place them among the highest stars of the firmament that if this author puts a stop to that scheme by dragging the new stars down from the heavens all the way to the elemental sphere, he will indeed be worthy of great exaltation himself; even of being transported to the stars, or at least of having his name perpetuated among them by fame.

But let us proceed with this first part, where he opposes the Copernican opinion; begin by setting forth his objections.

SIMP. We need not read them word for word, as they are very prolix. In reading them attentively many times I have, as you see, indicated in the margin the words which contain the meat of his arguments, and it will be sufficient to read these.

In the Copernican doctrine the criterion of philosophy is spoiled.

The first argument begins here: *Et primo, si opinio Copernici recipiatur, criterium naturalis philosophiae, ni prorsus tollatur, vehementer saltem labefactari videtur.* ("And first, if Copernicus's opinion is embraced, the criterion of science itself will be badly shaken if not completely overturned.") By which criterion he means, in agreement with philosophers of every school, that the senses and experience should be our guide in philosophizing. But in the Copernican position, the senses much deceive us when they visually show us, at close range and in a perfectly clear medium, the straight perpendicular descent of very heavy bodies. Despite all, according to Copernicus, vision deceives us in even so plain a matter and the motion is not straight at all, but mixed straight-and-circular.

SALV. This is the first argument adduced by Aristotle and Ptolemy and all their followers, which has been sufficiently replied to and shown to be a paralogism. It has been very clearly explained that such motion as is common to us and to the moving bodies is as if it did not exist. But since true conclusions meet with support from many things, I wish to add a few for the benefit of this philosopher. You, Simplicio, shall take his side, and answer my questions for him.

Motion in common is as if nonexistent.

First tell me what effect that stone has upon you when it falls from the summit of the tower, and what is the cause of your perceiving its motion? Because if nothing new or different acted upon you in its fall than in its rest upon the top of the tower, then you surely would not perceive its descent or distinguish its moving from its standing still.

SIMP. I am aware of its descent in relation to the tower because now I see it beside one mark on the tower, now at a lower one, and so on successively until I discover it united with the earth.

The argument taken from vertically falling bodies refuted in another way.

SALV. Then if the stone were dropped from the claws of a flying eagle and fell through mere invisible air, and you had no other visible and stable object to compare it with, you would not be aware of its motion?

SIMP. Even then this would be perceived by me; I should have to raise my head to see it when it was on high, and then as it fell I should have to lower it, and in a word to move it continually (or my eyes) to follow its motion.

How the motion of a falling body is recognized.

SALV. Now that is the correct answer. You know that the stone is at rest, then, when without moving your eyes a bit you can see it always right before you. And you know that it is moving when, in order not to lose it from sight, you must move your organs of vision — that is, your eyes. So whenever, without moving your eyes at all, you can see an object continually in the same aspect, you would always judge it to be motionless.

Motion of the eyes implies for us the motion of the object seen.

SIMP. I believe that would necessarily be so.

SALV. Now imagine yourself in a boat with your eyes fixed on a point of the sail yard. Do you think that because the boat is moving along briskly, you will have to move your eyes in order to keep your vision always on that point of the sail yard and to follow its motion?

SIMP. I am sure that I should not need to make any change at all; not just as to my vision, but if I had aimed a musket I should never have to move it a hairsbreadth to keep it aimed, no matter how the boat moved.

SALV. And this comes about because the motion which the ship confers upon the sail yard, it confers also upon you and upon your eyes, so that you need not move them a bit in order to gaze at the top of the sail yard, which consequently appears motionless to you. [And the rays of vision go from the eye to the sail yard just as if a cord were tied between the two ends of the boat. Now a hundred cords are tied at different fixed points, each of which keeps its place whether the ship moves or remains still.]

Now transfer this argument to the whirling of the earth and to the rock placed on top of the tower, whose motion you cannot discern because in common with the rock you possess from the earth that motion which is required for following the tower; you do not need to move your eyes. Next, if you add to the rock a downward motion which is peculiar to it and not shared by you, and which is mixed with this circular motion, the circular portion of the motion which is common to the stone and the eye continues to be imperceptible. The straight motion alone is sensible, for to follow that you must move your eyes downward.

I wish I could tell this philosopher, in order to remove him from error, to take with him a very deep vase filled with water some time when he goes sailing, having prepared in advance a ball of wax or some other material which would descend very slowly to the bottom — so that in a minute it would scarcely sink a yard. Then, making the boat go as fast as he could, so that it might travel more than a hundred yards in a minute, he should gently immerse this ball in the water and let it descend freely, carefully observing its motion. And from the first, he would see it going straight toward that point on the bottom of the vase to which it would tend if the boat were standing still. To his eye and in relation to the vase its motion would appear perfectly straight and perpendicular, and yet no one could deny that it was a compound of straight (down) and circular (around the watery element).

Now these things take place in motion which is not natural, and in materials with which we can experiment also in a state of rest or moving in the opposite direction, yet we can discover no difference in the appearances, and it seems that our senses are deceived. Then what can we be expected to detect as to the earth, which, whether it is in motion or at rest, has always been in the same state? And when is it that we are supposed to test by experiment whether there is any difference to be discovered among these events of local motion in their different states of motion and of rest, if the earth remains forever in one or the other of these two states?

SAGR. These arguments have somewhat quieted my stomach, which was a bit upset by those fishes and snails. The first has called to my mind the correction of an error which had such an

[margin note:] Experiment showing that motion in common is imperceptible.

appearance of truth that I think not one person in a thousand would have questioned it. When sailing to Syria, and having quite a good telescope which had been given me by our mutual friend, who had devised it not many days before, I proposed to the sailors that it would be of great benefit to navigation to make use of it in the foretop of the ship to spy out distant ships and identify them. The suggestion was approved, but it was argued that it would be difficult to use on account of the continual pitching of the ship, and especially at the top of the mast, where the agitation is so much greater; that it would be better to use it at the foot of the mast, where the movement is less than in any other place on the ship. I concurred in this view (for I do not wish to conceal my own mistake), and for a time I did not reply, nor do I know how to tell you just what it was that made me meditate further on the matter. Finally I recognized my foolishness (which may therefore be pardoned) in admitting what was false to have been true. It was false, I mean, that the great agitation of the foretop in comparison with that at the foot of the mast would necessarily make it harder to use the telescope for finding the object.

Subtle inquiry into the possible use of the telescope with equal facility at the top of a ship's mast and at the foot.

SALV. I should have taken sides with the sailors and with your first impression.

SIMP. I should have done so, too, and still would; nor do I believe that I would understand differently if I were to think about it for a century.

SAGR. Then, for once, I may be able to instruct both of you. And since proceeding by interrogations seems to me to shed much light upon things, in addition to the pleasure one may get out of pumping one's companion and making things drop from his lips which he never knew that he knew, I shall make use of that artifice. And first I assume that the ships, galleys, or other vessels which one seeks to discover and recognize are very distant, say 4, 6, 10, or 20 miles. For no glass is needed to recognize nearby ones, while the telescope can easily reveal the whole of a vessel at such a distance as 4 or 6 miles, and even much larger bulks. Now I ask how many and what kind of movements are made by the foretop as a result of the pitching of the ship?

SALV. I am picturing to myself a ship sailing eastward. First, in a tranquil sea, there would be no motion except this progress. Adding the agitation of the waves, there would be a motion

Different movements which depend upon the fluctuations of the ship.

Two changes
occur at the tele-
scope from rock-
ing of the ship.

which, alternately raising and lowering the stern and the prow, would make the foretop tilt forward and back. Other waves, tipping the ship to one side, would tilt the mast from right to left. Others might sometimes turn the ship and make it deflect its boom, let us say from directly east now to northeast and again to southeast. Still others, lifting the keel from below, might make the ship rise and fall without deflecting it. To sum up, it appears to me that there would be two kinds of movements: one kind which alters the angle of the telescope, and one which changes its alignment, so to speak, without changing its angle — that is, keeps the barrel of the instrument always parallel to itself.

SAGR. Next tell me this. Having first directed the telescope there toward that tower at Burano, about six miles away, if we should then move it through an angle to the right or left, or up or down by the breadth of a fingernail, what effect would this have upon its view of the tower?

SALV. It would make the tower disappear immediately from view; for such a tilting, small though it is, can mean a great many yards there.

SAGR. But if, without changing the angle (keeping the barrel always parallel to itself), we moved it 10 or 12 yards away to the left or right, or up or down, what effect would this have so far as the tower is concerned?

SALV. It would be absolutely imperceptible, because the space here and the space there are contained between parallel rays, and the changes made here and there must be equal; and since the space there revealed by the telescope could hold many such towers, this one would not be lost from view at all.

SAGR. Now going back to the ship, we may unquestionably affirm that a movement of the telescope to the right or left, or up or down, and forward or back even 20 or 25 yards, while keeping the telescope always parallel to itself, could not divert the visual ray from the observed point of the object any more than this same 25 yards. And since in a distance of 8 or 10 miles the scope of the instrument embraces a much larger space than the galley or other vessel seen, such a small change will not lose it to view. Thus the obstacle and the cause of losing the object can come only from a change made in the angle; for the deflection of the telescope due to the pitching of the ship up or down, or to right or left, cannot amount to very many yards.

Now suppose that you have two telescopes, one attached to the lower part of the ship's mast, and another not just to the round-top, but to the maintop or even the main topgallant where the pennant is hung, and that both are pointed at a vessel ten miles away. Tell me whether you believe that a greater change is made in the angle of the higher tube than in that of the lower, let the agitation of the ship be what it may. When a wave raises the prow, it may well make the highest point go back 30 or 40 yards more than the foot of the mast, and the tube of the upper tele-scope will be seen to be withdrawn by that amount while the lower one goes only a foot, but the angle changes just as much in the latter instrument as in the former. Likewise, a wave which comes from the side will move the higher tube to the right or left a hundred times as much as the lower one, but their angles either do not change or they change equally. Now changes to the right or left, forward or back, and up or down produce no noticeable obstacle in the sighting of distant objects, though an alteration of angle brings about a large one. It must therefore be admitted that the use of the telescope at the top of the mast is no more difficult than at the foot, for the changes of angle are the same in both places.

SALV. How carefully we must proceed before affirming or deny-ing a statement! I say again that anyone will be persuaded, upon hearing someone resolutely declare it, that because of the greater movement made at the top of the mast than at the foot, the use of the telescope should be far more difficult aloft than below. Accordingly I wish to pardon those philosophers who throw up their hands or fly into a rage at people who do not wish to admit that a cannon ball which is plainly seen to go perpendicularly down along a straight line must absolutely be moving in that way, but who want to have its motion be along an arc, ever more slanted and diagonal.

Well, let us leave them in their distress and listen to the other objections this present author makes against Copernicus.

SIMP. The author goes on to show that in the Copernican doc-trine the senses must be denied; even the grossest sensations. This would be the case if we, who feel the blowing of a slight breeze, were then not made to feel the impetus of a perpetual gale that drives with a velocity of more than 2,529 miles an hour.† For such is the distance which the center of the earth

travels in an hour in its annual motion along the circumference
of the *orbis magnus,* as he carefully calculates; and because, as
he says, it nevertheless appears to Copernicus, *Cum Terra move-
tur circumpositus aër; motus tamen eius, velocior licet ac rapidior
celerrimo quocumque vento, a nobis non sentiretur, sed summa
tum tranquillitas reputaretur, nisi alius motus accederet. Quid
est vero decipi sensum, nisi haec esset deceptio?* ("The surrounding air is moved with the earth; yet its motion, though swifter
than the most rapid wind, would not be perceived by us, but
would be considered quite tranquil unless some other motion
occurred. If this is not deception of the senses, then what is?")

SALV. This philosopher must believe that the earth which Copernicus makes go around the circumference of its orbit together
with its circumambient air is not this one which we inhabit, but
some other separate one; for this one of ours takes us along too,
with its own velocity and that of its surrounding air. What beating could we feel when fleeing with equal speed over the course
pursued by him who would whip us? This gentleman has forgotten that we also, and not just the earth and the air, are carried
around; and that consequently we are always touched by the
same part of the air, which cannot strike upon us.

SIMP. Not at all; look at the words which follow immediately:
Praeterea nos quoque rotamur ex circumductione Terrae, etc.
("Besides, we too would therefore be turned about as a result
of the earth's revolution, etc.")

SALV. Now I can neither help him nor excuse him. You must
pardon him for this, Simplicio, and help him out of it.

SIMP. Offhand, no satisfactory defense occurs to me at the
moment.

SALV. Well, think it over tonight and come to his defense on this
point tomorrow. Meanwhile let us hear his other objections.

SIMP. The same objection is continued, it being shown that in
Copernicus's view one must deny one's own senses. For this
principle by which we go around with the earth is either intrinsic
to us, or it is external to us, as a snatching along by the earth.
If it is the latter, then since we do not feel any such snatching
along, it must be said that the sense of touch does not feel its
own related object nor the impression of that upon our consciousness. But if the principle is intrinsic, then we shall not be feeling
a local motion deriving from our very selves, and we shall never
perceive a tendency perpetually attached to us.

SALV. So that this philosopher's objection emphasizes that the principle by which we move along with the earth, whether this is external or internal, ought to be felt by us in either case; and since we do not feel it, it is neither the one nor the other. Therefore we do not move, and neither does the earth. And I say that it could be either one without our feeling it. As to the possibility of its being external, the experiment aboard ship more than removes every difficulty. For we are able at will to make the ship move, and also to make it stand still, and go about observing with great accuracy whether or not we could detect whether it moves, by means of any difference which might be apprehended by the sense of touch. And seeing that as yet no such thing has been learned, how can it be any wonder if the same condition with regard to the earth remains unknown? The earth may have been carrying us forever without our ever having been able to devise any experiment with it at rest.

I know that you, Simplicio, have gone from Padua by boat many times, and, if you will admit the truth of the matter, you have never felt within yourself your participation in that motion except when the boat has been stopped by running aground or by striking some obstacle, when you and the other passengers, taken by surprise, have stumbled perilously. The terrestrial globe would have only to encounter some obstacle which would arrest it, and I assure you that you would become aware of the impetus which resides in you when you were thrown by it toward the stars.

It is true that you can perceive the motion of the boat by means of the other senses, accompanied by reasoning, as by vision when you are watching poles and buildings situated in the fields, which, being separated from the boat, appear to move in the opposite direction. If you want to be convinced of the terrestrial motion by such an experience, I may tell you to look at the stars, which for the same reason appear to you to move in the opposite direction.

Next, any surprise at not feeling this principle if it is internal is still less reasonable; if we do not feel a similar motion coming from the outside and frequently absent, why should we feel one when it immutably resides continually within us?

Now, has he anything further as to this first argument?

SIMP. There is this little complaint: *Ex hac itaque opinione necesse est diffidere nostris sensibus, ut penitus fallacibus vel stu-*

255 The Second Day

Our motion may be either internal or external without being known or felt by us.

Motion of a boat insensible by the sense of touch to those within it.

Motion of the boat is perceptible by vision coupled with reason.

Terrestrial motion may be known from the stars.

pidis in sensibilibus, etiam coniunctissimis, diiudicandis; quam ergo veritatem sperare possumus, a facultate adeo fallaci ortum trahentem? ("And from this opinion we must necessarily suspect our own senses as wholly fallible or stupid in judging sensible things which are very close at hand. Then what truth can we hope for, deriving its origin from so deceptive a faculty?")

SALV. Oh, I wish to derive still more useful and more certain precepts from it, learning to be more circumspect and less confident about that which the senses represent to us at a first impression, for they may easily deceive us. And I wish that this author would not put himself to such trouble trying to have us understand from our senses that this motion of falling bodies is simple straight motion and no other kind, nor get angry and complain because such a clear, obvious, and manifest thing should be called into question. For in this way he hints at believing that to those who say such motion is not straight at all, but rather circular, it seems they see the stone move visibly in an arc, since he calls upon their senses rather than their reason to clarify the effect. This is not the case, Simplicio; for just as I (who am impartial between these two opinions, and masquerade as Copernicus only as an actor in these plays of ours) have never seen nor ever expect to see the rock fall any way but perpendicularly, just so do I believe that it appears to the eyes of everyone else. It is therefore better to put aside the appearance, on which we all agree, and to use the power of reason either to confirm its reality or to reveal its fallacy.

SAGR. If I ever had a chance to meet this philosopher, who seems to me a cut above most of the followers of these doctrines, I should as a token of my esteem acquaint him with an event which he has surely seen many times, from which (in complete agreement with what we are saying) one may learn how easily anyone may be deceived by simple appearances, or let us say by the impressions of one's senses. This event is the appearance to those who travel along a street by night of being followed by the moon, with steps equal to theirs, when they see it go gliding along the eaves of the roofs. There it looks to them just as would a cat really running along the tiles and putting them behind it; an appearance which, if reason did not intervene, would only too obviously deceive the senses.

SIMP. To be sure, there are plenty of experiences which make evident the fallacies of the simple senses. Therefore postponing such sensations for the present, let us listen to the ensuing arguments which are taken *ex rerum natura*, so to speak.

The first is that the earth cannot move by its own nature in three widely differing movements without actually contradicting many manifest axioms. The first of these is that every effect depends upon some cause; the second, that nothing is self-created; from these it follows that the thing causing motion (*movente*) and the thing moved cannot be one and the same. This holds not only for things which are moved by an extrinsic and obvious mover, but the above principles imply also that the same holds for natural motion depending upon an intrinsic principle. Otherwise, since the moving thing (*movente*), as such, is a cause, and the thing moved, as such, is an effect, then the cause and the effect would be identical in all respects. Therefore a body does not move entirely of itself so that the whole is mover as well as moved, but there is required in the thing moved some way of distinguishing the efficient principle of motion from that which is moved with such motion.

The third axiom is that in things subject to sensation, one thing, in so far as it is one, produces but one effect. In an animal, to be sure, the soul (*anima*) does produce various operations, such as sight, hearing, smell, generation, etc., but it does so by means of different instruments; in a word, it may be seen that different actions in sensible objects derive from differences which exist in the causes.

Now if these axioms are combined, it will be quite evident that a simple body, such as the earth, will by its nature be unable to move with three widely differing motions at the same time. For by the assumptions made, the whole cannot move all by itself. Hence three principles must be distinguished for three motions in it; otherwise the same principle would be producing more than one motion. But if a body contained within itself three principles of natural motion, besides the part moved, it would not be a simple body, but one composed of three moving principles plus the part moved. Therefore if the earth is a simple body, it does not move with three motions.

Furthermore it will not move with any of the motions which Copernicus attributes to it, being obliged to move with only one motion; for it is obvious (for reasons given by Aristotle) that the

Arguments against the earth's motion taken *ex rerum natura*.

Three axioms assumed to be evident.

A simple body, such as the earth, cannot move with three diverse motions.

The earth cannot move with any of the motions attributed to it by Copernicus.

earth does move toward its center — as shown by particles of earth, which descend to the spherical surface of the earth at right angles.

SALV. There is much that might be said and considered with regard to the weaving of this argument. But since we can resolve it in a few words, I do not wish at the moment to enlarge upon it unnecessarily; the more so as the answer is put in my possession by the author himself when he says that various operations can be produced in an animal from a single principle. Therefore I answer him for the present that diverse movements in the earth are derived from a single principle in a similar way.

SIMP. This answer will not at all satisfy the author of the objection; in fact, it is completely overthrown by what he adds next in further substantiation of his attack, as you shall hear. He corroborates the argument, I mean, by one more axiom, which

is this: Nature is neither deficient nor excessive in that which is necessary. This is obvious to observers of natural things, especially of animals, in which, since they must make many move-

ments, nature has made their many joints and has knitted their parts suitably for motion — as at the knees and hips, so that animals may travel or lie down at their pleasure. Besides, in man, nature has made many joints and tendons at the elbow and at the hand, so that these may perform many motions. It is from

these things that the argument against the threefold motion of the earth is drawn. Either a body which is one and continuous without being tied or jointed at all can perform different movements, or it cannot do so without having joints. If it can do so without, then it is in vain that nature has made joints in animals, which is against the axiom. But if it cannot, then the earth (being a body which is one and continuous and without joints and tendons) by its nature cannot move with more than one motion. Now you see how ingeniously he controverts your reply, almost as if he had foreseen it.

SALV. Are you serious, or are you speaking ironically?

SIMP. I am giving you the very best that is in me.

SALV. Then you must think you have a good enough case to be able to defend this philosopher against additional counterattacks than those which he has made up. So be kind enough to answer

me on his behalf, since we cannot have him present.

In the first place you admit it as true that nature has made joints, tendons, and muscles in animals in order that they may move in many different manners. I deny this statement, and I tell you that the joints are made so that the animal can move one or more of its parts, keeping the rest stationary, and that as to kinds and differences of the movements, they are of one kind only — the circular. That is why you see the ends of all moving bones to be convex or concave, and some of these spherical; namely, those which have to move in every direction, as must the arm in the shoulder knot of an ensign when he is displaying the colors, or that of the falconer when bringing the hawk to his lure. And such is the elbow joint, upon which the hand turns round when boring with an auger. Others are circular in one direction only, and almost cylindrical, being used by a member which bends in only one way; such are the parts of the fingers, one above another, etc. But without more detailed counter-instances, the truth may be made known by means of a single general reason. This is that if a solid body moves while one of its extremities remains still without changing place, the motion cannot be anything but circular. And since, in animal motion, no member parts company with any other which is coterminous with it, such motion is necessarily circular.

SIMP. That is not the way I see it, for I observe animals moving in a hundred noncircular motions, all very different from each other; running and jumping, climbing up and down, swimming, and many other things.

SALV. Quite so; but these are secondary motions, dependent upon the primary motions of joints and flexures. As a consequence of bending the leg at the knee and the thigh at the hip, which are circular motions of the parts, comes the jumping or running, which are movements of the whole body and may be noncircular. Now since the terrestrial globe need not move one of its parts upon another stationary one, but any of its movements must belong to the whole body, there is no need of joints.

SIMP. It will be said on the other side that this might be the case if there were a question of but one motion; but, there being three quite different from one another, it is impossible for them to be accommodated in an unarticulated body.

SALV. I really think that that would be the answer of this philosopher. I now attack it from another side, and ask you whether you suppose that by means of joints and flexures the earth might

Motions of animals all of one sort.

The ends of all movable bones are rounded.

Necessity shown for ends of all movable bones being rounded, and for all motions of animals being circular.

Secondary motions of animals dependent upon the primary.

No joints required for the earth's motion.

be adapted to participation in three different circular motions?

What, no answer? Since you remain silent, I shall reply for your philosopher. He would certainly say yes, because otherwise it would have been superfluous and irrelevant to bring into consideration the fact that nature makes flexures in order that movable bodies may have a variety of motions, and therefore that the earth, having no flexures, may not have three motions attributed to it. For if he had believed that even flexures would not render it fit for such movement, he would have said without qualification that the globe could not have three motions.

Now this being the case, I wish to ask you (and if it were possible I should ask the philosopher and author of this argument through you) to be so kind as to show me how these joints would have to be arranged so that all three motions could be conveniently performed; and for your answer I shall allow you four months — no, six. In the meantime it seems to me that a single principle can cause more than one motion of the earth in exactly the same way in which I have just told you that a single principle, by means of various instruments, produces many and diverse motions in animals. As to joints, there is no need for them, the required movements being of the whole and not of some parts only. And since they must be circular, the simple spherical shape is the most beautiful joint one could ask for.

It is desired to know by means of what joints the terrestrial globe could move in three different ways.

A single principle can cause more than one motion in the earth.

SIMP. The most that one must allow you would be that a single movement might take place. But three different ones cannot be possible in my view, or in this author's, as he goes on to support his objection by writing: "Let us imagine with Copernicus that the earth moves by a property of its own and from an intrinsic principle from west to east in the plane of the ecliptic, and moreover that it revolves also by an intrinsic principle about its own center from east to west; and for a third motion that it tilts by a tendency of its own from north to south and vice versa." In a continuous body, not put together with joints and sections, can our feelings and judgment ever understand how one vague natural principle — one single propensity — might break down into different and almost contradictory motions? I do not believe that anyone exists who would say such a thing unless he had resolved to defend this position through thick and thin.

Another objection against the threefold motion of the earth.

SALV. Wait a moment. Find me this place in the book, and show it to me.

*Fingamus modo cum Copernico, Terram aliqua sua vi et ab
indito principio impelli ab occasu ad ortum in eclipticae plano,
tum rursus revolvi ab indito etiam principio circa suimet centrum
ab ortu in occasum, tertio deflecti rursus suopte nutu a septen-
trione in austrum et vicissim.*

I was questioning, Simplicio, whether you had not made a
mistake in quoting the author's words; but I see that he himself
is in error, and very seriously. I am grieved to learn that he set
himself up to dispute a position which he did not even understand, for these are not the movements which Copernicus attributes to the earth. Where did he get the idea that Copernicus made
the earth's annual motion to be along the ecliptic opposite to its
motion around its own center? He must never have read Copernicus's book, which says in a hundred places — even in the
opening chapters — that both these movements are in the same
direction; that is, from west to east. But without hearing this
from anyone else, couldn't he see for himself that the motions
which are attributed to the earth, one taken from the sun and
the other from the *primum mobile,* must necessarily be in the
same direction?

SIMP. Take care that you do not fall into error yourself, and
Copernicus along with you. Is not the daily motion of the *primum
mobile* from east to west? And on the other hand, isn't the sun's
motion along the ecliptic just the opposite, from west to east?
So how do you make contraries become agreements when transferred to the earth?

SAGR. Surely Simplicio has revealed the origin of the error of
this philosopher, who doubtless would have made this same
argument.

SALV. Then if it can be done, let us at least remove Simplicio
from error. Seeing the stars climb above the eastern horizon
upon rising, he will have no trouble understanding that if this
motion did not belong to the stars, the horizon would necessarily
have to be considered as going down in the opposite direction,
and in consequence that the earth would revolve upon itself
opposite to the apparent motion of the stars; that is, from west
to east in the order of the signs of the zodiac. Next, as to the
other motion, the sun being fixed in the center of the zodiac and
the earth moving around its circumference: In order to make the
sun appear to move through the zodiac in the order of its signs,

Serious error by
the assailant of
Copernicus.

Clever and sim-
ple objection
against
Copernicus.

The opponent's
error made man-
ifest by explain-
ing how the
annual and diur-
nal motions,
belonging to the
earth, are in the
same direction
and not contrary.

the earth would have to travel in that same order; for the sun always appears to occupy in the zodiac that sign opposite to the one in which the earth is found. Thus, the earth running through Aries, the sun will appear to be running through Libra; the earth passing through Taurus, the sun will appear in Scorpio; the earth at Gemini, the sun is at Sagittarius. This amounts to both of them moving in the same direction; that is, following the order of the signs, as the revolution of the earth about its own center was also made.

SIMP. I quite understand, and I do not know what to say in excuse for such an error.

From another, graver, error the opponent is shown to have made scant study of Copernicus.

SALV. Go easy, Simplicio, for there is another, still worse than this; it is his making the earth move diurnally around its own center from east to west. He does not understand that in that case the movement of the universe during twenty-four hours would appear to be made from west to east, just the opposite of what we see.

SIMP. Why, I am sure that I myself, who have scarcely learned the elements of spherical astronomy, would not have made so grave an error.

SALV. Judge, then, how much study this opponent may be supposed to have spent on the books of Copernicus, when he gets this basic and principal hypothesis backwards, upon which are founded all the dissents of Copernicus from the doctrine of Aristotle and Ptolemy.

Now as to the third motion† which this author assigns to the terrestrial globe as being Copernicus's idea, I do not know what he means by it. It is certainly not the one which Copernicus attributes to the earth along with the other two (annual and diurnal), for that has nothing to do with any tilting toward the south and north, but merely serves to keep the axis of the earth's

It is doubted that the opponent understood the third motion attributed by Copernicus to the earth.

diurnal revolution continually parallel to itself. So one must say that either the adversary has not understood this, or that he has pretended not to. Although this grave deficiency is enough to relieve us of any further obligation to occupy ourselves with the consideration of his objections, yet I should like to consider them anyway, as they are truly much worthier of evaluation than those of many other foolish opponents.

Getting back to the objection, then, I say that the two movements, annual and diurnal, are not contrary movements at all.

Rather, they are in the same direction and may therefore depend upon the same principle. The third movement follows as a consequence of the annual motion spontaneously and by itself, in such a way that you need not appeal to any internal or external principle for the cause from which (as I shall prove in due course) it comes into being.

SAGR. With common sense as a guide, I should also like to say something to this opponent. He wants to condemn Copernicus if I cannot resolve on the spot all doubts, and reply to all the objections that he makes — as if from my ignorance it necessarily followed that the doctrine were false. But if this means of condemning a writer appears judicious to him, then he should not think it unreasonable for me not to endorse Aristotle and Ptolemy, when this is the best he can do to resolve for me the same difficulties which I point out to him in their doctrines.

He asks me what the principles are by which the terrestrial globe makes its annual motion through the zodiac, and its diurnal motion around the equator upon itself. I say to him that they are similar to those by which Saturn moves through the zodiac in thirty years, and about its own center in the equinoctial plane in a much shorter time, as the disclosure and hiding of its collateral globest shows us. This is similar to something which he concedes without question; that the sun runs through the ecliptic in one year, and revolves parallel to the equator in less than one month, as its spots visibly show us. It is also similar to the principle by which the satellites of Jupiter traverse the zodiac in twelve years, and among themselves revolve around Jupiter in very small circles and very short times.

SIMP. This author would deny all these things as visual deceptions due to the lenses of the telescope.

SALV. Oh, that is asking too much for himself, when he will have it that the unaided eye cannot be deceived in judging the straight motion of falling heavy bodies, but that it is deluded in understanding these other movements when its power is perfected and increased thirty times. Let us tell him, then, that the earth participates in its plurality of motions in a similar way, perhaps the same way, as that in which a compass needle has one motion downward as a heavy object, and two circular motions — a horizontal one, and a vertical one along the meridian.

Now what else? Tell me, Simplicio, between which do you

The same objection is resolved by example of the similar movements of other celestial bodies.

Motion differs
more from rest
than straight
motion from
circular.

One may more
reasonably at-
tribute to the
earth two inter-
nal principles
for straight and
circular motion
than two for
motion and rest.

believe that this author would pose the greater disparity: between straight and circular motion, or between motion and rest?

SIMP. Undoubtedly between motion and rest. This is obvious,
for circular motion is not contrary to straight motion for Aristotle; he even concedes that they may mix, which motion and
rest cannot do.

SAGR. Then it is a less improbable proposition to put in one
natural body two internal principles, one for straight motion and
the other for circular, than two others, one for motion and the
other for rest. Now both positions are in agreement as to the
natural inclination which resides in the earth's parts to return
to their whole when separated from it by force. They differ
only as to the operation of the earth's whole; the former would
have this remain motionless from an internal principle, and the
latter would attribute circular motion to it. But by your concession and according to this philosopher, two principles, one for
motion and the other for rest, are incompatible, just as the effects
are incompatible; whereas this does not happen for the two
movements, straight and circular, which have no repugnance for
each other.

Motion of the
earth's parts in
returning to its
whole may be
circular.

SALV. Add furthermore that very probably the motion made by
a separated part of the earth while it is getting back to its whole
is also circular, as has been explained already. Hence in every
respect, so far as the present objection is concerned, movability
appears to be more acceptable than rest.

Now, Simplicio, proceed with whatever remains.

SIMP. The author strengthens the objection by pointing out
another absurdity, which is that the same motions would thus
be adapted to things of very different nature, whereas observation teaches us that the actions and motions of things of diverse
natures are diverse. And reason confirms this, for otherwise we
should have no way of comprehending and distinguishing their
natures, if they did not have their special motions and actions
which reveal their substances to our understanding.

Diversity of mo-
tions leads to a
knowledge of
diversity of
nature.

SAGR. Two or three times in this author's arguments I have
noticed that in order to prove that matters stand in such-and-
such a way, he makes use of the remark that in just this way do
they accommodate themselves to our comprehension, and that
otherwise we should have no knowledge of this or that detail; or
that the criterion of philosophizing would be ruined; as if nature

Nature first
made things in
her own way,
and then con-
structed human
reason able to
understand them.

had first made the brain of man, and then arranged everything to conform to the capacity of his intellect. But I should think rather that nature first made things in her own way, and then made human reason skillful enough to be able to understand, but only by hard work, some part of her secrets.

SALV. I, too, am of this opinion. But tell me, Simplicio, what are these diverse natures to which, against observation and reason, Copernicus assigned the same motions and actions?

SIMP. They are these: water and air (which are yet different in nature from earth), and all the things which are found in those elements, each of which has to have these three movements which Copernicus gives to the terrestrial globe. He goes on to prove geometrically how it is true that in Copernicus's view a cloud which is suspended in the air and hovers for a long time over our heads without changing place must necessarily have all three movements which the terrestrial globe has. Here is the proof, which you may read for yourself, for I cannot recite it from memory.

Copernicus mistakenly assigns the same operations to diverse natures.

SALV. I am not in any hurry to read it; I even think it superfluous to have put it there, since I am sure that none of the adherents of the earth's motion would deny it to him. So granting him his proof, let us speak of the objection. This seems to me to have no conclusive force against the Copernican position, since nothing is detracted from those motions and actions by which we come into cognition of essences, etc. Please tell me, Simplicio, can those properties in which certain things agree exactly serve to make known to us the diverse natures of such things?

SIMP. No indeed; rather the opposite, for an identity of actions and properties can argue nothing but an identity of natures.

From common events, diverse natures could not be known.

SALV. So that the diverse natures of water, earth, air, and other things that exist in those elements are not inferred by you from those actions in which all these elements and the things connected with them agree, but from other actions. Is that right?

SIMP. That is so.

SALV. Then whatever would leave to the elements all those motions, actions, and other properties by which their natures are distinguished would not take away our power to gain a knowledge of these, even though it removed those actions in which they unite in agreement and which are therefore of no use in distinguishing their natures.

SIMP. I think this reasoning is quite correct.

SALV. But is it not your opinion, and that of the author and of Aristotle and Ptolemy and all their followers, that earth, water, and air are equally of such a nature as to be constituted immovable about the center?

SIMP. That is taken as an irrefutable truth.

SALV. Then the argument for the different natures of these elements and elemental things is not taken from this common natural condition of rest with respect to the center, but must be learned by taking notice of other qualities which they do not have in common. Therefore whoever should take from the elements only this common state of rest, and leave them all their other actions, would not in the least obstruct the road which leads us to an awareness of their essences.

Now Copernicus takes from them nothing except this common rest, leaving to them weight or lightness; motion up or down, slow or fast; rarity and density; the qualities of heat, cold, dryness, moistness; and, in a word, everything else. Hence no such absurdity as this author imagines exists anywhere in the Copernican position. Agreement in an identical motion means neither more nor less than agreement in an identical state of rest, so far as any diversification or nondiversification of natures is concerned. Now tell me if he has other opposing arguments.

SIMP. There follows a fourth objection, taken once again from an observation of nature. It is that bodies of the same kind have motions which agree in kind, or else they agree in rest. But in Copernicus's theory, bodies agreeing in kind and quite similar to each other would have great discrepancies as to motion, or even be diametrically opposed. For stars, so very similar to one another, would nevertheless have such dissimilar motions that six planets† would perpetually go around, while the sun and the fixed stars would remain forever unmoved.

SALV. The form of this argumentation appears to me valid, but I believe that its content or its application is at fault, and if the author were to persist in this assumption the consequences would run directly counter to his. The method of argument is this:

Among world bodies, there are six which perpetually move; these are the six planets. Of the others (that is, the earth, the sun, and the fixed stars) the question is which move and which stand still. If the earth stands still, the sun and the fixed stars necessarily move, and it may also be that the sun and the fixed stars

A common motion suiting the elements means neither more nor less than a common rest suiting them.

Bodies of the same kind have motions which agree in kind.

Another argument against Copernicus.

From the earth being naturally dark and the sun and fixed stars light, it is argued that the former is movable and the latter are motionless.

are motionless if the earth is moving. This matter being in question, we inquire which ones may more suitably have motion attributed to them, and which ones rest.

Common sense says that motion ought to be deemed to belong to those which agree better in kind and in essence with the bodies which unquestionably do move, and rest to those which differ most from them. Eternal rest and perpetual motion being very different events, it is evident that the nature of an ever-moving body must be quite different from that of one which is always fixed. Let us therefore find out, when in doubt about motion and rest, whether by way of some other relevant condition we can investigate which — the earth, or the sun and the fixed stars — more resembles those bodies which are known to be movable.

Now behold how nature, favoring our needs and wishes, presents us with two striking conditions no less different than motion and rest; they are lightness and darkness — that is, being brilliant by nature or being obscure and totally lacking in light. Therefore bodies shining with internal and external splendor are very different in nature from bodies deprived of all light. Now the earth is deprived of light; most splendid in itself is the sun, and the fixed stars are no less so. The six moving planets entirely lack light, like the earth; therefore their essence resembles the earth and differs from the sun and the fixed stars. Hence the earth moves, and the sun and the stellar sphere are motionless.

SIMP. But the author will not concede that the six planets are dark, and will stand firm upon that denial; or else he will argue the great conformity in nature between the six planets and the sun and fixed stars, as well as the contrast between the latter and the earth, with respect to conditions other than those of darkness and light. Indeed, I now see that here in the fifth objection, which follows, there is set forth the great disparity between the earth and the heavenly bodies. He writes that there would be great confusion and trouble in the system of the universe and among its parts, according to the Copernican hypothesis, because of its placing among the heavenly bodies (immutable and incorruptible according to Aristotle, Tycho, and others); among bodies of such nobility by the admission of everyone (including Copernicus himself, who declares them to be ordered and arranged in the best possible manner and who

Another difference between the earth and the heavenly bodies, taken from purity and impurity.

removes from them any inconstancy of power); because, I say, of its placing among bodies as pure as Venus and Mars this sink of all corruptible material; that is, the earth, with the water, the air, and all their mixtures!

How much superior a distribution, and how much more suitable it is to nature — indeed, to God the Architect Himself — to separate the pure from the impure, the mortal from the immortal, as all other schools teach, showing us that impure and infirm materials are confined within the narrow arc of the moon's orbit, above which the celestial objects rise in an unbroken series!

SALV. It is true that the Copernican system creates disturbances in the Aristotelian universe, but we are dealing with our own real and actual universe.

Copernicus
causes disturb-
ances in the
universe of
Aristotle.

If a disparity in essence between the earth and the heavenly bodies is inferred by this author from the incorruptibility of the latter and the corruptibility of the former in Aristotle's sense, from which disparity he goes on to conclude that motion must exist in the sun and fixed stars, with the earth immovable, then he is wandering about in a paralogism and assuming what is in question. For Aristotle wants to infer the incorruptibility of heavenly bodies from their motion, and it is being debated whether this is theirs or the earth's. Of the folly of this rhetorical deduction, enough has already been said. What is more vapid than to say that the earth and the elements are banished and sequestered from the celestial sphere and confined within the lunar orbit? Is not the lunar orbit one of the celestial spheres, and according to their consensus is it not right in the center of them all? This is indeed a new method of separating the impure and sick from the sound — giving to the infected a place in the heart of the city! I should have thought that the leper house would be removed from there as far as possible.

Paralogism of
the author of
the *Anti-Tycho.*

It seems foolish
to say that
the earth is
outside the
heavens.

Copernicus admires the arrangement of the parts of the universe because of God's having placed the great luminary which must give off its mighty splendor to the whole temple right in the center of it, and not off to one side. As to the terrestrial globe being between Venus and Mars, let me say one word about that. You yourself, on behalf of this author, may attempt to remove it, but please let us not entangle these little flowers of rhetoric in the rigors of demonstration. Let us leave them rather to the orators, or better to the poets, who best know how to exalt by their graciousness the most vile and sometimes even pernicious things. Now if there is anything remaining for us to do, let us get on with it.

SIMP. Here is the sixth and last argument, in which he puts it down as an unlikely thing that a corruptible and evanescent body could have a perpetual regular motion. This he supports by the example of the animals, which, though they move with their natural motion, nevertheless get tired and must rest to restore their energy. And what is such motion compared to the motion of the earth, which is immense in comparison with theirs? Yet the earth is made to move in three discordant and distractingly different ways! Who would ever be able to assert such a thing, except someone who was sworn to its defense?

Argument taken from animals, which need rest even though their motion is natural.

Nor in this case is there any use in Copernicus saying that this motion, because it is natural to the earth and not constrained, works contrary effects to those of forced motions; and that things which are given impetus are destined to disintegrate and cannot long subsist, whereas those made by nature maintain themselves in their optimum arrangement. This reply, I say, is no good; it falls down before our answer. For the animal is a natural body too, not an artificial one; and its movement is natural, deriving from the soul; that is, from an intrinsic principle, while that motion is constrained whose principle is outside and to which the thing moved contributes nothing. Yet if the animal continues its motion long, it becomes exhausted and would even die if it obstinately tried to force itself on.

You see, therefore, how everywhere in nature traces are to be found which are contrary to the position of Copernicus, and never one in favor of it. And in order that I shall not have to resume the role of this opponent, hear what he has to say against Kepler (with whom he is in disagreement) in regard to what this Kepler has objected against those to whom it seemed an unsuitable or even an impossible thing to expand the stellar sphere as much as the Copernican position requires. Kepler objects to this by saying: *"Difficilius est accidens praeter modulum subiecti intendere, quam subiectum sine accidente augere: Copernicus igitur verisimilius facit, qui auget orbem stellarum fixarum absque motu, quam Ptolemaeus, qui auget motum fixarum immensa velocitate."* ("It is harder to stretch the property

Kepler's argument in favor of Copernicus.

The author of
the *Anti-Tycho*
objects against
Kepler.

Velocity
increases in
circular motion
as the diameter
of the circle
grows.

beyond the model of the thing than to augment the thing without the property. Copernicus therefore has more probability on his side, increasing the orb of the stars as fixed without motion, than does Ptolemy who augments the motion of the fixed stars by an immense velocity.") The author resolves this objection, marveling that Kepler was so misled as to say that the Ptolemaic hypothesis increases the motion beyond the model of the subject, for it appears to him that this is increased only in proportion to the model, and that in accordance with this latter the velocity of motion is augmented. He proves this by imagining a millstone which makes one revolution in twenty-four hours, which motion will be called very slow. Next he supposes its radius to be prolonged all the way to the sun; the velocity of its extremity will equal that of the sun; prolonging it to the stellar sphere, it will equal the velocity of the fixed stars. Yet at the circumference of the millstone it will be very slow. Next, applying this reflection about the millstone to the stellar sphere, let us imagine a point on the radius of that sphere as close to its center as the radius of the millstone. Then the same motion which is very rapid in the stellar sphere will be very slow at this point. The size of the body is what makes it become very fast from being very slow, and thus the velocity does not grow beyond the model of the subject, but rather it increases according to that and to its size, very differently from what Kepler thinks.

SALV. I do not believe that this author entertained so poor and low an opinion of Kepler as to be able to persuade himself that Kepler did not understand that the farthest point on a line drawn from the center out to the starry orb moves faster than a point on the same line no more than two yards from the center. Therefore he must have seen and comprehended perfectly well that what Kepler meant was that it was less unsuitable to increase an immovable body to an enormous size than to attribute an excessive velocity to a body already vast, paying attention to the proportionality (*modulo*) — that is to say, to the standard and example — of other natural bodies, in which it is seen that as the distance from the center increases, the velocity is decreased; that is, the period of rotation for them requires a longer time. But in a state of rest, which is incapable of being made greater or less, the size of the body makes no difference whatever. So that if the author's reply is to have any bearing upon Kepler's

Explanation
of the true
sense of Kepler's
words and a
defense of him.

The size or
smallness of the
body makes a
difference in
motion but
not in rest.

argument, this author will have to believe that it is all the same to the motive principle whether a very tiny or an immense body is moved for the same time, the increase of velocity being a direct consequence of the increase in size. But this is contrary to the architectonic rule of nature as observed in the model of the smaller spheres, just as we see in the planets (and most palpably in the satellites of Jupiter) that the smaller orbs revolve in the shorter times. For this reason Saturn's time of revolution is longer than the period of any lesser orb, being thirty years. Now to pass from this to a much larger sphere, and make that revolve in twenty-four hours, can truly be said to go beyond the rule of the model. So that if we consider the matter carefully, the author's answer does not go against the sense and idea of the argument, but against its expression and manner of speaking. And here also the author is wrong, nor can he deny having in a way perverted the sense of the words in order to charge Kepler with too crass an ignorance. But the imposture is so crude that with all his censure he has not been able to detract from the impression that Kepler has made upon the minds of the learned with his doctrine.

Then as to the objection against the perpetual motion of the earth, taken from the impossibility of its keeping on without becoming fatigued, since animals themselves that move naturally and from an internal principle get tired and have need of repose to relax their members . . .

SAGR. It seems to me that I hear Kepler answering him that there are also animals which refresh themselves from weariness by rolling on the ground, and that hence there is no need to fear that the earth will tire; it may even be reasonably said that it enjoys a perpetual and tranquil repose by keeping itself in an eternal rolling about.

SALV. Sagredo, you are too caustic and sarcastic. Let us put all joking aside, for we are dealing with serious matters.

SAGR. Excuse me, Salviati, but to me what I have just said is not so far from relevant as perhaps you make it out to be. For a movement that serves for repose and removes the weariness from a body tired of traveling may much more easily serve to ward it off, just as preventive remedies are easier than curative ones. And I am sure that if the motion of animals took place as does this one which is attributed to the earth, they would not weary

Cause of
animals tiring.

Motion of
animals should
be called
forced rather
than natural.

Power is not
dissipated
where none of
it is used.

Chiaramonti's
objection
rebounds
upon him.

There are
conclusive
arguments
for true but
not for false
propositions.

at all. For the fatigue of the animal body proceeds, to my thinking, from the employment of but one part in moving itself and
the rest of the body. Thus, for instance, in walking, only the
thighs and the legs are used to carry themselves and all the rest,
but on the other hand you see the movement of the heart to be
indefatigable, because it moves itself alone.

Besides, I don't know how true it is that the movement of
animals is natural rather than constrained. Rather, I believe it
can be truly said that the soul naturally moves the members of
the animal with a preternatural motion. For if motion upward
is preternatural to heavy bodies, the raising of such heavy bodies
as the thigh and the leg to walk cannot be done without constraint, and therefore not without tiring the mover. Climbing
up a ladder carries a heavy body upward against its natural
tendency, from which follows weariness because of the natural
repugnance of heaviness to such a motion. But if a movable body
has a motion to which it has no repugnance whatever, what
tiredness or diminution of force and of power need be feared on
the part of the mover? And why should power be dissipated
where it is not employed at all?

SIMP. It is against the contrary motions by which the terrestrial
globe is imagined to move that the author directs his objection.

SAGR. It has already been said that they are not contrary at all,
and that in this the author is much deceived, so that the strength
of his objection is turned against the objector himself when he
will have it that the *primum mobile* carries all the lower spheres
along, contrary to the motion which they are continually employing at the same time. Therefore it is the *primum mobile*
which ought to get tired, since besides moving itself it has to
take along many other spheres which moreover oppose it with
a contrary motion. Hence the last conclusion that the author
drew, saying that in going over the effects of nature, things favorable to the Aristotelian and Ptolemaic opinion are always
found and never any that do not contradict Copernicus, stands
in need of careful consideration. It is better to say that if one of
these positions is true and the other necessarily false, it is impossible for any reason, experiment, or correct argument to be
found to favor the false one, as none of these things can be
repugnant to the true position. Therefore a great disparity must
exist between the reasons and arguments that are adduced by

the one side and by the other for and against these two opinions, the force of which I leave you to judge for yourself, Simplicio.

SALV. Carried away by the nimbleness of your wit, Sagredo, you have taken the words out of my mouth just when I meant to say something in reply to this last argument of the author's; and although you have replied more than adequately, I wish to add anyway what I had more or less in mind.

He puts it down as a very improbable thing that an evanescent and corruptible body such as the earth could move perpetually with a regular motion, especially since we see animals finally exhaust themselves and stand in need of rest. And to him this improbability is increased by this motion being immeasurably greater in comparison with that of animals. Now I cannot understand why he should be disturbed at present about the speed of the earth, when that of the stellar sphere, which is so much greater, causes him no more considerable disturbance than does that which he ascribes to the velocity of a millstone performing only one revolution every twenty-four hours. If the velocity of rotation of the earth, by being in accord with the model of the millstone, implies no consequence of greater moment than that does, then the author can quit worrying about the exhaustion of the earth; for not even the most languid and sluggish animal — not even a chameleon, I say — would get exhausted from moving no more than five or six yards every twenty-four hours. But if he means to consider the velocity absolutely, and no longer on the model of this millstone, then inasmuch as the movable body must pass over a very great space in twenty-four hours, he should show himself so much the more reluctant to concede this to the starry sphere, which, with incomparably greater speed than that of the earth, must take along with it thousands of bodies, each much larger than the terrestrial globe.

Tiring is more to be feared for the stellar sphere than the terrestrial globe.

It would now remain for us to see the proof by which this author concludes that the new stars of 1572 and 1604 were sublunar in position, and not celestial, as the astronomers of that time were commonly persuaded; truly a great undertaking. But since these writings are new to me, and long by reason of so many calculations, I thought that it would be more expeditious for me to look them over as well as I can between this evening and tomorrow morning; and then tomorrow, returning to our accustomed discussions, I shall tell you what I have got out of them.

Then, if there is time enough, we shall discuss the annual movement attributed to the earth.

Meanwhile, if there is anything else you want to say — particularly you, Simplicio — about matters pertaining to this diurnal motion which has been so lengthily examined by me, there
is yet a little while left to us in which this can be discussed.

SIMP. I have nothing else to say, except that the discussions held
today certainly seem to me full of the most acute and ingenious
ideas adduced on the Copernican side in support of the earth's
motion. But I do not feel entirely persuaded to believe them;
for after all, the things which have been said prove nothing except that the reasons for the fixedness of the earth are not necessary reasons. But no demonstration on the opposing side is
thereby produced which necessarily convinces one and proves
the earth's mobility.

SALV. I have never taken it upon myself, Simplicio, to alter your
opinion; much less should I desire to pass a definite judgment on
such important litigation. My only intention has been, and will
still be in our next debate, to make it evident to you that those
who have believed that the very rapid motion every twenty-four
hours belongs to the earth alone, and not to the whole universe
with only the earth excepted, were not blindly persuaded of the
possibility and necessity of this. Rather, they had very well
observed, heard, and examined the reasons for the contrary
opinion, and did not airily wave them aside. With this same
intention, if such is your wish and Sagredo's, we can go on to
the consideration of that other movement attributed to the same
terrestrial globe, first by Aristarchus of Samos and later by
Nicholas Copernicus, which is, as I believe you well know, that
it revolves under the zodiac in the space of a year around the
sun, which is immovably placed in the center of the zodiac.

SIMP. The question is so great and noble that I shall listen to its
discussion with deep interest, expecting to hear everything that
can be said upon the subject. Following that, I shall go on by
myself at my leisure in the deepest reflections upon what has
been heard and what is to be heard. And if I gain nothing else,
it will be no small thing to be able to reason upon more solid
ground.

SAGR. Then in order not to weary Salviati further, let us put an
end to today's discussions, and tomorrow we shall take up the

discourse again according to our custom, hoping to hear great new things.

Simp. I shall leave the book on the new stars, but I am taking back this booklet of theses in order to look over once more what is there written against the annual motion, which will be the subject of tomorrow's discussion.

End of the Second Day

THE THIRD DAY

SAGREDO. I have been impatiently awaiting your arrival, that I might hear the novel views about the annual rotation of this globe of ours. This has made the hours seem very long to me last night and this morning, though I have not passed them idly. On the contrary, I have lain awake most of the night running over in my mind yesterday's arguments and considering the reasons adopted by each side in favor of these two opposing positions — the earlier one of Aristotle and Ptolemy, and this later one of Aristarchus and Copernicus. And truly it seems to me that whichever of these theories happens to be wrong, the arguments in its favor are so plausible that it deserves to be pardoned — so long as we pause at the ones produced by its original weighty authors. Yet because of its antiquity the Peripatetic opinion has had many followers, while the other has had but few, partly because of its difficulty and partly because of its novelty. And among the partisans of the former, especially in modern times, I seem to discern some who introduce very childish, not to say ridiculous, reasons in maintaining the opinion which appears to them to be true.

SALV. The same thing has struck me even more forcibly than you. I have heard such things put forth as I should blush to repeat — not so much to avoid discrediting their authors (whose names could always be withheld) as to refrain from detracting so greatly from the honor of the human race. In the long run my observations have convinced me that some men, reasoning preposterously, first establish some conclusion in their minds

Some men first fix in their minds the conclusion they believe, and then adapt their reasoning to it.

which, either because of its being their own or because of their having received it from some person who has their entire confidence, impresses them so deeply that one finds it impossible ever to get it out of their heads. Such arguments in support of their fixed idea as they hit upon themselves or hear set forth by others, no matter how simple and stupid these may be, gain their instant acceptance and applause. On the other hand whatever is brought forward against it, however ingenious and conclusive, they receive with disdain or with hot rage — if indeed it does not make them ill. Beside themselves with passion, some of them would not be backward even about scheming to suppress and silence their adversaries. I have had some experience of this myself.

SAGR. I know; such men do not deduce their conclusion from its premises or establish it by reason, but they accommodate (I should have said discommode and distort) the premises and reasons to a conclusion which for them is already established and nailed down. No good can come of dealing with such people, especially to the extent that their company may be not only unpleasant but dangerous. Therefore let us continue with our good Simplicio, who has long been known to me as a man of great ingenuity and entirely without malice. Besides, he is intimately familiar with the Peripatetic doctrine, and I am sure that whatever he does not think up in support of Aristotle's opinion is not likely to occur to anybody.

But here, all out of breath, comes the very person who has been wished for so long today. — We were just now maligning you.

SIMP. Please don't scold me; blame Neptune for my long delay. For in this morning's ebb he withdrew the waters in such a manner that the gondola in which I was riding, having entered an unlined canal not far from here, was left high and dry. I had to stay there over an hour awaiting the return of the tide. And while I was there, unable to get out of the boat (which had run aground almost instantly), I fell to observing an event which struck me as quite remarkable. As the water slackened, it might be seen to run very swiftly through various rivulets, the mud being exposed in many places. While I was watching this effect, I saw this motion along one stretch come to a halt, and without pausing a moment the same water would begin to return, the sea turning from retreat to advance without remaining stationary

Motion of the water between ebb and flow is not interrupted by rest.

for an instant. This is an effect which I have never happened to see before in all the time I have frequented Venice.

SAGR. Then you cannot often have happened to be stranded among little trickles. On account of their having scarcely any slope, the sinking or rising of the open sea by merely the thickness of a sheet of paper is enough to make the water flow and return a long distance through such rivulets. On some seacoasts the rising of the sea only a few yards makes the water spill over the plains for many thousands of acres.

SIMP. I know that well enough, but I should think that between the lowest point of the sinking and the first point of the rising, some perceptible interval of rest would be bound to intervene.

SAGR. It will appear so to you when you have in mind walls or pilings, upon which this change takes place vertically. But actually there is no state of rest.

SIMP. It would seem to me that these being two contrary motions, there would have to be some rest midway between them, in agreement with Aristotle's doctrine proving that *in puncto regressus mediat quies.*

SAGR. I remember the passage well, and I also recall that when I was studying philosophy I was not convinced by Aristotle's proof. Indeed, I have had many experiences to the contrary. I might mention them now, but I do not want to have us wander into any more abysses. We have met here to discuss our subject, if possible, without interrupting it as we have in the past two days.

SIMP. Still it will be good, if not to interrupt it, at least to extend it somewhat. For upon returning home yesterday evening I fell to rereading that booklet of theses, where I found some very convincing proofs against this annual motion which is attributed to the earth. And since I did not trust myself to quote them exactly, I have brought the booklet along with me.

SAGR. You have done well. But if we mean to take up our discussion again in accordance with yesterday's agreement, we must first hear what Salviati has to say about the book on the new stars. Then, without further interruptions, we may examine the annual motion.

Now, Salviati, what have you to say in regard to these stars? Have they really been drawn down from the heavens into these baser regions by virtue of the calculations made by this author whom Simplicio has produced?

SALV. Last night I undertook to study his procedures, and this morning I gave them another glance, wondering whether what I thought I had been reading the night before was really written there, or whether I was the victim of ghosts and fantastic imaginings of the night. To my great regret, I found actually written and printed there that which, for the sake of this philosopher's reputation, I should have wished had not been. It seems impossible to me that he does not realize the vanity of his enterprise, both because it is so obvious and because I remember having heard our friend the Academician praise him. It also seems to me very hard to believe that out of deference to others he could be persuaded to hold his own reputation in such low esteem as to be induced to publish a work from which nothing but censure could be expected from the learned.

SAGR. You might add that there will be rather less than one in a hundred of these, to offset those who will celebrate and exalt him over all the most learned men who exist now or ever have. A man able to sustain the Peripapetic inalterability of the heavens against a host of astronomers, and one who, to their greater shame, has done battle against them with their own weapons! And if there are half a dozen to a province who perceive his trivialities, what are they against the innumerable multitude who (being able neither to discover these nor to comprehend them) are taken in by all the shouting, and applaud the more the less they understand? And even the few who do understand scorn to make a reply to such worthless and inconclusive scribbles. With good reason, too; for those who do understand have no need of this, and upon those who do not understand it is wasted effort.

SALV. Silence would indeed be the most appropriate reprimand for their worthlessness, were there not other reasons which practically force one to repudiate them. One reason is that we Italians are making ourselves look like ignoramuses and are a laughing-stock for foreigners, especially for those who have broken with our religion; I could show you some very famous ones who joke about our Academician and the many mathematicians in Italy for letting the follies of a certain Lorenzini appear in print and be maintained as his views without contradiction. But this also might be overlooked in comparison with another and greater occasion for laughter that might be mentioned, which is the hyprocrisy of the learned toward the trifling of opponents of this stripe in matters which they do not understand.

SAGR. I could not ask for a better example of their petulance, or of the unhappy situation of a man like Copernicus, placed under the carping of those who do not understand even the rudiments of the position against which they have declared war.

SALV. You will be no less astonished at their manner of refuting the astronomers who declare the new stars to be above the orbits of the planets, and perhaps among the fixed stars themselves (*nel firmamento*).

SAGR. But how can you have examined this whole book in such a short time? It is certainly a large volume, and there must be numerous demonstrations in it.

SALV. I stopped after these first refutations of his in which, with twelve demonstrations founded upon the observations of twelve of the astronomers who thought that the new star of 1572 (which appeared in Cassiopeia) was in the firmament, he proves it on the contrary to have been sublunar. To do this he compares, two by two, the meridian altitudes taken by different observers in places of different latitude, proceeding in a manner which you will understand presently. And it seems to me that in examining this first procedure of his I have detected in this author a great inability to prove anything against the astronomers or in favor of the Peripatetic philosophers, and that indeed he only confirms their opinion more conclusively. Therefore I did not want to devote myself with equal patience to the examination of his other methods; having given them a superficial glance, I am positive that the inconclusiveness which pervades his first refutation would exist in the others likewise. And the fact is (as you will soon see) that a very few words suffice to refute this work, although it is built up with so many laborious calculations, as you have perceived.

Therefore you shall hear how I proceeded. The author, I say, in order to attack his adversaries with their own weapons, takes a large number of the observations which they themselves have made, these authors being twelve or thirteen in number.[†] On a part of these he bases his calculations, and he deduces such stars to have been below the moon. Now since I am very fond of proceeding by interrogation, and since the author is not here himself, you, Simplicio, shall reply to the queries I am going to make, and say whatever you believe he would say.

Assuming that we are dealing with the nova of 1572 appearing

Method followed
by Chiaramonti
in refuting the
astronomers,
and by Salviati
in refuting him.

in Cassiopeia, tell me, Simplicio, whether you think it might have been in different places at the same time. That is, could it be amidst the elements and also be among the planetary orbits, and in addition be above these among the fixed stars, as well as being infinitely higher?

SIMP. Doubtless one must say that it was located in a single place, at a unique and determinate distance from the earth.

SALV. Then if the observations made by the astronomers were correct, and if the calculations made by this author were not erroneous, both the former and the latter would necessarily have to yield exactly the same distance; isn't that right?

SIMP. So far as I can see it would necessarily be so, nor do I believe that the author would contradict this.

SALV. But if, of many computations, not even two came out in agreement, what would you think of that?

SIMP. I should judge that all were fallacious, either through some fault of the computer or some defect on the part of the observers. At best I might say that a single one, and no more, might be correct; but I should not know which one to choose.

SALV. But would you want to deduce a questionable conclusion and establish it as true, from a false basis? Surely not. Now this author's calculations are such that not one of them agrees with any other; you see, then, how much faith you can put in them.

SIMP. If that is how matters stand, it is truly a serious defect.

SAGR. I want to help Simplicio and his author out by saying to you, Salviati, that your case would indeed be conclusive if the author had undertaken to find out definitely how far the star was from the earth. But I do not believe that that was his intent; he wished only to show that the star was sublunar. Now if, from the observations mentioned and from all the calculations made on these, the height of the star can always be inferred to have been less than that of the moon, this would suffice the author to convict of the crassest ignorance all those astronomers who, whether they erred in geometry or in arithmetic, could not deduce the true conclusions from their own observations.

SALV. Then I had better turn my attention to you, Sagredo, since you so cunningly sustain the author's doctrine. And let us see whether I can also persuade Simplicio (although he is unskilled at calculations and proofs) that this author's demonstrations are inconclusive to say the least. Consider first that both he and all the astronomers he is in conflict with agree that the new star had no motion of its own, but merely went around with the diurnal motion of the *primum mobile*. But they disagree about its place, the astronomers putting it in the celestial regions (that is, above the moon) and perhaps among the fixed stars, while he judges it to be near the earth; that is, under the arc of the moon's orbit. And since the site of the new star of which we are speaking was toward the north and at no great distance from the pole, so that for us northerners it never set, it was a simple matter to take its meridian altitudes by means of astronomical instruments—its minimal below the pole as well as its maximal above the pole. By combining these, when the observations were made at different places on the earth and at different distances from the north (that is, at places differing among themselves as to polar elevation), the distance of the star could be reasoned out. For if it was placed in the firmament among the other fixed stars, its meridian altitudes when taken at different elevations of the pole would have to differ among themselves in the same way as did these polar elevations. Thus, for example, if the altitude of the star above the horizon had been thirty degrees when taken at a place where the polar elevation was, say, forty-five degrees, then the altitude of the star ought to be increased four or five degrees in those more northerly lands in which the pole is four or five degrees higher. But if the distance of the star from the earth was very small in comparison with that of the firmament, then its meridian altitudes should have increased noticeably more than the polar elevations as the pole was approached. From such a greater increase—that is, from the excess of the increase of the star's elevation over the increase of the polar altitude, which is called a difference of parallax—the distance of the star from the center of the earth may be quickly calculated by a clear and certain method.

Minimum and maximum altitudes of a new star will differ no more than the polar elevations, if the new star is in the firmament.

Now this author takes the observations made by thirteen astronomers at different polar elevations, and comparing a part of these (which he selects) he calculates, by using twelve pairings, that the height of the new star was always below the moon. But he achieves this by expecting such gross ignorance on the part of everyone into whose hands his book might fall that it quite turns my stomach. I can hardly see how the other astron-

omers contain themselves in silence. Especially Kepler, against
whom this author particularly declaims; he would not be one to
hold his tongue, unless he considered the matter beneath his
notice.

Now for your information I have copied on these pages the
conclusions that he deduces from his twelve investigations. Of
these the first is from the two observations

1. of Maurolycus and Hainzel, from which it is deduced that the star is distant from the center by
   less than three terrestrial radii, the difference of
   parallax being 4° 42′ 30″ . . . .   3 radii;
2. and he calculates from the observations of Hainzel *and Schuler,* with a parallax of 8′ 30″, and
   infers that its distance from the center is more
   than . . . .   25 radii;
3. and upon the observations of Tycho and of
   Hainzel, with a parallax of 10 minutes; and the
   distance from the center is inferred to be a little
   less than . . . .   19 radii;
4. and upon the observations of Tycho and of the
   Landgrave of Hesse, with parallax of 14 minutes, and he renders the distance from the center
   about . . . .   10 radii;
5. and on the observations of Hainzel and of Gemma, with a parallax of 42′ 30″, by which the distance is implied to be about . . . . . .   4 radii;
6. and on the observations of the Landgrave and
   of Camerarius, with parallax of 8 minutes, and
   the distance is found to be around . . . . .   4 radii;
7. and upon the observations of Tycho and of
   Hagek, with a parallax of 6 minutes, and infers
   a distance of . . . .   32 radii;
8. and with the observations of Hagek and of Ursinus, with a parallax of 43 minutes, and he
   takes the distance of the star from the surface
   of the earth at . . . . 1/2 radius;
9. and on the observations of the Landgrave and
   of Busch, with a parallax of 15 minutes, he gives
   the distance from the surface of the earth as 1/48 radius;

10. and upon the observations of Maurolycus and of
Muñoz, with a parallax of *4° 30'*, yielding a distance from the surface of the earth of   .   .   .   1/5 radius;
11. and with the observations of Muñoz and of
Gemma, with a parallax of 55 minutes, there is
produced a distance from the center of about   .   13 radii;
12. and with the observations of Muñoz and of Ursinus, with a parallax of 1° 36', the distance
from the center is found to be less than   .   .   .   7 radii.

These are twelve investigations which the author has made at
his own choice from among the multitude which, as he says, could
be made with the combinations of the observations of these thirteen observers; the twelve selected are, one may believe, those
most favorable to his case.

SAGR. But I should like to know whether among all the other
investigations, omitted by this author, there were any in his dis-
favor; that is, any from the calculation of which it would be
inferred that the new star was above the moon. It seems to me
at first glance that this may reasonably be asked. For I see these
results differing so much among themselves that some of them
give me distances of the new star from the earth which are four,
six, ten, a hundred, a thousand, and fifteen hundred times as
great as others, so that I may well suspect that among those not
calculated there might be some in favor of the opposite side. This
seems to me so much the more credible in that I do not suppose
these astronomical observers would lack the intelligence and skill
for such computations, which I think do not depend upon the
most abstruse things in the world. Indeed, when among just these
twelve researches there are some which would place the star
only a few miles from the earth, and others which make it but
little short of the distance to the moon, it would seem almost
miraculous to me if none were to be found which favored the
other side and put the star at least twenty yards beyond the lunar
orbit. And what would be still more absurd is for all these astronomers to be so blind as not to discover so obvious a mistake of
their own.

SALV. Well, prepare now to hear, with unbounded astonishment,
to what excesses of confidence in one's own authority and the
foolishness of other people one may be carried by a desire to
argue and to show oneself more intelligent than others.

Among the researches which the author has omitted, there are some which place the new star not merely beyond the moon, but even above the fixed stars. And these are not just a few, but the majority, as you see here upon this page where I have set them down.

SAGR. But what does the author say about these? Or perhaps he has not considered them?

SALV. All too much has he considered them; he says that those observations are erroneous upon which such calculations are based as would put the star infinitely distant, and that these cannot be reconciled.

SIMP. Well, that certainly looks to me like a feeble evasion, since with just as much right the other side might say that these from which he deduces that the star is in the elemental regions are in error.

SALV. Oh, Simplicio, if I should succeed in convincing you of the artfulness—though it is no great artistry—of this author, I should rouse you to wonder—and also to indignation—when you discovered how he, covering his cunning with the veil of your naïveté and that of other mere philosophers, tries to insinuate himself into your good graces by gratifying your ear and puffing up your ambition, pretending to have convicted and silenced these trifling astronomers who wanted to assail the ineradicable inalterability of the Peripatetic heavens, and what is more, to have struck them dumb and overpowered them with their own weapons. I shall make every effort to do this. Meanwhile you, Sagredo, will excuse Simplicio and me if we bore you unduly while I, with a superfluous course of words (superfluous, I mean, to your swift apprehension), go on trying to make clear something which it is best should not remain hidden from and unknown to him.

SAGR. I shall hear your discourse not only without boredom, but with pleasure. If only all the Peripatetic philosophers might do the same, so that they might find out in that way just what their obligations are to this protector of theirs!

SALV. Tell me, Simplicio, assuming the new star to lie to the north and in the meridian circle, whether you are indeed convinced that for a person who should travel half a day toward the north star, this new star would keep rising above the horizon just as much as would the polestar, if the new star were truly located among the fixed stars. And that on the other hand if it were considerably below them — that is, closer to the earth — it would appear to rise more than that polestar, and the closer it was to the earth, the more it would do so.

Simp. I think I thoroughly understand this, as a token of which I shall try to make a mathematical diagram for it.

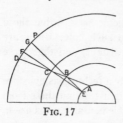

FIG. 17

In this large circle I mark the pole P, and in these two lower circles I shall indicate two stars seen from a point on the earth A, these two stars being B and C, seen along the same line ABC against a fixed star D. Then, as I move along the earth toward a point E, the two stars will appear to separate from the fixed star D and approach the pole P — the lower one, B, moving more, appearing to me as at G; and C somewhat less, appearing as at F. But the fixed star D will have kept the same distance from the pole.

Salv. I see that you understand quite well. I think you understand also how, the star B being lower than C, the angle formed by the visual rays leaving from the two places A and E and meeting at C (that is, the angle ACE) is narrower, or let us say more acute, than the angle formed at B by the rays AB and EB.

Simp. That is easily seen.

Salv. And also, since the earth is very small or practically imperceptible with respect to the firmament, and consequently the distance AE which can be traversed upon the earth is also very short in comparison with the immense length of the lines EG and EF from the earth to the firmament, you can understand that the star C might be raised higher and higher from the earth, so that the angle formed at it by rays leaving these same points A and E would become extremely acute — as if this angle were absolutely imperceptible and nonexistent.

Simp. This also I understand perfectly.

Salv. Now, Simplicio, you must know that astronomers and mathematicians have discovered infallible rules of geometry and arithmetic, by means of which, using the sizes of these angles B and C and their difference, and taking into account the distance between the two places A and E, one may determine the distance of the most sublime bodies within one foot, whenever the said distance and angles are taken precisely.

SIMP. Then if the rules depending upon geometry and arithmetic are correct, all the fallacies and errors that might arise in attempting to determine the altitudes of new stars, or comets, or the like, would have to depend upon improper measurement of the distance AE or the angles B and C. And hence all the differences that are seen in these twelve estimates depend not upon any defect in the rules of calculation, but upon errors made in determining those angles and distances by instrumental observations.

SALV. Exactly so; there is no doubt about that. Now you must carefully note that in moving the star from B to C, by which the angle is made always more acute, the ray EBG continually becomes more distant from that part of the ray ABD which is underneath the angle. This is shown by the line ECF, whose lower part EC is farther from the part AC than is EB. But it can never happen that by any lengthening, however great, the lines AD and EF would be totally severed, since they must ultimately come together at the star. They could be said to be separated and reduced to parallelness only if the lengthening were infinite, which is out of the question. But mark well that since the distance of the firmament may be regarded as infinite in relation to the smallness of the earth (as already mentioned), the angle included between the rays drawn from the points A and E and ending at a fixed star is to be considered as null, and such rays are to be considered parallel lines. Hence we conclude that the new star may be declared to have been among the fixed stars only if comparisons of the observations made at various places, upon calculation, imply this angle to have been null and the lines to have been parallel. But if the angle was of any perceptible size, the new star must necessarily have been beneath the fixed stars; even beneath the moon, if the angle ABE was greater than that formed at the center of the moon.

SIMP. Then the distance of the moon is not so great that such an angle remains imperceptible at it?

SALV. No indeed; this angle is perceptible not only at the moon, but even at the sun.

SIMP. In that case such an angle might be observed for the new star without its having been beneath the sun, let alone the moon.

SALV. So it might be, and so it is in the present instance, as you shall see in due course — that is, when I have cleared the road in such a way that you too, although ignorant of astronomical calculations, may be satisfied in your own mind how much more this author intended writing to please the Peripatetics by veiling and distorting various things than to establish the truth by bringing it out in its naked frankness. Therefore let us move on.

From what has been said up to this point, I believe that you know that the distance of the new star can never be made so great that this angle so often mentioned would entirely vanish, and the rays from observers at A and E would become parallel lines. This amounts to your completely understanding that if calculation should imply that angle to be entirely null or those lines truly parallel, based upon the observations, then we should surely know those observations to be mistaken in at least some small degree. And if the calculations should give us these same lines as being not merely separated to equidistance (that is, as having become parallel), but as having passed beyond the limit and become wider above than below, then it would have to be definitely concluded that the observations had been made with very little accuracy, and were quite erroneous, leading us to an obvious impossibility.

Next, you must believe me and take it as certainly true that two straight lines which leave from two given points upon another straight line are wider above than below whenever the angles included between them on that straight line are greater than two right angles. If they were equal to two right angles, these lines would be parallel; if they were less than two right angles, then the lines would converge, and if prolonged would undoubtedly form a triangle.

SIMP. I know this without taking your word for it. I am not so devoid of geometry as not to know a proposition which I have read in Aristotle a thousand times; namely, that the three angles of every triangle are equal to two right angles. So if I take the triangle ABE in my diagram, assuming that the line EA is straight, I know very well that its three angles A, E, and B are equal to two right angles, and that consequently E and A alone are less than two right angles by the angle B. Whence, widening the lines AB and EB (keeping them fixed in the points A and E) until the angle contained by them in the direction of B vanishes, the two angles at the base would remain equal to two right angles, and these lines would be reduced to parallelness. And if this widening were continued, the angles at points E and A would become greater than two right angles.

SALV. You are an Archimedes, and you have saved me spending more words in explaining to you that whenever the calculations imply that the two angles A and E exceed two right angles, the observations are to be taken as unquestionably mistaken. It is this which I so much desired to have you completely understand, and I was worried about not being able to explain it in such a way that a pure philosopher and Peripatetic would get a firm grip on it. Now we may proceed with the rest.

Taking up again what you granted to me a short time ago, that the new star could not be in more than one place, then whenever the calculations made from the observations of these astronomers do not agree in putting it in the same place, there must be errors in the observations; that is, either in taking the elevation of the pole or the altitude of the star, or both. Now since in many estimates, made from the combinations of the observations two at a time, there are few which place the star in the same position, only these few can be free from error; the rest are certainly mistaken.

SAGR. Then one would have to trust these few alone more than all the rest put together. And since you say that there are few of these which agree, and I see two among these twelve which put the distance of the star from the center of the earth at four radii (the fifth and sixth of them), then the star is more likely to have been elemental than celestial.

SALV. Not so, for if you will look carefully, it does not say here that the distance is exactly four radii, but about four radii. And you see that those distances differ between themselves by hundreds of miles. Look here: this fifth one, you see, which is 13,389 miles, exceeds this sixth one of 13,100 miles by nearly 300 miles.

SAGR. Then which are these few which agree in placing the star in the same position?

SALV. There are five investigations, to the disgrace of this author, which all place it among the fixed stars, as you may see in this other note where I have recorded many more combinations. But I am going to concede to this author more than he would perhaps demand of me — that, to be brief, there is some error in every combination of these observations. This I believe to be absolutely unavoidable, for the observations used in every investigation being four in number (that is, two different polar elevations and two different altitudes for the star, made by different observers in different places and with different instruments), anybody who knows anything about the matter will say that it cannot be that no error will have fallen in among the four. Especially when we know that in taking a single polar elevation with the same instrument in the same place and by the same observer (who may have made it many times), there will be a variance of a minute or so; even of many minutes, as may be seen in plenty of places in this same book.

These things granted, I ask you, Simplicio, whether you think this author took these thirteen observers for clever men, intelligent and dextrous in handling their instruments, or for inexpert bunglers?

SIMP. He must have considered them very acute and intelligent, for if he had thought them unfit for their work he would have been condemning his own book as inconclusive, being based upon assumptions full of errors. And he would have made us out as much too simple, if he had thought he could persuade us by means of their inexpertness to take a false proposition for true.

SALV. Then these observers being capable, and having erred for all that, and their errors needing to be corrected in order for us to get the best possible information from their observations, it will be appropriate for us to apply the minimum amendments and smallest corrections that we can — just enough to remove the observations from impossibility and restore them to possibility. So that, for example, if one can modify an obvious error and a patent impossibility in one of their observations by adding or subtracting two or three minutes, rendering it possible by such a correction, then one ought not to adjust it by adding or subtracting fifteen, twenty, or fifty minutes.

SIMP. I do not believe that the author would deny this; for granted that these were wise and expert men, one must believe that they would be more likely to err little than much.

SALV. Next note this. Of the various locations where the star is placed, some are obviously impossible and others are possible. It is absolutely impossible that it was infinitely higher than the fixed stars, for there is no such place in the universe; and if there were, a star placed there would be invisible to us. Also it is impossible that the star went creeping along the surface of the earth, much less that it was inside the very body of the earth. The possible places include those which are in question, there being nothing repugnant to our minds in a visible starlike object being above the moon, any more than beneath it.

Now when one attempts to deduce what its true place was, by observations and calculations made with as much accuracy as human diligence can achieve, one finds that the majority of these calculations place it an infinite distance beyond the fixed stars, whereas some have it close to the earth's surface, and some even beneath the surface. And of the others which give it places that are not impossible, none are in agreement among themselves. Thus it is proper to call all the observations erroneous, so that if we wish all this labor to bear any fruit, we must be reduced to correcting and amending all the observations.

SIMP. But the author will say that one ought to make no use at all of those observations which imply the star to have been in impossible positions, these being infinitely mistaken and fallacious; that one should accept only those which put it in places which are not impossible. He would say that only among the latter, using the most probable and most numerous data, should one seek, if not for the exact and specific position (that is, its true distance from the center of the earth), at least to find out whether it was among the elements or among the celestial bodies.

SALV. The reasoning you have just given is exactly what the author has put forth in favor of his case, but with too unreasonable a disadvantage to his opponents; and this is the principal point which has made me marvel above all at the excessive confidence he has placed no less in his own authority than in the blindness and carelessness of the astronomers. I shall speak on this, and do you answer on behalf of the author.

First I ask you whether astronomers, in observing with their instruments and seeking, for example, the degree of elevation of the star above the horizon, may deviate from the truth by excess as well as by defect; that is, erroneously deduce sometimes that it is higher than is correct, and sometimes lower? Or must the errors be always of one kind, so that when they err they are always mistaken by an excess, or always by a defect and never by an excess?

SIMP. I do not doubt that they are equally prone to err in one direction and the other.

SALV. I believe the author would say the same. Now, of these two kinds of error, which are opposite and into which the observers of the new star may equally have run, one kind when applied to the calculations will render the star higher than it should be, and the other lower. And since we are already agreed that all the observations are erroneous, what is the reason for this author wanting us to accept those which show the star to have been close as being more congruous with the truth than those others which show it to have been exceedingly remote?

SIMP. Judging by what I have got from what has been said up to this point, I do not see that the author rejects those observations and estimates which might make the star more distant than the moon, or even than the sun, but only those which would make it more than an infinite distance away, as you yourself have put the matter. This distance you also reject as impossible, and he accordingly passes over such observations as being convicted of infinite falsehood and impossibility. Thus it seems to me that if you want to refute the author you ought to produce more exact investigations, or more numerous ones, or by more careful observers, which place the star at such-and-such a distance above the moon or above the sun, at a place where it is entirely possible for it to be — just as he produces these twelve which all place the star beneath the moon in places which exist in the universe and are possible for the star.

SALV. Oh, but Simplicio, right here is your equivocation and the author's — yours in one regard, and his in another. I see from your way of talking that you have the idea that the anomalies (*esorbitanze*) created in establishing the distance of the star increase in proportion to the instrumental errors made in the observations, and conversely that from the size of the anomalies one may deduce the size of the errors. Thus if it is said that from such observations the distance of the star is implied to be infinite, you believe the errors of observation must necessarily have been infinite, and therefore not subject to correction and accordingly to be rejected. The case is quite otherwise, my dear Simplicio. On account of your not having understood how matters do stand, I excuse you, as one untrained in such matters, but I cannot cloak the author's error under the same veil. He, pretending not to know this, and persuading himself that we would really not understand it, hoped to make use of our ignorance for boosting the stock of his doctrine among the multitude of the ill-informed. Therefore, for the information of those who are more credulous than well-informed, and to rescue you from error, know that it may be (and it happens more often than not) that an observation which gives you the star at the distance of Saturn, for example, with the addition or subtraction of a single minute of elevation to that taken by the instrument, will send the star to an infinite distance, and thus take it from the possible to the impossible. Conversely, in these calculations made from the observations which would put the star infinitely distant, the addition or subtraction of one single minute would often restore it to a possible location. And while I say one minute, a correction of one-half that, or one-sixth, or less may suffice.

Now fix it well in mind that at very remote distances like that of Saturn or of the fixed stars, the most trifling errors made by the observer with his instrument will change the location from finite and possible to infinite and impossible. It does not happen thus with distances that are sublunar, and close to the earth, where it may happen that an observation which implies the star to be, for instance, four radii distant, may be increased or decreased not merely by one minute but by ten, or a hundred, or even more, without the calculation rendering the star not only not infinitely distant but not even farther than the moon. From this you may see that the size of the instrumental errors, so to speak, must not be reckoned from the outcome of the calculation, but according to the number of degrees and minutes actually counted on the instrument. Those observations must be called the more exact, or the less in error, which by the addition or subtraction of the fewest minutes restore the star to a possible position. And among the possible places, the actual place must be believed to be that in which there concur the greatest number of distances, calculated on the most exact observations.

SIMP. I am not so sure of what you say, nor do I myself understand how it could happen that in the largest distances a greater anomaly could result from an error of a single minute than could result from ten or a hundred in small distances. But I should be glad to learn.

SALV. You may see it, if not theoretically at least practically, from this brief abstract which I have made of all the combinations and of part of the estimates left out by this author, which I have calculated and entered upon this same sheet.

SAGR. Then from yesterday until now, in the mere eighteen hours which have passed, you must have done nothing but calculate without stopping to eat or sleep.

SALV. No, I have done both. I make such calculations very speedily, and the truth of the matter is that I was much astonished to see how this author goes to such lengths, and puts in so many computations which are not in the least necessary to the question which he is examining. And for a full knowledge of this, as well as to make it quickly apparent that from the observations of the astronomers which this author uses it may be deduced with more probability that the new star was above the moon (and even above the planets, among the fixed stars or higher), I have copied on this page all the observations noted by this same author, made by thirteen astronomers, setting down the polar elevations and the meridian altitudes of the star, both the minima below the pole and the maxima above it. They are as follows:

Tycho.

Altitude of the pole	55° 58′	
Altitude of the star	84° 0′ maximum	
	27° 57′ minimum	

And these are from
the first paper, but
from the second the
minimum is 27° 45′

Hainzel.

Polar altitude	48° 22′		
Altitude of the star	76° 34′	20° 9′ 40″	
	76° 33′ 45″	20° 9′ 30″	
	76° 35′	20° 9′ 20″	

Peucer and Schuler.		The Landgrave.	
Polar altitude	51° 54′	Polar altitude	51° 18′
Altitude of the star	79° 56′	Altitude of the star	79° 30′
	23° 33′		23° 3′

Camerarius.

Polar altitude	52° 24′	
Altitude of the star	80° 30′	24° 28′
	80° 27′	24° 20′
	80° 26′	24° 17′

	Hagek.			Ursinus.	
Polar altitude	48°	22′	Polar altitude	49°	24′
Altitude of the star	20°	15′	Altitude of the star	79°	
				22°	

	Muñoz.			Maurolycus.	
Polar altitude	39°	30′	Polar altitude	38°	30′
Altitude of the star	67°	30′	Altitude of the star	62°	
	11°	30′			

	Gemma.			Busch.	
Polar altitude	50°	50′	Polar altitude	51°	10′
Altitude of the star	79°	45′	Altitude of the star	79°	20′
				22°	40′

	Reinhold.	
Polar altitude	51°	18′
Altitude of the star	79°	30′
	23°	2′

Now, in order to see my entire method, let us begin with these five calculations which were omitted by the author — perhaps because they went against him, since they place the star above the moon by many terrestrial radii. This is the first of them, calculated on observations by the Landgrave of Hesse and by Tycho, who by the author's own admission are among the most exact of the observers. In this first one, I shall explain the order which I follow in my researches, giving you information which will serve for all the others, since they go according to the same rule, varying in nothing except the data given. The data consist of the number of degrees of polar elevation and of altitude of the new star above the horizon, from which one seeks its distance from the center of the earth in terms of terrestrial radii. In this matter it is of no consequence how many miles are involved; solving for the distances in miles between the places from which the observations were made, as this author does, is a waste of time and effort. I do not know why he has done it, especially when he ultimately reconverts miles into terrestrial radii.

SIMP. Maybe he did so in order to find the distance of the star down to smaller measures and such fractions of them as a few inches. Those of us who do not understand your arithmetical rules are amazed to hear the results when we read, for instance: "Therefore the comet, or the new star, was distant from the center of the earth three hundred seventy-three thousand, eight hundred and seven miles, and two hundred eleven four-thousand-and-ninety-sevenths ($373,807 \frac{211}{4097}$)." From such painstaking precision as this, in which these minutiae are noted, we get the impression that it would be simply impossible for you, who take account in your calculations of a mere inch, to deceive us in the end by a hundred miles.

SALV. Your reasoning and your excuse for him would be appropriate if, in a distance of thousands of miles, one yard more or less were of any moment, and if the assumptions which we take as true were so certain as to assure us that we would in the end deduce an indubitable truth. But, as you see here in the author's twelve computations, the distances of the star which are deduced differ from one another (and are therefore wide of the truth) by many hundreds and thousands of miles. Now when I am quite sure that what I seek must necessarily differ from correctness by hundreds of miles, why should I vex myself with calculations lest I miss one inch?

But let us get down to the operations, which I perform in the following way. Tycho, as is seen in the note, observed the star at a polar altitude of 55° 58', and the polar altitude of the Landgrave was 51° 18'. The height of the star at the meridian, as taken by Tycho, was 27° 45'; the Landgrave found it to be 23° 3'. These altitudes are set down together in this way:

Tycho	Pole 55° 58'	Star 27° 45'
The Landgrave	Pole 51° 18'	Star 23° 3'

This done, I subtract the lesser from the greater, and there remain these differences, as below:

4° 40'	4° 42'

Parallax	2'

where the difference in polar altitudes, 4° 40', is less than the difference in altitudes of the star, 4° 42'; and therefore there is a difference in parallax of 2 minutes.

These things determined, take the author's own figure in which the point B is the position of the Landgrave, D is that of Tycho, C the position of the star, A the center of the earth, ABE is the vertical line at the Landgrave's station, ADF is that at Tycho's,

and the angle BCD is the parallactic difference.

Since the angle BAD included between the two verticals is equal to the difference of the polar altitudes, it will be 4° 40′, and I note it separately here. I then find the chord of this from a table of arcs and chords, and set that down next to it; it is 8,142 parts where the radius AB is 100,000. Then I easily find the angle BDC, because half the angle BAD, which is 2° 20′, added to a right angle, gives the angle BDF as 92° 20′. Add to this the angle CDF, which is the deviation from the vertical of

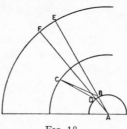

FIG. 18

the greater altitude of the star, in this case 62° 15′, giving the size of the angle BDC as 154° 45′.† This I set down together with its sine, taken from the table, which is 42,657, and under this I note the angle of parallax BCD, 2 minutes, with its sine, 58.

Angle BAD	4° 40′,	its chord, 8,142 parts where the radius AB is 100,000.

	BDF	92° 20′		
	BDC	154° 45′ ⎫	sines	42657
	BCD	0° 2′ ⎭		58

	58	42657	8142
		8142	
		85314	
		170628	
		42657	
		341256	

58)	347313794 (59
571	
5	

And since in the triangle BCD, the side DB is to the side BC as the sine of the opposite angle BCD is to the sine of the other opposite angle BDC, then if the line BD were 58, BC would be 42,657. Now since the chord DB is 8,142 when the radius is 100,000, and we are trying to find out how many of these same

100,000 parts make up BC, let us say, by the Rule of Three: If, when BD is 58, BC is 42,657, then if the same DB were 8,142, how much would BC be?

Therefore I multiply the second term by the third, and get 347,3*13*,294; this must be divided by the first, or by 58; the quotient will be the number of parts in the line BC when the radius is 100,000 parts. And to find how many radii BA the same line BC would contain, the same quotient would have to be again divided by 100,000, and we shall have the number of radii included in BC. Now the number 347,313,294 divided by 58 is 5,988,160¼, as may be seen below:

$$\overline{5988160\tfrac{1}{4}}$$
$$58\,)\,347313294$$
$$5717941$$
$$54\ 3$$

and this divided by 100,000 gives us $59\frac{88,160}{100,000}$:

$$1\,|\,00000\quad|\quad59\,|\,88160.$$

We could somewhat shorten the operations, dividing the first product† found (that is, 347,313,294) by the product of the two numbers, 58 and 100,000, thus:

$$5800000\,)\,3473\ 13294\,(\ 59$$
$$571$$
$$5$$

and this likewise gives us $59\frac{5,113,294}{5,800,000}$. That many radii are contained in the line BC, and adding one for the line AB will give us a little less than 61 radii for the two lines ABC. Therefore the distance from the center A to the star C is over 60 radii, which puts it above the moon by more than 27 radii as Ptolemy reckons, and by more than 8 according to Copernicus, assuming that the distance of the moon from the center of the earth according to the account of Copernicus himself is, as this author states, 52 radii.†

By this kind of investigation, from the observations of Camerarius and Muñoz I find the star to have been situated at a similar distance — that is to say, more than 60 radii. Here are these observations, followed by the computation.

Polar Altitude of:		Altitude of the star:	
Camerarius	52° 24'		24° 28'
Muñoz	39° 30'		11° 30'
Difference	12° 54'	Difference	12° 58'
			12° 54'

Parallactic difference (angle BCD) 0° 4'

$$\text{Angles} \begin{cases} \text{BAD} & 12°\ 54', \text{ and its chord } 22466 \\ \text{BDC} & 161°\ 59' \\ \text{BCD} & 0°\ 4' \end{cases} \text{sines} \begin{cases} 30930 \\ 116 \end{cases}$$

Rule of Three

22466

116	30930	22466

673980
202194
67398

Distance BC:

116) 6948⟨73380⟩ (59   59 radii, almost 60.
1144
10

The investigation below is based upon two observations by Tycho and Muñoz, from which the star is calculated to have been 478 or more radii distant from the center of the earth.

Polar Altitude of:		Altitude of the star:	
Tycho	55° 58'		84° 0'
Muñoz	39° 30'		67° 30'
Difference	16° 28'	Difference	16° 30'
			16° 28'

Parallactic difference (angle BCD) 0° 2'

$$\text{Angles} \begin{cases} \text{BAD} & 16°\ 28', \text{ and its chord } 28640 \\ \text{BDC} & 104°\ 14' \\ \text{BCD} & 0°\ 2' \end{cases} \text{sines} \begin{cases} 96930 \\ 58 \end{cases}$$

Rule of Three

58	96930	28640
	28640	

3877200
58158
77544
19386

58) 27760⟨73200⟩ (478
4506
53

This following investigation gives the star as more than 358 radii distant from the center.

Polar Altitude	{ Peucer	51° 54'	Altitude of the star	{	79° 56'
	Muñoz	39° 30'			67° 30'
		12° 24'			12° 26'
					12° 24'
					0° 2'

$$\text{Angles} \begin{cases} \text{BAD} & 12° \ 24', \text{ chord } 21600 \\ \text{BDC} & 106° \ 16' \\ \text{BCD} & 0° \ \ 2' \end{cases} \text{sines} \begin{cases} 95996 \\ 58 \end{cases}$$

Rule of Three

58   95996   21600
            21600
            ─────────
            57597600
            95996
            191992

58) 20735̷1̷3̷0̷0̷0̷ (357
      3339
      42

From this other investigation the star is found to be more than 716 radii distant from the center.

Polar Altitude	} The Landgrave	51° 18'	Altitude of the star	}	79° 30'
	Hainzel	48° 22'			76° 33' 45"
		2° 56'			2° 56' 15"
					2° 56' 0"
					0° 0' 15"

$$\text{Angles} \begin{cases} \text{BAD} & 2° \ 56', \text{ chord } 5120 \\ \text{BDC} & 101° \ 58' \\ \text{BCD} & 0° \ \ 0' \ 15'' \end{cases} \text{sines} \begin{cases} 97845† \\ 7 \end{cases}$$

Rule of Three

7   97845   5120
           5120
           ─────────
           1956900
           97845
           489225

7) 5009̷6̷6̷4̷0̷0̷ (715
     134

These are, as you see, five investigations which range the star well above the moon. Now I want you to consider what I told you a little while ago; namely, that at great distances a change — or I should say a correction — of a very few minutes will move a star through an immense distance. For example, in the first of the above investigations, where the calculation puts the star 60 radii away from the center with a parallax of 2 minutes, those who wish to maintain that it was among the fixed stars need only correct the observations by two minutes or less, for then the parallax vanishes or becomes so small as to place the star at an immense distance, such as everyone takes that of the firmament to be. In the second investigation, an amendment of less than 4 minutes does the like. In the third and fourth, as in the first, 2 minutes only will also place the star among the fixed stars. In the last, a quarter of a minute — 15 seconds — will give the same result.

But you will not find it so for the sublunar altitudes. For imagine any distance you please, and try to amend the investigations made by the author to adjust them so that they all correspond with that definite distance, and you will discover how much greater are the corrections which you will have to make. SAGR. It would not do any harm at all for our complete understanding of this if we were to see an example of what you are saying.

SALV. Decide at your pleasure what the given sublunar distance shall be at which the star is to be located, for with little difficulty we can assure ourselves whether corrections like those which we have seen to be sufficient to put it back among the fixed stars would move it to the place decided upon by yourselves.

SAGR. In order to pick the distance most favorable to the author, let us assume it to be that which is the greatest among all his twelve investigations. For if there is a dispute about it between him and the astronomers, and the latter assert the star to have been above the moon while he places it below, then even the smallest amount by which he proves it to be below will give him the victory.

SALV. Let us accordingly take the seventh investigation, made upon the observations of Tycho and Thaddeus Hagek, by which the author finds the star to have been 32 radii distant from the center, this being the distance which is most favorable to his side.† And to give him every advantage, I wish us to place it at the distance least favorable to the astronomers, which means putting it beyond the firmament.

These things assumed, then, let us find out what corrections would be necessary to apply to the balance of his eleven investigations so as to raise the star up to the distance of 32 radii. We shall commence with the first, computed on the observations of Hainzel and Maurolycus, in which the author finds the distance from the center to be about 3 radii, with a parallax of 4° 42′ 30″; now let us see whether it would be carried up to 32 radii by cutting this down to a mere 20 minutes. Here are the operations, which are very brief and quite exact. I multiply the sine of the angle BDC by the chord BD, and divide the product (ignoring the last five digits) by the sine of the parallax. This gives 28½ radii; so not even by a correction made by taking away 4° 22′ 30″ from 4° 42′ 30″ is the star elevated to 32 radii; this correction, for Simplicio's information, is one of 262½ minutes.

Hainzel	Pole	48° 22′		Star	76° 34′ 30″
Maurolycus	Pole	38° 30′		Star	62°
		9° 52′			14° 34′ 30″
					9° 52′
			Parallax		4° 42′ 30″

BAD	9° 52′	chord	17200
BDC	108° 21′ 30″	sine	94910
BCD	0° 20′	sine	582

$$94910$$
$$17200$$
$$\overline{}$$
$$18982000$$
$$66437$$
$$9491$$
$$582\,)\,16324\cancel{87000}\,(\,28$$
$$4688$$
$$2$$

In the second calculation,† made upon observations by Hainzel and Schuler, with a parallax of 8′ 30″, the star is found to be at a height of about 25 radii, as seen in the following operations:

BD	chord	6166

BDC }
BCD } sines { 97987
{ 247

```
    97987
     6166
   ------
  587922
  587922
   97987
  587922
  -----------------
247 ) 604187842 ( 24
    1103
      11
```

And reducing the parallax of 8′ 30″ to 7′, whose sine is 204, the star is raised to around 30 radii. Therefore a correction of 1′ 30″ is not enough.

```
204 ) 604187842 ( 29
      1965
        12
```

Now let us see what correction is needed for the third investigation, made upon the observations of Hainzel and Tycho, which puts the star approximately 19 radii high, with a parallax of 10 minutes. The usual angles and their sines and chord are shown as found by this author, and they imply the star to be about 19 radii distant, as in the author's calculations. Hence in order to raise it, the parallax must be reduced according to the rule which he, too, observes in the ninth computation. Meanwhile let us assume the parallax to be 6 minutes, the sine of which is 175. Having made this division, we find that the star is less than 31 radii distant. Therefore a correction of 4 minutes is too small for the author's needs.

```
          ( BAD    7° 36′   chord  13254
Angles   { BDC  155° 52′   sine   40886
          ( BCD    0° 10′   sine     291
            13254
            40886
           -------
            79524
           106032
           106032
            53016
   -----------------------        --------------
291 ) 541903044 ( 18        175 ) 5419 ( 30
     2501                         16
       18
```

Let us go on with the fourth investigation and the remaining ones by the same rule, using the chords and sines as found by the author himself. In this one the parallax is 14 minutes, and the height established is less than 10 radii. Reducing the parallax from 14 minutes to 4 minutes, you see that the star is not raised up even to 31 radii in any case, so that a correction of 10 minutes out of 14 is not sufficient.

BD	chord	8142	43235
BDC	sine	43234	8142
BCD	sine	407	86470
			172940
			43235
			345880

$$116 \ ) \ 3520\cancel{1}\cancel{9}\cancel{3}\cancel{7}\cancel{0} \ (\ 30$$
$$4$$

In the author's fifth calculation we have the sines and chord as below:

BD	chord	4034†	97998
BDC	sine	97998	4034
BCD	sine	1236	391992
			293994
			391992

$$145 \ ) \ 3953\cancel{2}\cancel{3}\cancel{9}\cancel{3}\cancel{2} \ (\ 27$$
$$1058$$
$$3$$

and the parallax is 42′ 30″, which implies an altitude of about 4 radii for the star. Correcting the parallax by reducing it from 42′ 30″ to merely 5′ does not suffice to raise it up even to 28 radii, so an amendment of 37′ 30″ is too little.

Here are the chord, the sines, and the parallax in the sixth computation:

BD		chord	1920	40248
BDC		sine	40248	1920
BCD	8′	sine	233	804960
				362232
				40248

$$29 \ ) \ 7727\cancel{0}\cancel{1}\cancel{6}\cancel{0} \ (\ 26$$
$$198$$
$$1$$

and the star is found to be about 4 radii above the earth. Let us see what this becomes by reducing the parallax from 8 minutes to only one. Look at the calculation, with the star not even raised to 27 radii; hence it is insufficient to correct this by 7 minutes out of 8.

In the eighth calculation the chord, the sines, and the parallax, as you see, are these:

BD	chord	1804		36643
BDC	sine	36643†		1804
BCD	sine	29		146572
				293144
				36643

29) 6610̸3̸9̸7̸7̸ (22
    83
     2

and from this the author calculates the height of the star at 1½ radii, with a parallax of 43 minutes, which when reduced to 1 minute still leaves the star less than 24 radii distant. So a correction of 42 minutes is inadequate.

Now let us look at the ninth. Here are the chord, the sines, and the parallax — which is 15 minutes. From these the author calculates that the star is separated from the surface of the earth by less than one forty-seventh of a radius. But this is an error of calculation, for as we shall see in a moment, it really comes out as more than one-fifth. See here: it is about $90/_{436}$, which is greater than ⅕.

BD	chord	232		39046
BDC	sine	39046		232
BCD	sine	436		78092
				117138
				78092

436) 90̸5̸8̸6̸7̸2̸

What the author next remarks is quite true — that to correct the observations it is not sufficient to reduce the parallactic difference either to a single minute, nor even to the eighth part of a minute. But I can tell you that a difference as small as the tenth part of a minute would not restore the altitude of the star to 32 radii; for the sine of one-tenth of one minute (i.e., of 6 seconds) is 3. This, if divided into 90 according to our rule — or I should say, if we divide 9,058,762 by 300,000 — becomes $30\frac{58,672}{100,000}$;† that is, a little more than 30½ radii.

The tenth gives the height of the star as one-fifth of one radius, with these angles and sines, and a parallax of 4° 30′. This could, I see, be reduced from 4½ degrees to 2 minutes without promoting the star up to 29 radii.

$$
\begin{array}{lllr}
\text{BD} & & \text{chord} & 1746 \\
\text{BDC} & & \text{sine} & 92050 \\
\text{BCD} & 4°\ 30′ & \text{sine} & 7846
\end{array}
\qquad
\begin{array}{r}
1746 \\
92050 \\
\hline
87300 \\
3492 \\
15714
\end{array}
$$

$$58\ )\ 1607\cancel{1930}0\ (\ 27$$
$$441$$
$$4$$

The eleventh makes the star some 13 radii away for this author, with a parallax of 55 minutes. Let us see where it will take the star if we reduce this to 20 minutes. Here is the computation; it elevates the star to a little less than 33 radii, so the correction would be somewhat less than 35 minutes out of 55.

$$
\begin{array}{lllr}
\text{BD} & & \text{chord} & 19748 \\
\text{BDC} & & \text{sine} & 96166 \\
\text{BCD} & 55′ & \text{sine} & 1600
\end{array}
\qquad
\begin{array}{r}
96166 \\
19748 \\
\hline
769328 \\
384664 \\
673162 \\
865494 \\
96166
\end{array}
$$

$$582\ )\ 18990\cancel{86106}\ (\ 32$$
$$1536$$
$$36$$

The twelfth, with a parallax of 1° 36′, implies the star to be less than 6 radii high. Reducing the parallax to 20 minutes takes the star to a distance of less than 30 radii; therefore a correction of 1° 16′ is not enough.

BD		chord	17257	17258
BDC		sine	96150	96150
BCD	1° 36′	sine	2792	862900

$$\begin{array}{r} 17258 \\ 103548 \\ 155322 \end{array}$$

582) 16593̸3̸6̸7̸0̸0̸ (28
       4957
         29

These are the corrections of the parallaxes in the ten estimates by the author to replace the star at an altitude of 32 radii:

Degrees	Minutes	Seconds		Degrees	Minutes	Seconds
4	22	30	out of	4	42	30
0	4	0	out of	0	10	0
0	10	0	out of	0	14	0
0	37	30	out of	0	42	30
0	7	0	out of	0	8	0
0	42	0	out of	0	43	0
0	14	50	out of	0	15	0
4	28	0	out of	4	30	0
0	35	0	out of	0	55	0
1	16	0	out of	1	36	0
9	216			9	296	
x60=	540			x60=	540	
	756				836	

From this it is seen that in order to move the star to an altitude of 32 radii we must subtract 756 from the total of 836 minutes of parallax, reducing this to 80 — and even these corrections are insufficient.

Hence you may see (as I noted almost at once) that if the author should decide he wanted to take the distance of 32 radii for the true height of the star, then the correction of the above ten estimates (ten, because the second calculation we made was also very high, and restored the altitude to 32 radii with only 2 minutes of correction), in order to make them all restore the star to that distance, would require such a reduction of parallaxes as to amount to more than 756 minutes in all the subtractions together. But in the five which I calculated and which imply the star to have been beyond the moon, 10¼ minutes correction alone is enough to adjust them so that all place the star in the firmament. Now, in addition to these, there are five more investigations which imply the star to be precisely among the fixed stars without any correction, making ten computations which agree in placing it in the firmament with merely the correction of five of them by 10¼ minutes, as we have seen; whereas in order to adjust ten of the author's computations, amendments of 756 minutes out of 836 are required to raise the star to a height of 32 radii. That is, if you want the star to have a distance of 32 radii, it is necessary to subtract 756 minutes from the total of 836, and even that correction is not sufficient.

Now for the investigations which render the star devoid of parallax directly and without any correction, and which thus place it in the firmament and even in the most distant part of this (in a word, as high as the very pole), here are these five:

	Polar Altitude	Star Altitude
Camerarius	52° 24′	80° 26′
Peucer	51° 54′	79° 56′
	0° 30′	0° 30′
The Landgrave	51° 18′	79° 30′
Hainzel	48° 22′	76° 34′
	2° 56′	2° 56′
Tycho	55° 48′	84°
Peucer	51° 54′	79° 56′
	4° 4′	4° 4′
Reinhold	51° 18′	79° 30′
Hainzel	48° 22′	76° 34′
	2° 56′	2° 56′
Camerarius	52° 24′	24° 17′
Hagek	48° 22′	20° 15′
	4° 2′	4° 2′

Of the remaining combinations that can be made of observations taken by all these astronomers, those which imply the star to be infinitely high are much more numerous — about thirty more — than those which upon calculation place the star beneath the moon. Now, as we agreed, it is plausible that the observers are more likely to have erred little than much, and it is obvious that the corrections to be applied to observations which give the star as at an infinite distance will, in drawing it down, bring it first and with least amendment into the firmament rather than below the moon. Hence everything supports the opinion of those who place it among the fixed stars. Moreover, the corrections needed for this amendment are much smaller than those by which the star may be moved from an improbable proximity up to an altitude more favorable to this author, as has been seen from the previous examples.

Among those of impossible proximity are the three by which the star seems to be separated from the center of the earth by a distance of less than one radius, making it move around subterraneously, so to speak. Such are the combinations in which the polar altitude of one of the observers is greater than that of the other, while the elevation of the star as taken by the former is less than that taken by the latter; they are the combinations recorded below. The first is that of the Landgrave combined with Gemma's, where the Landgrave's polar altitude is 51° 18′, greater than the polar altitude of Gemma, which is 50° 50′, while the altitude of the star for the Landgrave, 79° 30′, is less than that of the star for Gemma, 79° 45′.

	Polar Altitude	Star Altitude
The Landgrave	51° 18′	79° 30′
Gemma	50° 50′	79° 45′

The other two are those below:

Busch	51° 10′	79° 20′
Gemma	50° 50′	79° 45′
Reinhold	51° 18′	79° 30′
Gemma	50° 50′	79° 45′

From what I have shown you up to this point, you can see how unfavorable to the author's case is this first method of his for investigating the distance of the star and proving it to be sublunar, and how much more clearly and with how much greater probability it is implied that the distance of the star placed it in the most remote heavens.

SIMP. The ineffectiveness of this author's proofs thus far seems to me to have been very clearly exposed. But I see that all this takes up only a few pages of his book, and it may be that other arguments of his are more conclusive than these first ones.

SALV. Rather, they can only be less valid, if we take what has gone before as a sample of what is left. For it is obvious that the uncertainty and inconclusiveness of the former are clearly the result of errors committed in instrumental observations, which he assumed to permit polar elevations and the altitude of the star to be precisely taken, whereas in fact all of them may easily be wrong. Astronomers have had centuries in which to take the elevation of the pole at their leisure, and the meridian altitudes of a star are the simplest ones to observe, as they are very definite; moreover, they allow the observer plenty of time to proceed with them, for they do not perceptibly change in a brief interval as do altitudes remote from the meridian.

Now if this is the case (and it most certainly is), what faith can we have in calculations founded upon observations which are more numerous, more difficult to make, more capricious in their variation, and on top of all this are made with less convenient and more unreliable instruments? From just the glance that I have given to the ensuing proofs, the observations are taken upon the altitudes of stars in the various vertical circles, which are known by the Arabic term *azimuths*. In such observations, one makes use of instruments that are movable not only in vertical circles, but also in horizontal ones at the same time, so that one must have observed, at the same time that the altitude was taken in the vertical one, the distance of the star from the meridian in the horizontal one. Moreover the operation must be repeated after a considerable interval of time, and careful track must be kept of the elapsed time, trusting either to·clocks† or to other observations of the stars.

It is that kind of a web of observations that he next goes about comparing with another one like it, made by a different observer in a different country with different instruments and at different times. From this, the author attempts to deduce the altitude of the star and its horizontal latitudes at the same moments of time as the first observations; and ultimately he bases his calculations upon such adjustments. Now I leave it to you to judge how much confidence may be placed in deductions made from such methods of investigation.

Besides, I have no doubt that if anyone wished to suffer through such long calculations he would find, just as before, that there were more which favored the opposing side than the au-

thor's. But this does not seem to me worth the trouble for a
matter which is not of prime interest to us in any case.
SAGR. I share your opinion in this. But if the matter is surrounded
with so much confusion, uncertainty, and error, how does it
happen that so many astronomers have so confidently declared
the new star to have been very remote?

SALV. Either of two sorts of observations, both very simple, easy,
and correct, would be enough to assure them of the star being
located in the firmament, or at least a long way beyond the moon.
One of these is the equality — or very slight disparity — of its
distances from the pole when at its lowest point on the meridian
and at its highest. The other is that it remained always at the
same distance from certain surrounding fixed stars; especially
x Cassiopeiae, from which it was less than one and one-half
degrees distant. From these two things it may unquestionably be
deduced that parallax was either entirely lacking, or was so small
that the most cursory calculation proves the star to have been a
great distance from the earth.

SAGR. But didn't the author know about these things? And if so,
what did he have to say in his own defense?

SALV. When a person finds no defense to be of any avail against
his mistake and produces a frivolous excuse, people say that
he is reaching for ropes from the sky. This author grasps not at
ropes, but at spiderwebs from the sky, as you will plainly see
upon examining these two points just mentioned.

First, as to what is shown by the observed polar distances one
by one, I have noted these down in these brief calculations. For
complete understanding, I should first inform you that if the
new star, or some other phenomenon, is close to the earth and is
turning in the diurnal motion about the pole, it will show itself
more distant from the pole when it is below the pole on the
meridian than when above it. This is seen in the next diagram, in
which the point T denotes the center of the earth, O the place
of the observer; the arc of the firmament is marked VPC, the
pole, P. The phenomenon, moving on the circle FS, is seen at
one time beneath the pole along the ray OFC, and at another
above it along the ray OSD. Hence its places as seen against the
firmament are D and C, but the true places with respect to the
center T are B and A, equally distant from the pole. From this
it is obvious that one apparent place of the phenomenon S (that

is, the point D) is closer to the pole than the other apparent place, C, seen along the ray OFC. And this is the first point to be noted.

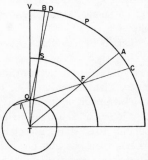

FIG. 19

In the second place you must notice that the amount by which its apparent lower distance from the pole exceeds its apparent upper distance from the pole is greater than is its lower parallax. By this I mean that the excess of the arc CP (lower apparent distance) over the arc PD (upper apparent distance) is greater than the arc CA (which is the lower parallax). This is easily deduced, since the arc CP must exceed the arc PD more than it does PB, PB being greater than PD. But PB is equal to PA, and the excess of CP over PA is the arc CA. Therefore the excess of the arc CP over the arc PD is greater than the arc CA, which is the parallax of the phenomenon assumed to be at F; and that is what was required to be known. And to give every advantage to the author, we shall assume the parallax of the star at F to be the entire excess of the arc CP (that is, the distance below the pole) over the arc PD (the upper distance).

Now I come to the examination of what is implied by the observations of all the astronomers cited by the author, among which there is not one that does not work out against him and his purposes. First let us take this one by Busch, which finds the distance of the star from the pole when it is above to be 28° 10', and when underneath, 28° 30', so that the excess is 20 minutes, which (to the author's advantage) we shall take as if it were the parallax of the star at F; that is, the angle TFO. Then the distance from the zenith, that is, the arc CV, is 67° 20'. These two things found, produce the line CO and let fall the perpendicular TI upon it, and let us consider the triangle TOI, of which the angle I is a right angle. And angle IOT is known, from being opposite angle VOC, the distance of the star from the zenith. Furthermore the angle F is known, triangle TIF being a right triangle; and this is taken to be the parallax. Hence we set down here the two angles IOT and IFT, and take the sines of these, which are as you see them noted.

Now since in the triangle IOT the sine of angle IOT gives TI as 92,276 where TO as a whole is 100,000, and moreover in triangle IFT the sine of angle IFT gives TI as 582 when TF as a whole is 100,000, let us say by the Rule of Three: If TI is 582, TF is 100,000; but if TI were 92,276, what would TF be?

We multiply 92,276 by 100,000 and get 9,227,600,000, and this is to be divided by 582, which comes out, as you see, 15,854,982; and that would be the length of TF if the length of TO were 100,000. So to find out how many lines TO there are in TF, we divide 15,854,982 by 100,000, and there would be approximately 158½; that is how many radii distant the star F will be from the center T. And to shorten the operations, seeing that the product of multiplying 92,276 by 100,000 has to be divided first by 582, and then the quotient by 100,000, we can get the same result by dividing the sine 92,276 by the sine 582, without any multiplication of 92,276 by 100,000. This is seen below, where 92,276 divided by 582 is this same 158½ approximately. Thus let us keep it in mind that merely the division of TI considered as sine of the angle TOI, by TI considered as sine of the angle IFT, gives us the required distance TF in terms of the radius TO.

Angles $\begin{Bmatrix} \text{IOT} & 67° & 20' \\ \text{IFT} & & 20' \end{Bmatrix}$ sines $\begin{Bmatrix} 92276 \\ 582 \end{Bmatrix}$

				15854982
			582)	9227600000
TI	TF	TI	TF	3407002246
582	100000	92276	?	49297867
				325414
			100000)	15854982

582) 92276 (158
34070
492
3

Now see what the observations of Peucer give us. In these, the distance underneath the pole is 28° 21′, and the distance above is 28° 2′; the difference is 19 minutes, and the distance from the zenith is 66° 27′. From these data the distance of the star from the center is deduced to be almost 166 radii.

Third
Day

Angles $\left\{\begin{array}{ll}\text{IAC} & 66° \ 27' \\ \text{IEC} & 19'\end{array}\right\}$ sines $\left\{\begin{array}{l} 91672 \\ 553\end{array}\right.$

$$553 \ ) \ 91672 \ (\ 165\ ^{427}/_{553}$$
$$36397$$
$$312$$
$$4$$

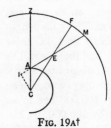

Fig. 19A†

Here is what is shown by taking those observations of Tycho which are most favorable to the opponent; that is, the lower distance from the pole is 28° 13′, and the upper 28° 2′, leaving the entire difference of 11 minutes as if it were all parallax. The distance from the zenith is 62° 15′. And below is the calculation; the distance of the star from the center is found to be 276 $^9/_{16}$ radii.

Angles $\left\{\begin{array}{ll}\text{IAC} & 62° \ 15' \\ \text{IEC} & 11'\end{array}\right\}$ sines $\left\{\begin{array}{l} 88500 \\ 320\end{array}\right.$

$$320 \ ) \ 88500 \ (\ 276\ ^9/_{16}$$
$$2418$$
$$21$$

The observations of Reinhold, which follow, yield a distance of the star from the center as 793 radii.

Angles $\left\{\begin{array}{ll}\text{IAC} & 66° \ 58' \\ \text{IEC} & 4'\end{array}\right\}$ sines $\left\{\begin{array}{l} 92026 \\ 116\end{array}\right.$

$$116 \ ) \ 92026 \ (\ 793\ ^{38}/_{116}$$
$$10888$$
$$33$$

From the following observations of the Landgrave, the distance of the star from the center is 1,057 radii.

Angles $\left\{\begin{array}{ll}\text{IAC} & 66° \ 57' \\ \text{IEC} & 3'\end{array}\right\}$ sines $\left\{\begin{array}{l} 92012 \\ 87\end{array}\right.$

$$87 \ ) \ 92012 \ (\ 1057\ ^{53}/_{87}$$
$$5663$$
$$5$$

Taking from Camerarius those two of his observations which are most favorable to the author, we find the distance of the star from the center to be 3,143 radii.

Angles $\left\{ \begin{array}{ll} \text{IAC} & 65°\ 43' \\ \text{IEC} & 1' \end{array} \right\}$ sines $\left\{ \begin{array}{l} 91152 \\ 29 \end{array} \right.$

$$29 \overline{)\ 91152\ }(\ 3143$$
$$4295$$
$$1$$

The observations of Muñoz give no parallax, and therefore place the new star among the highest fixed stars. Those of Hainzel make it infinitely distant, but with the amendment of one-half minute they would place it among the fixed stars; the same would be implied by those of Ursinus with a correction of 12 minutes. There are no distances above and below the pole given by the other astronomers, so nothing can be deduced there. You see now how all these observations agree against the author, by placing the star in the highest celestial regions.

SAGR. But what is his defense against so patent a contradiction?

SALV. One of those weakest filaments: He says that the parallaxes have become diminished because of refraction which, operating contrary to them, elevates the phenomena whereas parallaxes lower them. Now just how much use this miserable refuge is, you may judge from the fact that if refraction had as great an effect as some astronomers have suggested in recent times, the most that it could do to raise the true position of a phenomenon already twenty-three or twenty-four degrees above the horizon would be to diminish the parallax about three minutes of arc. This adjustment is much too small to pull the star down below the moon, and in some cases it gives him less advantage than does our concession in admitting that the entire excess of the distance below the pole as compared with that above it is due to parallax. And this advantage is a much clearer and more palpable thing than the effect of refraction, the amount of which I question, and not without reason.

Moreover, I would ask this author whether he believes that the astronomers whose observations he uses would have known of this effect, and whether they would have taken it into consideration. If they knew of it and considered it, one may reasonably believe that they took it into account in assigning the true elevation of the star, reducing the degrees of altitude shown on their instruments to the extent required by alterations due to refraction, so that the distances they announced were correct and exact, and not merely apparent and false. But if he believes that such authors did not reflect upon the matter of refraction, he ought to confess that they likewise erred in their determinations of all those things which cannot be completely adjusted without allowance for refraction, among which is the exact determination of the polar altitude. This is commonly derived from the two meridian altitudes of certain fixed stars which are always visible. Such altitudes would be altered by refraction in precisely the same way as would that of the new star, so that the polar altitude deduced from these would be defective, and would share in the same defect that this author ascribes to the altitudes assigned to the new star; that is, both the former and the latter would be higher than actual, with equal error. Now such an error, so far as it concerns our present subject, does not prejudice it in any way. For since we need no more than to know the difference between the two distances of the new star from the pole when seen above the pole and below it, we can plainly see that this difference would remain the same, assuming a common alteration due to refraction for the star and for the pole which would affect both the former and the latter.

The author's argument would be of some importance, though not much, if he had ascertained that the height of the pole had been assigned precisely and corrected for the error due to refraction, against which error the same astronomers had then neglected to guard themselves in assigning the altitude of the new star. But he has not assured us of this, and perhaps could not have done so, nor perhaps (and this is more likely) was such a precaution ignored by the observers.

SAGR. This objection seems to me to be more than adequately nullified. But tell me how he frees himself from that of the star having always kept the same distance from the surrounding fixed stars.

SALV. He grasps similarly at two threads, still weaker than the first, one of which is still tied to refraction but even less firmly. For he says that refraction, altering the true site of the new star and making it appear higher, operates so as to make uncertain its apparent distances from the neighboring fixed stars with which it is compared. I cannot sufficiently marvel at the way he pretends not to know that the same refraction operates in the same way upon the new star as upon the old ones near to it, raising them both equally, whence the interval between them remains unaltered.

The other refuge he takes is still more miserable and contains something of the ridiculous, being based upon the possible occurrence of errors in the instrumental observations themselves because the observer is not able to place the center of the pupil of his eye at the pivot of the sextant (an instrument used for observing intervals between two stars). Holding it out from that point by the distance of the pupil from some bone or other of the cheek where he rests the head of the instrument, he thus forms at his eye a more acute angle than that formed by the sides of the sextant. And the angle formed by the rays also differs in itself when one looks at stars not much elevated above the horizon, and then later looks at the same stars when situated at a great altitude. A different angle is made, says he, as one continues to elevate the instrument with one's head held fixed.

But if, in raising the sextant, the neck is bent back and the head is raised together with the instrument, the angle would remain the same. Hence the author's remark assumes that in using the instrument the observers did not raise their heads as required, which is not very likely. But even supposing that this did happen, I leave it to you to judge what difference there could be between the vertex angles of two isosceles triangles of which the sides of one are each about four yards long, and those of the other four yards less the diameter of a lentil. Surely there can be no difference greater than this between the lengths of the two visual rays when a line falls vertically from the center of the pupil on the plane of the limbs of the sextant, this line being no more than a thumb's breadth in length, and the length of the same rays when, raising the sextant without elevating the head along with it, this line no longer falls perpendicularly on the said plane, but is inclined to it, making the angle in the direction of the scale somewhat acute.

But to free the author once and for all from his unhappy and beggarly excuses, let him know (since it is clear that he has not had much practice in the use of astronomical instruments) that

along each side of a sextant or quadrant there are placed two
sights, one at its center and the other at the opposite end, which
are raised an inch or more from the plane of the limbs, and the
line of vision is made to pass through the tops of those sights, the
eye being held quite a way from the instrument — a span or two,
maybe more — so that neither the pupil nor the cheekbone nor
any other part of the person touches the instrument or rests upon
it. Nor are these instruments held or raised by the arm, especially
when they are large, as they generally are; weighing tens, hundreds, or even thousands of pounds, they are supported upon
most solid bases. And thus the whole objection vanishes.

These are the author's subterfuges, which even if soundly
constructed would not guarantee him the hundredth part of a
single minute; yet he thinks he can make us believe that he has
with their help offset a difference of more than a hundred minutes. I mean that no perceptible difference was noted in the distance between a fixed star and the new star during all their cir-
culations, whereas if the nova had been as near as the moon, such
a difference ought to have made itself quite conspicuous even to
the naked eye without any instruments at all. When compared
with ϰ Cassiopeiae, which was within one and one-half degrees of
the new star, it should have strayed by more than two lunar
diameters, as the more intelligent astronomers of those days were
well aware.

SAGR. This is as if I were watching some unfortunate farmer who,
after having all his expected harvest beaten down and destroyed
by a tempest, goes about with pallid and downcast face, gathering up such poor gleanings as would not serve to feed a chicken
for one day.

SALV. Truly, it was with too scant a store of ammunition that
this author rose up against the assailers of the sky's inalterability, and it is with chains too fragile that he has attempted to
pull the new star down from Cassiopeia in the highest heavens
to these base and elemental regions. Now, since the great difference between the arguments of the astronomers and of this
opponent of theirs seems to me to have been very clearly demonstrated, we may as well leave this point and return to our main
subject. We shall next consider the annual movement generally
attributed to the sun, but then, first by Aristarchus of Samos and
later by Copernicus, removed from the sun and transferred to

the earth. Against this position I know that Simplicio comes strongly armed, in particular with the sword and buckler of his booklet of theses or mathematical disquisitions. It will be good to commence by producing the objections from this booklet.

SIMP. If you don't mind, I am going to leave those for the last, since they were the most recently discovered.

SALV. Then you had better take up in order, in accordance with our previous procedure, the contrary arguments by Aristotle and the other ancients. I also shall do so, in order that nothing shall be left out or escape careful consideration and examination. Likewise Sagredo, with his quick wit, shall interpose his thoughts as the spirit moves him.

SAGR. I shall do so with my customary lack of tact; and since you have asked for this, you will be obliged to pardon it.

SALV. This favor will oblige me to thank and not to pardon you. But now let Simplicio begin to set forth those objections which restrain him from believing that the earth, like the other planets, may revolve about a fixed center.

SIMP. The first and greatest difficulty is the repugnance and incompatibility between being at the center and being distant from it. For if the terrestrial globe must move in a year around the circumference of a circle — that is, around the zodiac — it is impossible for it at the same time to be in the center of the zodiac. But the earth is at that center, as is proved in many ways by Aristotle, Ptolemy, and others.

SALV. Very well argued. There can be no doubt that anyone who wants to have the earth move along the circumference of a circle must first prove that it is not at the center of that circle. The next thing is for us to see whether the earth is or is not at that center around which I say it turns, and in which you say it is situated. And prior to this, it is necessary that we declare ourselves as to whether or not you and I have the same concept of this center. Therefore tell me what and where this center is that you mean.

SIMP. I mean by "center," that of the universe; that of the world; that of the stellar sphere; that of the heavens.

SALV. I might very reasonably dispute whether there is in nature such a center, seeing that neither you nor anyone else has so far proved whether the universe is finite and has a shape, or whether it is infinite and unbounded.† Still, conceding to you for the

It has not been proved by anyone so far whether the universe is finite or infinite.

Aristotle's proofs
that the universe
is finite all
collapse upon
denial that it
is movable.

Aristotle makes
center of the
universe that
point around
which all
celestial spheres
rotate.

It is asked
which of two
propositions
repugnant to
his doctrine
Aristotle would
admit, if forced
to accept one
of them.

moment that it is finite and of bounded spherical shape, and therefore has its center, it remains to be seen how credible it is that the earth rather than some other body is to be found at that center.

SIMP. Aristotle gives a hundred proofs that the universe is finite, bounded, and spherical.

SALV. Which are later all reduced to one, and that one to none at all. For if I deny him his assumption that the universe is movable all his proofs fall to the ground, since he proves it to be finite and bounded only if the universe is movable. But in order not to multiply our disputes, I shall concede to you for the time being that the universe is finite, spherical, and has a center. And since such a shape and center are deduced from mobility, it will be the more reasonable for us to proceed from this same circular motion of world bodies to a detailed investigation of the proper position of the center. Even Aristotle himself reasoned about and decided this in the same way, making that point the center of the universe about which all the celestial spheres revolve, and at which he believed the terrestrial globe to be situated. Now tell me, Simplicio: if Aristotle had found himself forced by the most palpable experiences to rearrange in part this order and disposition of the universe, and to confess himself to have been mistaken about one of these two propositions — that is, mistaken either about putting the earth in the center, or about saying that the celestial spheres move around such a center — which of these admissions do you think that he would choose?

SIMP. I think that if that should happen, the Peripatetics . . .

SALV. I am not asking the Peripatetics; I am asking Aristotle himself. As for the former, I know very well what they would reply. They, as most reverent and most humble slaves of Aristotle, would deny all the experiences and observations in the world, and would even refuse to look at them† in order not to have to admit them, and they would say that the universe remains just as Aristotle has written; not as nature would have it. For take away the prop of his authority, and with what would you have them appear in the field? So now tell me what you think Aristotle himself would do.

SIMP. Really, I cannot make up my mind which of these two difficulties he would have regarded as the lesser.

SALV. Please, do not apply this term "difficulty" to something that may necessarily be so; wishing to put the earth in the center of the celestial revolutions was a "difficulty." But since you do not know to which side he would have leaned, and considering him as I do a man of brilliant intellect, let us set about examining which of the two choices is the more reasonable, and let us take that as the one which Aristotle would have embraced. So, resuming our reasoning once more from the beginning, let us assume out of respect for Aristotle that the universe (of the magnitude of which we have no sensible information beyond the fixed stars), like anything that is spherical in shape and moves circularly, has necessarily a center for its shape and for its motion. Being certain, moreover, that within the stellar sphere there are many orbs one inside another, with their stars which also move circularly, our question is this: Which is it more reasonable to believe and to say; that these included orbs move around the same center as the universe does, or around some other one which is removed from that? Now you, Simplicio, say what you think about this matter.

The Third Day

SIMP. If we could stop with this one assumption and were sure of not running into something else that would disturb us, I should think it would be much more reasonable to say that the container and the things it contained all moved around one common center rather than different ones.

It is more suitable for the container and the contained to move around the same center than diverse ones.

SALV. Now if it is true that the center of the universe is that point around which all the orbs and world bodies (that is, the planets) move, it is quite certain that not the earth, but the sun, is to be found at the center of the universe. Hence, as for this first general conception, the central place is the sun's, and the earth is to be found as far away from the center as it is from the sun.

If the center of the world is the point around which the planets move, the sun and not the earth is located there.

SIMP. How do you deduce that it is not the earth, but the sun, which is at the center of the revolutions of the planets?

SALV. This is deduced from most obvious and therefore most powerfully convincing observations. The most palpable of these, which excludes the earth from the center and places the sun there, is that we find all the planets closer to the earth at one time and farther from it at another. The differences are so great that Venus, for example, is six times as distant from us at its farthest as at its closest, and Mars soars nearly eight times as high in the one state as in the other. You may thus see whether Aristotle was not some trifle deceived in believing that they were always equally distant from us.

Observations from which it is deduced that the sun and not the earth is at the center of the celestial revolutions.

Changes of
shape in Venus
argue its motion
to be around
the sun.

The moon
cannot be
separated from
the earth.

The annual
motion of the
earth, mixing
with the motions
of the other
planets, produces
apparent
anomalies.

SIMP. But what are the signs that they move around the sun?

SALV. This is reasoned out from finding the three outer planets —
Mars, Jupiter, and Saturn — always quite close to the earth
when they are in opposition to the sun, and very distant when
they are in conjunction with it. This approach and recession is
of such moment that Mars when close looks sixty times as large
as when it is most distant. Next, it is certain that Venus and
Mercury must revolve around the sun, because of their never
moving far away from it, and because of their being seen now
beyond it and now on this side of it, as Venus's changes of shape
conclusively prove. As to the moon, it is true that this can never
separate from the earth in any way, for reasons that will be set
forth more specifically as we proceed.

SAGR. I have hopes of hearing still more remarkable things arising from this annual motion of the earth than were those which
depended upon its diurnal rotation.

SALV. You will not be disappointed, for as to the action of the
diurnal motion upon celestial bodies, it was not and could not
be anything different from what would appear if the universe
were to rush speedily in the opposite direction. But this annual
motion, mixing with the individual motions of all the planets,
produces a great many oddities which in the past have baffled
all the greatest men in the world.

Now returning to these first general conceptions, I repeat that
the center of the celestial rotation for the five planets, Saturn,
Jupiter, Mars, Venus, and Mercury, is the sun; this will hold
for the earth too, if we are successful in placing that in the
heavens. Then as to the moon, it has a circular motion around
the earth, from which as I have already said it cannot be separated; but this does not keep it from going around the sun
along with the earth in its annual movement.

SIMP. I am not yet convinced of this arrangement at all. Perhaps
I should understand it better from the drawing of a diagram,
which might make it easier to discuss.

SALV. That shall be done. But for your greater satisfaction and
your astonishment, too, I want you to draw it yourself. You will
see that however firmly you may believe yourself not to understand it, you do so perfectly, and just by answering my questions
you will describe it exactly. So take a sheet of paper and the
compasses; let this page be the enormous expanse of the uni-

verse, in which you have to distribute and arrange its parts as reason shall direct you. And first, since you are sure without my telling you that the earth is located in this universe, mark some point at your pleasure where you intend this to be located, and designate it by means of some letter.

SIMP. Let this be the place of the terrestrial globe, marked A.

SALV. Very well. I know in the second place that you are aware that this earth is not inside the body of the sun, nor even contiguous to it, but is distant from it by a certain space. Therefore assign to the sun some other place of your choosing, as far from the earth as you like, and designate that also.

SIMP. Here I have done it; let this be the sun's position, marked O.

System of the universe sketched from the appearances.

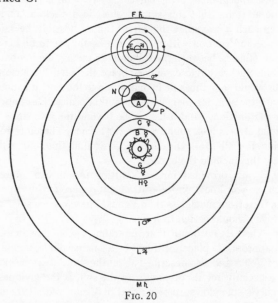

FIG. 20

SALV. These two established, I want you to think about placing Venus in such a way that its position and movement can conform to what sensible experience shows us about it. Hence you must call to mind, either from past discussions or from your own observations, what you know happens with this star. Then assign it whatever place seems suitable for it to you.

Venus is very
large at its
evening con-
junction, and
very small at the
morning one.

SIMP. I shall assume that those appearances are correct which you have related and which I have read also in the booklet of theses; that is, that this star never recedes from the sun beyond a certain definite interval of forty degrees or so; hence it not only never reaches opposition to the sun, but not even quadrature, nor so much as a sextile aspect. Moreover, I shall assume that it displays itself to us about forty times as large at one time than at another; greater when, being retrograde, it is approaching evening conjunction with the sun, and very small when it is moving forward toward morning conjunction, and furthermore that when it appears very large, it reveals itself in a horned shape, and when it looks very small it appears perfectly round.

These appearances being correct, I say, I do not see how to escape affirming that this star revolves in a circle around the sun, in such a way that this circle cannot possibly be said to embrace and contain within itself the earth, nor to be beneath the sun (that is, between the sun and the earth), nor yet beyond the sun. Such a circle cannot embrace the earth because then Venus would sometimes be in opposition to the sun; it cannot be beneath the sun, for then Venus would appear sickle-shaped at both conjunctions; and it cannot be beyond the sun, since then it would always look round and never horned. Therefore for its lodging I shall draw the circle CH around the sun, without having this include the earth.

SALV. Venus provided for, it is fitting to consider Mercury, which, as you know, keeping itself always around the sun, recedes therefrom much less than Venus. Therefore consider what place you should assign to it.

Mercury is
concluded to
revolve around
the sun, inside
the orbit
of Venus.

SIMP. There is no doubt that, imitating Venus as it does, the most appropriate place for it will be a smaller circle, within this one of Venus and also described about the sun. A reason for this, and especially for its proximity to the sun, is the vividness of Mercury's splendor surpassing that of Venus and all the other planets. Hence on this basis we may draw its circle here and mark it with the letters BG.

SALV. Next, where shall we put Mars?

Mars necessarily
includes within
its orbit the
earth and also
the sun.

SIMP. Mars, since it does come into opposition with the sun, must embrace the earth with its circle. And I see that it must also embrace the sun; for, coming into conjunction with the sun, if it did not pass beyond it but fell short of it, it would appear horned as Venus and the moon do. But it always looks round; therefore its circle must include the sun as well as the earth. And since I remember your having said that when it is in opposition to the sun it looks sixty times as large as when in conjunction, it seems to me that this phenomenon will be well provided for by a circle around the sun embracing the earth, which I draw here and mark DI. When Mars is at the point D, it is very near the earth and in opposition to the sun, but when it is at the point I, it is in conjunction with the sun and very distant from the earth.

And since the same appearances are observed with regard to Jupiter and Saturn (although with less variation in Jupiter than in Mars, and with still less in Saturn than in Jupiter), it seems clear to me that we can also accommodate these two planets very neatly with two circles, still around the sun. This first one, for Jupiter, I mark EL; the other, higher, for Saturn, is called FM.

SALV. So far you have comported yourself uncommonly well. And since, as you see, the approach and recession of the three outer planets is measured by double the distance between the earth and the sun, this makes a greater variation in Mars than in Jupiter because the circle DI of Mars is smaller than the circle EL of Jupiter. Similarly, EL here is smaller than the circle FM of Saturn, so the variation is still less in Saturn than in Jupiter, and this corresponds exactly to the appearances. It now remains for you to think about a place for the moon.

SIMP. Following the same method (which seems to me very convincing), since we see the moon come into conjunction and opposition with the sun, it must be admitted that its circle embraces the earth. But it must not embrace the sun also, or else when it was in conjunction it would not look horned but always round and full of light. Besides, it would never cause an eclipse of the sun for us, as it frequently does, by getting in between us and the sun. Thus one must assign to it a circle around the earth, which shall be this one, NP, in such a way that when at P it appears to us here on the earth A as in conjunction with the sun, which sometimes it will eclipse in this position. Placed at N, it is seen in opposition to the sun, and in that position it may fall under the earth's shadow and be eclipsed.

SALV. Now what shall we do, Simplicio, with the fixed stars? Do we want to sprinkle them through the immense abyss of the universe, at various distances from any predetermined point, or

Probable
situation of
the fixed stars.

What should be
considered the
sphere of the
universe.

place them on a spherical surface extending around a center of their own so that each of them will be the same distance from that center?

SIMP. I had rather take a middle course, and assign to them an orb described around a definite center and included between two spherical surfaces — a very distant concave one, and another closer and convex, between which are placed at various altitudes the innumerable host of stars. This might be called the universal sphere, containing within it the spheres of the planets which we have already designated.

SALV. Well, Simplicio, what we have been doing all this while is arranging the world bodies according to the Copernican distribution, and this has now been done by your own hand. Moreover, you have assigned their proper movements to them all except the sun, the earth, and the stellar sphere. To Mercury and Venus you have attributed a circular motion around the sun without embracing the earth. Around the same sun you have caused the three outer planets, Mars, Jupiter, and Saturn, to move, embracing the earth within their circles. Next, the moon cannot move in any way except around the earth and without embracing the sun. And in all these movements you likewise agree with Copernicus himself. It now remains to apportion three things among the sun, the earth, and the stellar sphere: the state of rest, which appears to belong to the earth; the annual motion through the zodiac, which appears to belong to the sun; and the diurnal movement, which appears to belong to the stellar sphere, with all the rest of the universe sharing in it except the earth. And since it is true that all the planetary orbs (I mean Mercury, Venus, Mars, Jupiter, and Saturn) move around the sun as a center, it seems most reasonable for the state of rest to belong to the sun rather than to the earth — just as it does for the center of any movable sphere to remain fixed, rather than some other point of it remote from the center.

Rest, the annual
motion, and the
diurnal one
must be
distributed
among the sun,
the earth, and
the firmament.

It appears more
reasonable for
the center of a
movable sphere
to be fixed, than
any other part.

Next as to the earth, which is placed in the midst of moving objects — I mean between Venus and Mars, one of which makes its revolution in nine months and the other in two years — a motion requiring one year may be attributed to it much more elegantly than a state of rest, leaving the latter for the sun. And such being the case, it necessarily follows that the diurnal motion, too, belongs to the earth. For if the sun stood still, and the

If the annual
motion be given
to the earth, the
diurnal is
fittingly assigned
to it also.

earth did not revolve upon itself but merely had the annual movement around the sun, our year would consist of no more than one day and one night; that is, six months of day and six months of night, as was remarked once previously.

See, then, how neatly the precipitous motion of each twenty-four hours is taken away from the universe, and how the fixed stars (which are so many suns) agree with our sun in enjoying perpetual rest. See also what great simplicity is to be found in this rough sketch, yielding the reasons for so many weighty phenomena in the heavenly bodies.

SAGR. I see this very well indeed. But just as you deduce from this simplicity a large probability of truth in this system, others may on the contrary make the opposite deduction from it. If this very ancient arrangement of the Pythagoreans is so well accommodated to the appearances, they may ask (and not unreasonably) why it has found so few followers in the course of centuries; why it has been refuted by Aristotle himself, and why even Copernicus is not having any better luck with it in these latter days.

SALV. Sagredo, if you had suffered even a few times, as I have so often, from hearing the sort of follies that are designed to make the common people contumacious and unwilling to listen to this innovation (let alone assent to it), then I think your astonishment at finding so few men holding this opinion would dwindle a good deal. It seems to me that we can have little regard for imbeciles who take it as a conclusive proof in confirmation of the earth's motionlessness, holding them firmly in this belief, when they observe that they cannot dine today at Constantinople and sup in Japan, or for those who are positive that the earth is too heavy to climb up over the sun and then fall headlong back down again. There is no need to bother about such men as these, whose name is legion, or to take notice of their fooleries. Neither need we try to convert men who define by generalizing and cannot make room for distinctions, just in order to have such fellows for our company in very subtle and delicate doctrines. Besides, with all the proofs in the world what would you expect to accomplish in the minds of people who are too stupid to recognize their own limitations?

Utterly childish reasons suffice to keep imbeciles believing in the fixity of the earth.

No, Sagredo, my surprise is very different from yours. You wonder that there are so few followers of the Pythagorean opin-

Showing how
improbable
the opinion of
Copernicus is.

In Aristarchus
and Copernicus,
reason and
argument
prevailed over
sensory evidence.

ion, whereas I am astonished that there have been any up to this day who have embraced and followed it. Nor can I ever sufficiently admire the outstanding acumen of those who have taken hold of this opinion and accepted it as true; they have through sheer force of intellect done such violence to their own senses as to prefer what reason told them over that which sensible experience plainly showed them to the contrary. For the arguments against the whirling of the earth which we have already examined are very plausible, as we have seen; and the fact that the Ptolemiacs and Aristotelians and all their disciples took them to be conclusive is indeed a strong argument of their effectiveness. But the experiences which overtly contradict the annual movement are indeed so much greater in their apparent force that, I repeat, there is no limit to my astonishment when I reflect that Aristarchus and Copernicus were able to make reason so conquer sense that, in defiance of the latter, the former became mistress of their belief.

SAGR. Then we are about to encounter still further strong attacks against this annual movement?

SALV. We are, and such obvious and sensible ones that were it not for the existence of a superior and better sense than natural and common sense to join forces with reason, I much question whether I, too, should not have been much more recalcitrant toward the Copernican system than I have been since a clearer light than usual has illuminated me.

SAGR. Well, then, Salviati, let us get down to cases, as they say; for every word spent otherwise seems to me to be wasted.

SALV. I am at your service . . .

[SIMP. Gentlemen, please give me a chance to restore harmony to my mind, which I now find very much upset by certain matters which Salviati has just touched upon. Then, when this storm has subsided, I shall be able to listen to your theories more profitably. For there is no use forming an image in a wavy mirror, as the Latin poet has told us so graciously by writing:

> . . . *nuper me in littore vidi,*
> *Cum placidum ventis staret mare.*

SALV. You are quite right; tell us your difficulties.

SIMP. Those who deny the diurnal motion to the earth because they do not see themselves being transported to Persia or Japan have been called by you just as dull-witted as those who oppose the annual motion because of the repugnance they feel against admitting that the vast and ponderous bulk of the terrestrial globe can raise itself on high and then descend to the depths, as it would have to do if it revolved about the sun annually. Now I, without blushing to be numbered among such simpletons, feel in my own mind this very repugnance as to the second point against the annual motion, the more so when I see how much resistance is made to motion even over a plain by, I shall not say a mountain, but a mere stone; and even the former would be but the tiniest fraction of an Alpine range. Therefore I beg you not to scorn such objections entirely, but to solve them; and not for me alone, but also for others to whom they seem quite real. For I think it is very difficult for some people, simple though they may be, to recognize and admit that they are simple just because they know themselves to be so regarded.

SAGR. Indeed, the simpler they are, the more nearly impossible it will be to convince them of their own shortcomings. And on this account I think that it is good to resolve this and all similar objections, not only that Simplicio should be satisfied, but also for other reasons no less important. For it is clear that there are plenty of people who are well versed in philosophy and the other sciences but who, either through lack of astronomy or mathematics or some other discipline which would sharpen their minds for the penetration of truth, adhere to silly doctrines like these. That is why the situation of poor Copernicus seems to me lamentable; he could expect only censure for his views and could not let them fall into the hands of anyone who, being unable to comprehend his arguments (which are very subtle and therefore difficult to master), would be convinced of their falsity on account of some superficial appearances, and would go about declaring them to be wrong and full of error. If people cannot be convinced by the arguments, which are quite abstruse, it is good to make sure that they recognize the vapidity of these objections. From such knowledge comes moderation in their judgment and condemnation of the doctrine which at present they consider erroneous. Accordingly I shall raise two other objections against the diurnal motion, which not so long ago were to be heard put forward by important men of letters, and after that we shall look into the annual motion.

The first was that if it were true that the sun and other stars did not rise over the eastern horizon, but the eastern side of the earth sank beneath them while they remained motionless, then it would follow that after a short time the mountains, sinking downward with the rotation of the terrestrial globe, would get into such a position that whereas a little earlier one would have had to climb steeply to their peaks, a few hours later one would have to stoop and descend in order to get there.

The other was that if the diurnal motion belonged to the earth, it would have to be so rapid that anyone placed at the bottom of a well would not for a moment be able to see a star which was directly above him, being able to see it only during the very brief instant in which the earth traverses two or three yards, this being the width of ,the well. Yet experiment shows that the apparent passage of such a star in going over the well takes quite a while— a necessary argument that the mouth of the well does not move with that rapidity which is required for the diurnal movement. Hence the earth is motionless.

SIMP. Of these two arguments, the second really does seem persuasive to me; but as to the first, I think I could clear that up myself. For I consider it the same thing for the terrestrial globe to move about its own center and carry a mountain eastward with it, as for the globe to stand still while the mountain was detached at the base and drawn along the earth. And I do not see that carrying the mountain over the earth's surface is an operation any different from sailing a ship over the surface of the sea. So if the objection of the mountain were valid, it would follow likewise that as the ship continued its voyage and became several degrees distant from our ports, we should have to climb its mast not merely in order to ascend, but to move about in a plane, or eventually even to descend. Now this does not happen, nor have I ever heard of any sailor, even among those who have circumnavigated the globe, who had found any difference in such actions (or any others performed on board ship) because of the ship being in one place rather than another.

SALV. You argue very well, and if it had ever entered the mind of the author of this objection to consider how this neighboring eastern mountain of his would, if the terrestrial globe revolved, be found in a couple of hours to have been carried by that motion to where Mt. Olympus, for example, or Mt. Carmel is now located, he would have seen that by his own line of reasoning he would be obliged to believe and admit that in order to get to the top of the latter mountains one would in fact have to descend. Such people have the same kind of mind as do those who deny the antipodes on the grounds that one cannot walk with his head down and his feet attached to the ceiling; they produce ideas that are true and that they completely understand, but they do not find it easy to deduce the simplest solutions for their difficulties. I mean, they understand very well that to gravitate or to descend is to approach the center of the terrestrial globe, and that to ascend is to depart from that; but they fail to understand that our antipodes have no trouble at all in sustaining themselves or in walking because they are just like us, having the soles of their feet toward the center of the earth and their heads toward the sky.

SAGR. Yet we know that men who are profoundly ingenious in other fields are blind to such ideas. This confirms what I have just said; it is good to remove every objection, even the feeblest. Therefore the matter of the well should also be answered.

SALV. This second argument does indeed have some elusive appearance of cogency. Nevertheless, I think it certain that if one were to interrogate the very person to whom it occurs, to the end that he might express himself better by explaining just what results ought to follow if one assumes the diurnal rotation of the earth, but which appear to him not to take place; then, I say, I believe that he would get all tangled up in explaining his question and its consequences — perhaps no less than he would disentangle it by thinking it over.

SIMP. To be perfectly frank, I am sure that that is what would happen, although I too find myself right now in this same confusion. For at first glance it seems to me that the argument is binding, but on the other hand I am beginning to realize that other troubles would arise if the reasoning were to continue along the same line. For this extremely rapid course, which ought to be perceived in the star if the motion belonged to the earth, should also be discovered in it if the motion were its own — even more so, since it would have to be thousands of times as fast in the star as in the earth. On the other hand, the star must be lost to sight by passing the mouth of the well, which would be only a couple of yards in diameter, if the well goes along with the earth more than two million yards per hour. Indeed, this seems to be such a transitory glimpse that one cannot even imagine it; yet from the bottom of a well a star is seen for quite a long time. So I should like to be put in the clear about this matter.

SALV. Now I am strongly confirmed in my belief about the confusion of the author of this objection, seeing that you too, Simplicio, becloud what you mean and do not really grasp what you should be saying. I deduce this principally from your omitting a distinction which is a principal point in this matter. So tell me whether in carrying out this experiment (I mean this one of the star passing over the mouth of the well) you would make any distinction between the well being deeper or shallower; that is, between the observer being farther from or closer to its mouth. For I have not heard you make any mention of this.

SIMP. The fact is that I had not thought about it, but your question has awakened my mind to it, and hints to me that such a distinction must be quite necessary. Already I begin to see that in order to determine the time of the passage, the depth of the well may perhaps make no less difference than its width.

SALV. Still, I rather question whether the width makes any difference to us, or very much.

SIMP. Why, it seems to me that having to travel 10 yards of breadth takes ten times as long as to pass 1 yard. I am sure that a boat 10 yards long will pass beyond my view long before a galley 100 yards long will do so.

SALV. So, we still persist in that inveterate idea of not moving unless our legs carry us.

What you are saying is true, my dear Simplicio, if the object you see is in motion while you remain stationary to observe it. But if you are in a well when the well and you together are carried along by the rotation of the earth, don't you see that not in an hour, nor in a thousand, nor in all eternity will you ever be overtaken by the mouth of the well? The manner in which the moving or nonmoving of the earth acts upon you in such a situation can be recognized not from the mouth of the well, but from some other separate object not sharing the same state of motion — or I should say, of rest.

SIMP. So far so good; but assume that I, being in the well, am carried together with it by the diurnal motion, and that the star seen by me is motionless. The opening of the well (which alone allows my sight to pass beyond) being not more than three yards, out of so many millions of yards in the balance of the terrestrial surface which are hindering my view, how can the time of my seeing be a perceptible fraction of that of my not seeing?

SALV. You are still falling into the same quibble, and in fact you will need someone to help you out of it. It is not the width of the well, Simplicio, which measures the time of visibility of the star, since in that case you would see it perpetually, as the well would give passage to your vision perpetually. No, the measure of this time must be obtained from that fraction of the motionless heavens which remains visible through the opening of the well.

SIMP. Is not that part of the sky which I perceive the same fraction of the entire heavenly sphere as the mouth of the well is of the terrestrial sphere?

SALV. I want you to answer that for yourself. Tell me whether the mouth of the well is always the same fraction of the earth's surface.

SIMP. There is no doubt that it is always the same.

SALV. And how about the part of the sky which is seen by the person in the well? Is that always the same fraction of the whole celestial sphere?

SIMP. Now I am beginning to sweep the darkness from my mind, and to understand what you hinted to me a little while ago — that the depth of the well has something to do with this matter. For I do not question that the more distant the eye is from the mouth of the well, the smaller will be the part of the sky which it will perceive, and consequently the sooner this will have been passed and become lost to view by whoever is looking at it from the bottom of the well.

SALV. But is there any place in the well from which he would perceive exactly that fraction of the celestial sphere which the mouth of the well is of the earth's surface?

SIMP. It seems to me that if the well were excavated to the center of the earth, perhaps from there one might see a part of the sky which would be to it as the well is to the earth. But leaving the center and rising toward the surface, an ever larger part of the sky would be revealed.

SALV. And finally, placing the eye at the mouth of the well, it would perceive one-half the sky, or very little less, which would take twelve hours in passing, assuming that we were at the equator.]

A while ago I sketched for you an outline of the Copernican system, against the truth of which the planet Mars launches a ferocious attack. For if it were true that the distances of Mars from the earth varied as much from minimum to maximum as twice the distance from the earth to the sun, then when it is closest to us its disc would have to look sixty times as large as when it is most distant. Yet no such difference is to be seen. Rather, when it is in opposition to the sun and close to us, it shows itself as only four or five times as large as when, at conjunction, it becomes hidden behind the rays of the sun.

Another and greater difficulty is made for us by Venus, which, if it circulates around the sun as Copernicus says, would be now beyond it and now on this side of it, receding from and approaching toward us by as much as the diameter of the circle it describes. Then when it is beneath the sun and very close to us, its disc ought to appear to us a little less than forty times as large as when it is beyond the sun and near conjunction. Yet the difference is almost imperceptible.

Add to these another difficulty; for if the body of Venus is intrinsically dark, and like the moon it shines only by illumination from the sun, which seems reasonable, then it ought to appear horned when it is beneath the sun, as the moon does when it is likewise near the sun — a phenomenon which does not make itself evident in Venus. For that reason, Copernicus declared that Venus was either luminous in itself or that its substance was such that it could drink in the solar light and transmit this through its entire thickness in order that it might look resplendent to us. In this manner Copernicus pardoned Venus its unchanging shape, but he said nothing about its small variation in size; much less of the requirements of Mars. I believe this was because he was unable to rescue to his own satisfaction an appearance so contradictory to his view; yet being persuaded by so many other reasons, he maintained that view and held it to be true.

Venus, according
to Copernicus, is
either luminous
by itself or is
of transparent
material.

Copernicus is
silent about
the inadequate
variation in size
of Venus and
of Mars.

The moon much
disturbs the
orderliness of the
other planets.

Besides these things, to have all the planets move around together with the earth, the sun being the center of their rotations, then the moon alone disturbing this order and having its own motion around the earth (going around the sun in a year together with the earth and the whole elemental sphere) seems in some way to upset the whole order and to render it improbable and false.

These are the difficulties which make me wonder at Aristarchus and Copernicus. They could not have helped noticing them, without having been able to resolve them; nevertheless they were confident of that which reason told them must be so in the light of many other remarkable observations. Thus they confidently affirmed that the structure of the universe could have no other form that that which they had described. Then there are other very serious but beautiful problems which are not easy for ordinary minds to resolve, but which were seen through and explained by Copernicus; these we shall put off until we have answered the objections of people who show themselves hostile to this position.

Coming now to the explanations and replies to the three grave objections mentioned, I say that the first two are not only not contrary to the Copernican system, but that they absolutely favor it, and greatly. For both Mars and Venus do show themselves variable in the assigned proportions, and Venus does appear horned when beneath the sun, and changes her shape in exactly the same way as the moon.

Answer to the first three objections against the Copernican system.

SAGR. But if this was concealed from Copernicus, how is it revealed to you?

SALV. These things can be comprehended only through the sense of sight, which nature has not granted so perfect to men that they can succeed in discerning such distinctions. Rather, the very instrument of seeing introduces a hindrance of its own. But in our time it has pleased God to concede to human ingenuity an invention so wonderful as to have the power of increasing vision four, six, ten, twenty, thirty, and forty times, and an infinite number of objects which were invisible, either because of distance or extreme minuteness, have become visible by means of the telescope.

SAGR. But Venus and Mars are not objects which are invisible because of any distance or small size. We perceive these by simple natural vision. Why, then, do we not discern the differences in their sizes and shapes?

SALV. In this the impediment of our eyes plays a large part, as I have just hinted to you. On account of that, bright distant objects are not represented to us as simple and plain, but are festooned with adventitious and alien rays which are so long and dense that the bare bodies are shown as expanded ten, twenty,

Why Venus and Mars appear not to vary in size as much as they should.

a hundred, or a thousand times as much as would appear to us if the little radiant crown which is not theirs were removed.

SAGR. Now I recall having read something of the sort, but I don't remember whether it was in the *Solar Letters* or in *Il Saggiatore* by our friend. It would be a good thing, in order to refresh my memory as well as to inform Simplicio, who perhaps has not read those works, to explain to us in more detail how the matter stands. For I should think that a knowledge of this would be most essential to an understanding of what is now under discussion.

SIMP. Everything that Salviati is presently setting forth is truly new to me. Frankly, I had no interest in reading those books, nor up till now have I put any faith in the newly introduced optical device. Instead, following in the footsteps of other Peripatetic philosophers of my group, I have considered as fallacies and deceptions of the lenses those things which other people have admired as stupendous achievements. If I have been in error, I shall be glad to be lifted out of it; and, charmed by the other new things I have heard from you, I shall listen most attentively to the rest.

Telescopic operations reputed fallacious by the Peripatetics.

SALV. The confidence which men of that stamp have in their own acumen is as unreasonable as the small regard they have for the judgments of others. It is a remarkable thing that they should think themselves better able to judge such an instrument without ever having tested it, than those who have made thousands and thousands of experiments with it and make them every day. But let us forget about such headstrong people, who cannot even be censured without doing them more honor than they deserve.

Getting back to our purpose, I say that shining objects, either because their light is refracted in the moisture that covers the pupil, or because it is reflected from the edges of the eyelids and these reflected rays are diffused over the pupil, or for some other reason, appear to our eyes as if surrounded by new rays. Hence these bodies look much larger than they would if they were seen by us deprived of such irradiations. This enlargement is made in greater and greater proportion as such luminous objects become smaller and smaller, in exactly such a manner as if we were to suppose a growth of shining hair, say four inches long, to be added around a circle four inches in diameter, which would increase its apparent size nine times; but . . .

Brilliant objects are seen surrounded by adventitious rays.

Reason why luminous bodies look more enlarged the smaller they are.

SIMP. I think you meant to say "three times," since four inches added on each side of a circle four inches in diameter would amount to tripling its magnitude and not to enlarging it nine times.

SALV. A little geometry, Simplicio; it is true that the diameter increases three times, but the surface (which is what we are talking about) grows nine times. For the surfaces of circles, Simplicio, are to each other as the squares of their diameters, and a circle four inches in diameter has to another of twelve inches the same ratio which the square of four has to the square of twelve; that is, 16 to 144. Therefore it will be nine times as large, not three. This is for your information, Simplicio.

Now, to continue, if we add this coiffure of four inches to a circle of only two inches in diameter, the diameter of the crown will be ten inches and the ratio of the circle to the bare body will be as 100 to 4 (for such are the squares of 10 and of 2), so the enlargement would be twenty-five times. And finally, the four inches of hair added to a tiny circle of one inch in diameter would enlarge this eighty-one times. Thus the increase is continually made larger and larger proportionately, according as the real objects which are increased become smaller and smaller.

SAGR. The question which gave Simplicio trouble did not really bother me, but there are some other things about which I desire a clearer explanation. In particular I should like to learn the basis upon which you affirm such a growth to be always equal in all visible objects.

SALV. I have already partly explained by saying that only luminous objects increase; not dark ones. Now I shall add the rest. Of shining objects, those which are brightest in light make the greatest and strongest reflections upon our pupils, thereby showing themselves as much more enlarged than those less bright. And so as not to go on too long about this detail, let us resort to what is shown us by our greatest teacher; this evening, when the sky is well darkened, let us look at Jupiter; we shall see it very radiant and large. Then let us cause our vision to pass through a tube, or even through a tiny opening which we may leave between the palm of our hand and our fingers, clenching the fist and bringing it to the eye; or through a hole made by a fine needle in a card. We shall see the disc of Jupiter deprived of rays and so very small that we shall indeed judge it to be even

Areas of surfaces increase in proportion to the squares of the lines in them.

The more vivid the light of objects, the more they appear to be enlarged.

A simple experiment showing the enlargement of stars to be caused by adventitious rays.

less than one-sixtieth of what had previously appeared to us to be a great torch when seen with the naked eye. Afterwards, we may look at the Dog Star, a very beautiful star and larger than any other fixed star. To the naked eye it looks to be not much smaller than Jupiter, but upon taking away its headdress in the manner described above, its disc will be seen to be so small that one would judge it to be no more than one-twentieth the size of Jupiter. Indeed, a person lacking perfect vision will be able to find it only with great difficulty, from which it may reasonably be inferred that this star is one which has a great deal more luminosity than Jupiter, and makes larger irradiations.

The sun and
moon are but
little enlarged.

Next, the irradiations of the sun and of the moon are as nothing because of the size of these bodies, which by themselves take up so much room in our eye as to leave no place for adventitious rays, so that their discs are seen as shorn and bounded.

Showing by an
obvious experi-
ment that the
most brilliant
bodies are much
more irradiated
than those
less lucid.

We may assure ourselves of the same fact by another experiment which I have made many times — assure ourselves, I mean, that the resplendent bodies of more vivid illumination give out many more rays than those which have only a pale light. I have often seen Jupiter and Venus together, twenty-five or thirty degrees from the sun, the sky being very dark. Venus would appear eight or even ten times as large as Jupiter when looked at with the naked eye. But seen afterward through a telescope, Jupiter's disc would be seen to be actually four or more times as large as Venus. Yet the liveliness of Venus's brilliance was incomparably greater than the pale light of Jupiter, which comes about only because Jupiter is very distant from the sun and from us, while Venus is close to us and to the sun.

These things having been explained, it will not be difficult to understand how it might be that Mars, when in opposition to the sun and therefore seven or more times as close to the earth as when it is near conjunction, looks to us scarcely four or five times as large in the former state as in the latter. Nothing but irradiation is the cause of this. For if we deprive it of the adventitious rays we shall find it enlarged in exactly the proper ratio. And to remove its head of hair from it, the telescope is the unique and supreme means. Enlarging its disc nine hundred or a thousand times, it causes this to be seen bare and bounded like that of the moon, and in the two positions varying in exactly the proper proportion.

Next in Venus, which at its evening conjunction when it is beneath the sun ought to look almost forty times as large as in its morning conjunction, and is seen as not even doubled, it happens in addition to the effects of irradiation that it is sickle-shaped, and its horns, besides being very thin, receive the sun's light obliquely and therefore very weakly. So that because it is small and feeble, it makes its irradiations less ample and lively than when it shows itself to us with its entire hemisphere lighted. But the telescope plainly shows us its horns to be as bounded and distinct as those of the moon, and they are seen to belong to a very large circle, in a ratio almost forty times as great as the same disc when it is beyond the sun, toward the end of its morning appearances.

SAGR. O Nicholas Copernicus, what a pleasure it would have been for you to see this part of your system confirmed by so clear an experiment!

SALV. Yes, but how much less would his sublime intellect be celebrated among the learned! For as I said before, we may see that with reason as his guide he resolutely continued to affirm what sensible experience seemed to contradict. I cannot get over my amazement that he was constantly willing to persist in saying that Venus might go around the sun and be more than six times as far from us at one time as at another, and still look always equal, when it should have appeared forty times larger.

Copernicus persuaded by reason against sensible experience.

SAGR. I believe then that in Jupiter, Saturn, and Mercury one ought also to see differences of size corresponding exactly to their varying distances.

SALV. In the two outer planets I have observed this with precision in almost every one of the past twenty-two years. In Mercury no observations of importance can be made, since it does not allow itself to be seen except at its maximum angles with the sun, in which the inequalities of its distances from the earth are imperceptible. Hence such differences are unobservable, and so are its changes of shape, which must certainly take place as in Venus. But when we do see it, it would necessarily show itself to us in the shape of a semicircle, just as Venus does at its maximum angles, though its disc is so small and its brilliance so lively that the power of the telescope is not sufficient to strip off its hair so that it may appear completely shorn.

Mercury does not admit of clear observations.

It remains for us to remove what would seem to be a great

objection to the motion of the earth. This is that though all the planets turn about the sun, the earth alone is not solitary like the others, but goes together in the company of the moon and the whole elemental sphere around the sun in one year, while at the same time the moon moves around the earth every month. Here one must once more exclaim over and exalt the admirable perspicacity of Copernicus, and simultaneously regret his misfortune at not being alive in our day. For now Jupiter removes this apparent anomaly of the earth and moon moving conjointly. We see Jupiter, like another earth, going around the sun in twelve years accompanied not by one but by four moons, together with everything that may be contained within the orbits of its four satellites.

SAGR. And what is the reason for your calling the four Jovian planets "moons"?

SALV. That is what they would appear to be to anyone who saw them from Jupiter. For they are dark in themselves, and receive their light from the sun; this is obvious from their being eclipsed when they enter into the cone of Jupiter's shadow. And since only that hemisphere of theirs is illuminated which faces the sun, they always look entirely illuminated to us who are outside their orbits and closer to the sun; but to anyone on Jupiter they would look completely lighted only when they were at the highest points of their circles. In the lowest part — that is, when between Jupiter and the sun — they would appear horned from Jupiter. In a word, they would make for Jovians the same changes of shape which the moon makes for us Terrestrials.

Now you see how admirably these three notes harmonize with the Copernican system, when at first they seemed so discordant with it. From this, Simplicio will be much better able to see with what great probability one may conclude that not the earth, but the sun, is the center of rotation of the planets. And since this amounts to placing the earth among the world bodies which indubitably move about the sun (above Mercury and Venus but beneath Saturn, Jupiter, and Mars), why will it not likewise be probable, or perhaps even necessary, to admit that it also goes around?

SIMP. These events are so large and so conspicuous that it is impossible for Ptolemy and his followers not to have had knowledge of them. And having had, they must also have found a way

to give reasons sufficient to account for such sensible appearances; congruous and probable reasons, since they have been
accepted for so long by so many people.

SALV. You argue well, but you must know that the principal
activity of pure astronomers is to give reasons just for the appearances of celestial bodies, and to fit to these and to the motions of the stars such a structure and arrangement of circles
that the resulting calculated motions correspond with those same
appearances. They are not much worried about admitting anomalies which might in fact be troublesome in other respects. Copernicus himself writes, in his first studies, of having rectified
astronomical science upon the old Ptolemaic assumptions, and
corrected the motions of the planets in such a way that the computations corresponded much better with the appearances, and
vice versa. But this was still taking them separately, planet by
planet. He goes on to say that when he wanted to put together
the whole fabric from all individual constructions, there resulted
a monstrous chimera composed of mutually disproportionate
members, incompatible as a whole. Thus however well the
astronomer might be satisfied merely as a calculator, there was
no satisfaction and peace for the astronomer as a scientist. And
since he very well understood that although the celestial appearances might be saved by means of assumptions essentially false
in nature, it would be very much better if he could derive them
from true suppositions, he set himself to inquiring diligently
whether any one among the famous men of antiquity had attributed to the universe a different structure from that of Ptolemy's
which is commonly accepted. Finding that some of the Pythagoreans had in particular attributed the diurnal rotation to the
earth, and others the annual revolution as well, he began to
examine under these two new suppositions the appearances and
peculiarities of the planetary motions, all of which he had readily
at hand. And seeing that the whole then corresponded to its parts
with wonderful simplicity, he embraced this new arrangement,
and in it he found peace of mind.

SIMP. But what anomalies are there in the Ptolemaic arrangement which are not matched by greater ones in the Copernican?

SALV. The illnesses are in Ptolemy, and the cures for them in
Copernicus. First of all, do not all philosophical schools hold it
to be a great impropriety for a body having a natural circular

Main task of
astronomers is
to give reasons
for the
appearances.

Copernicus
restored
astronomy on
the Ptolemaic
assumptions.

What motivated
Copernicus to
establish
his system.

Inconveniences
which exist in
Ptolemy's
system.

movement to move irregularly with respect to its own center and regularly around another point? Yet Ptolemy's structure is composed of such uneven movements, while in the Copernican system each movement is equable around its own center. With Ptolemy it is necessary to assign to the celestial bodies contrary movements, and make everything move from east to west and at the same time from west to east, whereas with Copernicus all celestial revolutions are in one direction, from west to east. And what are we to say of the apparent movement of a planet, so uneven that it not only goes fast at one time and slow at another, but sometimes stops entirely and even goes backward a long way after doing so? To save these appearances, Ptolemy introduces vast epicycles, adapting them one by one to each planet, with certain rules about incongruous motions — all of which can be done away with by one very simple motion of the earth. Do you not think it extremely absurd, Simplicio, that in Ptolemy's construction where all planets are assigned their own orbits, one above another, it should be necessary to say that Mars, placed above the sun's sphere, often falls so far that it breaks through the sun's orb, descends below this and gets closer to the earth than the body of the sun is, and then a little later soars immeasurably above it? Yet these and other anomalies are cured by a single and simple annual movement of the earth.

SAGR. I should like to arrive at a better understanding of how these stoppings, retrograde motions, and advances, which have always seemed to me highly improbable, come about in the Copernican system.

A strong argument for Copernicus is his removal of the stoppings and retrograde motions of the planets.

SALV. Sagredo, you will see them come about in such a way that the theory of this alone ought to be enough to gain assent for the rest of the doctrine from anyone who is neither stubborn nor unteachable. I tell you, then, that no change occurs in the movement of Saturn in thirty years, in that of Jupiter in twelve, that of Mars in two, Venus in nine months, or in that of Mercury in about eighty days. The annual movement of the earth alone,

The annual motion of the earth alone produces great inequalities of motion in the five planets.

between Mars and Venus, causes all the apparent irregularities of the five stars named. For an easy and full understanding of this, I wish to draw you a picture of it. Now suppose the sun to be located in the center O, around which we shall designate the orbit described by the earth with its annual movement, BGM. The circle described by Jupiter (for example) in 12 years will

be *BGM* here, and in the stellar sphere we shall take the circle
of the zodiac to be *PUA*. In addition, in the earth's annual orbit
we shall take a few equal arcs, BC, CD, DE, EF, FG, GH, HI,

Demonstration
that the
irregularity of
the three outer
planets comes
from the
annual motion
of the earth.

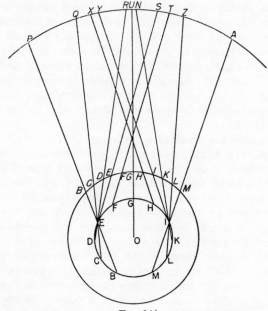

FIG. 21†

IK, KL, and LM, and in the circle of Jupiter we shall indicate
these other arcs passed over in the same times in which the earth
is passing through these. These are *BC, CD, DE, EF, FG, GH,*
HI, IK, KL, and *LM,* which will be proportionately smaller than
those noted on the earth's orbit, as the motion of Jupiter through
the zodiac is slower than the annual celestial motion.

Now suppose that when the earth is at B, Jupiter is at *B;* then
it will appear to us as being in the zodiac at *P,* along the straight
line B*B*P. Next let the earth move from B to C and Jupiter from
B to *C* in the same time; to us, Jupiter will appear to have arrived
at *Q* in the zodiac, having advanced in the order of the signs
from *P* to *Q.* The earth then passing to D and Jupiter to *D,* it
will be seen in the zodiac at *R;* and from E, Jupiter being at *E,*

it will appear in the zodiac at *S*, still advancing. But now when the earth begins to get directly between Jupiter and the sun (having arrived at F and Jupiter at *F)*, to us Jupiter will appear to be ready to commence returning backward through the zodiac, for during the time in which the earth will have passed through the arc EF, Jupiter will have been slowed down between the points *S* and *T*, and will look to us almost stationary. Later the earth coming to G, Jupiter at *G* (in opposition to the sun) will be seen in the zodiac at *U*, turned far back through the whole arc *TU* in the zodiac; but in reality, following always its uniform course, it has advanced not only in its own circle but in the zodiac too, with respect to the center of the zodiac and to the sun which is located there.

The earth and Jupiter then continuing their movements, when the earth is at H and Jupiter is at *H*, it will be seen as having returned far back through the zodiac by the whole arc *UX;* but the earth having arrived at I and Jupiter at *I*, it will apparently have moved in the zodiac by only the small space *XY*, and will there appear stationary. Then when the earth shall have progressed to K and Jupiter to *K*, Jupiter will have advanced through the arc *YN*, in the zodiac; and, continuing its course, from L the earth will see Jupiter at *L* in the point *Z*. Finally, Jupiter at *M* will be seen from the earth at M to have passed to *A*, still advancing. And its whole apparent retrograde motion in the zodiac will be as much as the arc *TX*, made by Jupiter while it is passing in its own circle through the arc *FH*, the earth going through FH in its orbit.

Now what is said here of Jupiter is to be understood of Saturn and Mars also. In Saturn these retrogressions are somewhat more frequent than in Jupiter, because its motion is slower than Jupiter's, so that the earth overtakes it in a shorter time. In Mars they are rarer, its motion being faster than that of Jupiter, so that the earth spends more time in catching up with it.

Next, as to Venus and Mercury, whose circles are included within that of the earth, stoppings and retrograde motions appear in them also, due not to any motion that really exists in them, but to the annual motion of the earth. This is acutely demonstrated by Copernicus, enlisting the aid of Apollonius of Perga, in chapter 35 of Book V in his *Revolutions*.

You see, gentlemen, with what ease and simplicity the annual

Retrograde
movements more
frequent in
Saturn, less so
in Jupiter, and
still less in
Mars; and why.

Retrogressions
of Venus and
Mercury demon-
strated by
Appolonius and
Copernicus.

motion — if made by the earth — lends itself to supplying reasons for the apparent anomalies which are observed in the movements of the five planets, Saturn, Jupiter, Mars, Venus, and Mercury. It removes them all and reduces these movements to equable and regular motions; and it was Nicholas Copernicus who first clarified for us the reasons for this marvelous effect.

But another effect, no less wonderful than this, and containing a knot perhaps even more difficult to untie, forces the human intellect to admit this annual rotation and to grant it to our terrestrial globe. This is a new and unprecedented theory touching the sun itself. For the sun has shown itself unwilling to stand alone in evading the confirmation of so important a conclusion, and instead wants to be the greatest witness of all to this, beyond exception. So now hear this new and mighty marvel.

The original discoverer and observer of the solar spots† (as indeed of all the other novelties in the skies) was our Lincean Academician; he discovered them in 1610, while he was still lecturer in mathematics at the University of Padua. He spoke about them to many people here in Venice, some of whom are yet living, and a year later he showed them to many gentlemen at Rome, as he tells in the first of his *Letters to Mark Welser, Prefect of Augsburg.* He was the first to affirm, against the opinions of those who were too timid or too solicitous about the inalterability of the heavens, that such spots were of a material which was produced and dissolved within a brief time. As to their place, they were contiguous to the body of the sun and revolved about it, or rather completed their rotations by being on the very globe of the sun, which revolves upon its own center in the space of nearly one month. At the beginning he judged this motion to be made by the sun about an axis at right angles to the plane of the ecliptic, since the arcs described by these spots on the sun's disc appeared to our eyes as straight lines parallel to the plane of the ecliptic. These, however, became altered in places by various wandering and irregular accidental movements to which they are subjected. In this way they change place chaotically and without any order among themselves — several now gathering together, and then again dispersing; some dividing into many, and greatly changing their shapes, which are for the most part very extraordinary. And although such inconstant mutations would partly alter the original periodic course of these spots, our

Annual motion of the earth most apt for explaining the peculiarities of the five planets.

The sun itself bears witness that the annual motion belongs to the earth.

The Lincean Academician was first to discover sunspots and all other celestial novelties.

History of the Academician's progress over a long period in observations of sunspots.

friend would not for that reason change his opinion so as to believe that for such deviations there were certain fixed and essential causes; he continued to believe that all the apparent alterations stemmed from some accidental mutations, exactly as would happen for someone who might observe from afar the motion of our clouds. These would appear to move with a very rapid and constant motion, carried around every twenty-four hours by the whirling of the earth (if such a motion belonged to it) along circles parallel to the equator, but somewhat varied by incidental movements caused in them by winds, which drive them casually in all directions.

It happened at this time that Welser sent to him some letters which had been written to him under the pseudonym "Apelle" on the subject of these spots, and urgently requested our friend to say frankly what he thought of these letters and to add his own opinion about the nature of such spots. This request he complied with in his three *Letters,* first showing how vain and foolish were the ideas of Apelle, next revealing his own opinions, and then predicting that no doubt Apelle, becoming better informed as time went on, would come around to his views, as indeed happened. And since it seemed to our Academician (just as it seemed to others who were informed about the natural facts) that in his *Letters* he had looked into and demonstrated everything that human reason could attain to in such matters, if not everything that human curiosity might seek and desire, he interrupted for a time his continual observations, being occupied with other studies. It was only in order to gratify some friend that he would, from time to time, make a few observations with him.

Now several years later, being with me at my Villa delle Selve and being enticed by a particularly clear and protracted serenity of the heavens, he happened to find one of those solitary sunspots which are very large and thick, and at my request he made observations of its entire journey, carefully noting down its places from day to day when the sun was on the meridian. We preceived that its passage was not exactly in a straight line, but a somewhat bent one; and it occurred to us to make other observations from time to time. We were strongly encouraged to do this by an idea which suddenly struck the mind of my guest and which he imparted to me in the following words:

"Filippo, it seems to me that the road is open for us into a matter of great consequence. For if the axis around which the sun revolves is not perpendicular to the plane of the ecliptic, but is somewhat inclined to this — as the curved path just observed suggests to me — then we shall have a more solid and convincing theory of the sun and earth than has ever yet been offered by anybody."

Excited by so rich a promise, I begged him to disclose his idea to me plainly, and he replied: "If the annual motion belongs to the earth — along the ecliptic and around the sun — and if the sun is situated in the center of this ecliptic, and if it turns upon itself not around the axis of the ecliptic (which would be the axis of the earth's annual motion) but around a tilted axis, then extraordinary changes would have to be seen by us in the apparent movements of the solar spots, provided we assume that the axis of the sun remains perpetually and unchangingly at the same tilt with the same orientation toward the same point in the universe. In the first place the earth, traveling around the sun with the annual motion and carrying us with it, would cause the passage of the spots to appear to us to be sometimes along straight lines, but only twice a year; at all other times they would appear to make perceptibly curved arcs. In the second place, the curvature of such arcs during one half of the year will appear to us as being tilted opposite to what appears in the other half. That is, for six months the convexity of the arcs will be toward the upper part of the solar disc, and for the other six months toward its lower part. Third, since the spots commence to appear and to our eyes are born, so to speak, at the left side of the solar disc, and then proceed so as to disappear and set at the right-hand side, the eastern points (that is, the first appearances) will for six months be lower than the points of occultation opposite to them. During the other six months, the contrary will take place; that is, the spots originating at points more elevated, and descending therefrom, they will disappear at the lowest points in their courses. Only on two days in the year will these points of rising and setting be balanced, after which the paths of the spots will begin to tilt by small degrees, as on a scale. And from day to day this tilt will become larger, attaining its greatest obliquity in three months and commencing from that point to diminish, being reduced once more to equilibrium in that much

The concept which suddenly occurred to the Lincean Academician concerning the great consequence which followed from the motions of sunspots.

Strange changes in the movements of sunspots foreseen by the Academician if the annual motion were the earth's.

time again. For a fourth remarkable thing, it will happen that the day of maximum obliquity will be the same as that on which the passage will be made in a straight line; and on the day of equilibration, the arc of the journey will appear to be most curved of all. At other times, accordingly as the tilting diminishes and proceeds toward equilibrium, the curvature of the arcs of passage will increase."

SAGR. I know it is bad manners to interrupt your discourse, Salviati, but I think it would be no worse to let you pour out more words when, as people say, they are falling on empty air. For to speak frankly, I cannot at present form any distinct idea for any of the conclusions you have announced. Yet as taken thus generally and confusedly, they suggest to me matters of remarkable consequence, so I should like somehow to be put in better possession of them.

SALV. The same thing happened to me that is happening to you, when these bare words were given to me by my guest. He then assisted my understanding by representing the facts for me upon a material instrument, which was nothing but an astronomical sphere, making use of some of its circles — though a different use from that which they ordinarily serve. Now I shall remedy the absence of a sphere by making diagrams on paper as they are required. To represent the first event which I related — that the passage of the spots could only twice a year appear to be made along straight lines — let us imagine this point O to be the center of the earth's orbit (or let us say of the ecliptic) and likewise of the globe of the sun itself, of which (considering the great distance between it and the earth) we may suppose that we Terres-

First event to be
discovered in the
motion of the
sunspots, and the
consequent
explanation of
all others.

trials can see one-half. So let us describe this circle ABCD around the center O; this will represent for us the extreme boundary which separates and divides the hemisphere of the sun which we can see from

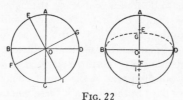

FIG. 22

that which is hidden. Now since our eyes, like the center of the earth, are supposed to be in the plane of the ecliptic, in which the center of the sun lies likewise, if we represent to ourselves the solar body as cut by the plane of the ecliptic, the section will appear to our eyes as a straight line. Let this be BOD, and suppose this to be perpendicular to AOC, which will be the axis of the ecliptic and of the annual motion of the terrestrial globe.

Now let us suppose the solar body to revolve upon itself without its center moving. Let it revolve not around the axis AOC, which is perpendicular to the plane of the ecliptic, but around some other one, tilted a certain amount, which shall be EOI here; and let this fixed and immutable axis perpetually remain at the same inclination and point toward the same part of the firmament and of the universe. Since in the revolution of the solar globe each point of its surface, except the poles, describes the circumference of a circle, greater or lesser according as it is located more or less distant from the poles, let us take the point F equidistant from them, and denote the diameter FOG, which will be perpendicular to the axis EI and will be the diameter of the great circle described about the poles E and I.

Now assume that the earth, carrying us along with it, is at such a point on the ecliptic that the solar hemisphere which is visible to us is bounded by the circle ABCD; this, passing through the poles A and C (as it always does) passes also through the poles E and I. It is obvious that the great circle whose diameter is FG will be vertical to the circle ABCD, to which that ray is perpendicular which reaches our eyes from the center O. Therefore the same ray falls in the plane of the circle whose diameter is FG and whose circumference therefore appears to us as a straight line, and is the same as FG. Whenever a spot is at the point F, then, and is carried by the sun's rotation, it will mark on the surface of the sun the circumference of that circle which appears to us as a straight line. Hence its passage will appear straight, and so will the movements of the other spots which describe smaller circles in the same revolution, since all of these are parallel to the great circle, our eye being placed at an immense distance from them.

Next, if you consider that after six months the earth will have run through half its orbit and will be situated opposite that solar hemisphere which is now hidden from us, so that the boundary of the part seen by us will still be the same circle ABCD passing through the poles E and I, you will understand that the same thing will occur in the course of the spots. That is, all of them will appear to be made in straight lines. But since this will not happen unless the boundary passes through the poles E and I, and since this boundary changes from moment to moment because of the annual motion of the earth, its passage through the fixed poles E and I is momentary, and consequently the time of the motion when these spots will appear linear is momentary.

From what has been said thus far, it may be understood how the motion of spots appearing first at side F and proceeding toward G gives an ascending passage from left to right. But assuming the earth to be diametrically opposite, the spots will indeed appear to the left of the observer, near G, but their paths will descend toward F at the right.

Now let us imagine the earth to be one quadrant removed from its present place, and let us denote in this second figure the boundary ABCD, and as before the axis AC, through which the plane of our meridian† would pass. In this plane would also be the axis of the sun, with one of its poles toward us in the visible hemisphere, which pole we shall represent by the point E, and with the other pole, I, falling in the hidden hemisphere. The axis EI tilting thus with its upper part E toward us, the great circle described by the rotation of the sun will be BFDG here, whose visible half, that is, BFD, will no longer appear as a straight line (because of the poles E and I not being on the circumference ABCD), but will appear to us curved, with its convex part toward the bottom, C. And it is obvious that the same will hold for all lesser circles parallel to the great circle BFD. It should also be understood that when the earth is diametrically opposite to this position, so that the other hemisphere of the sun is seen which is now hidden, one will see the same part DGB of the great circle curved, with its convex part toward the top, A; and the courses of the spots in this location will be first along the arc BFD and then along the other, DGB. And their first appearances and final disappearances, made near the points B and D, will be balanced, the former being neither more nor less elevated than the latter.

Now let us put the earth in such a place along the ecliptic that neither the boundary ABCD nor the meridian AC passes through the poles of the axis EI, as I show you by drawing this third figure, where the visible pole E falls between the arc of the boundary AB and the meridian section AC; the diameter of the great circle will be FOG, the visible semicircle being FNG and the hidden one GSF. The former has its convex part N curved toward the bottom part, and the latter curves with its summit S toward the upper part of the sun. The entrance and the exit of the spots (that is, the points F and G) will not be balanced as

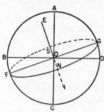

FIG. 23

B and D were before, but F will be lower and G higher — though indeed with less difference than in the first diagram — also, the arc FNG will be curved, but not as much as BFD in the preceding case. Hence in this arrangement the paths of the spots will ascend from F on the left to G on the right, and will be curved lines. Supposing the earth to be placed at the point diametrically opposite, so that the hemisphere of the sun here hidden will become visible and will be bounded by the same rim ABCD, obviously the course of the spots will be perceived to be along the arc GSF, commencing at the upper point G (which will still be to the left of the observer) and going toward the boundary at the point F, descending toward the right.

Once what I have explained is understood, I do not believe that any difficulty will remain in seeing how, from the passage of the boundary of the solar hemisphere through the poles of the sun's rotation or through points near or far from these, all the diversities of the apparent courses of the spots originate, so that the more distant these poles are from this boundary, the more curved the said courses will be, and the less oblique. At the maximum distance, which occurs when the said poles are at the meridian section, the curvature becomes greatest and the obliquity least (that is, the latter is reduced to equilibrium, as the second diagram shows). On the other hand, when the poles are at the boundary, as the first figure shows, the tilt is a maximum and the curvature is a minimum (is reduced to straightness). As the boundary leaves the poles, the curvature begins to become perceptible, increasing as it goes, while the tilt or inclination becomes less.

These are the strange changes which my guest told me would appear from one time to another in the courses of the sunspots, if it were true that the annual movement belonged to the earth, while the sun, being located at the center of the ecliptic, rotated upon an axis that was not perpendicular but was tilted to the plane of the ecliptic.

SAGR. I am quite convinced of these consequences, and believe that they will become better fixed in my mind when I examine them by placing a globe at this tilt and then looking at it from various angles.

It now remains for you to tell us what happened afterward with regard to the outcome of these conjectured results.

The events ob-
served corre-
sponded to the
predictions.

SALV. It came about that, continuing to make very careful observations for many, many months, and noting with consummate accuracy the paths of various spots at different times of the year, we found the results to accord exactly with the predictions.

SAGR. Simplicio, if what Salviati is telling us here is true (and it would be improper for us to doubt his word), the Ptolemaics and the Aristotelians will need most solid arguments, great theories, and sound experiments to offset so weighty a discovery and to save their opinions from ultimate defeat.

SIMP. Step gently, my friend; perhaps you have not got so far as you think you have. For although I have not entirely mastered the content of Salviati's discourse, still, when I consider the form of the argument, I cannot see that my logic teaches me that this mode of reasoning necessarily forces me to any conclusion in favor of the Copernican hypothesis; that is, of the stability of the sun in the center of the zodiac and the mobility of the earth

Although the
annual motion
attributed to the
earth corre-
sponds to the
appearances of
the solar spots,
it does not there-
fore follow that
conversely from
the appearances
of the spots one
must infer the
annual motion to
be the earth's.

around its circumference. For while it is true that assuming the rotation of the sun and the circulation of the earth, such-and-such pecularities must necessarily be perceived in the sunspots, it does not therefore necessarily follow that, arguing from the converse, from perceiving these oddities in the spots one must necessarily conclude that the earth does move around the circumference of the zodiac while the sun is posted in its center. For who is there to assure me that such peculiarities might not also be seen in a sun moving along the ecliptic, by inhabitants of an earth stationary in its center? Unless you first demonstrate to me that such an appearance cannot be accounted for when the sun is made movable and the earth fixed, I shall not change my opinion, nor believe that the sun moves and the earth remains at rest.

SAGR. Simplicio is behaving bravely, and he battles very cleverly to sustain the Aristotelian and Ptolemaic side. To tell the truth, it seems to me that conversing with Salviati even for such a short time has considerably increased his capacity to reason rigorously — an effect which I hear that this has had on other people, too. Now as to this inquiry and decision (as to whether it is possible to give an adequate cause for the visible peculiarities of the movements of the sunspots while leaving the earth motionless and keeping the sun in motion), I hope that Salviati will open his thoughts to us. For it is certainly reasonable to believe that he has reflected upon it and has deduced as much along that line as is possible.

SALV. I have thought about it many times, and have also talked it over with my friend and guest, and as to that which can be adduced by philosophers and astronomers in defense of the ancient system, we are sure of one thing. This is that the true pure Peripatetics, laughing at anyone who employs himself in what (to their thinking) are empty fooleries, will pretend that all these appearances are vain illusions of the lenses, and will thus free themselves with little trouble from the obligation of thinking any more about it. But as for scientific astronomers, after having given very careful thought to what might be said on this matter, we have not found under the ancient system any reply adequate to harmonize the course of the spots with human reason. I shall tell you what occurred to us, and you may make whatever use of it your own discretion tells you to.

Pure Peripatetic philosophers laugh at sunspots and their appearances as illusions of the telescope lenses.

Assuming that the visible motions of the sunspots are as we have declared above, and assuming the earth to be immovable in the center of the ecliptic, on whose circumference the center of the sun is placed, it is necessary that all the diversity which is perceived in these movements shall have causes residing in the motions of the solar body. In the first place this must revolve upon itself and carry along with it the spots, which have been assumed and even demonstrated to adhere to the surface of the sun. Secondly, it will be necessary to say that the axis of the sun's rotation is not parallel to the axis of the ecliptic, which amounts to saying that it is not perpendicular to the plane of the ecliptic. For if it were, the passages of these spots would appear to us to be made in straight lines, and parallel to the ecliptic. Therefore this axis is tilted, since the courses for the most part appear to be made along curved lines.

If the earth is immovable at the center of the zodiac, one must attribute to the sun four different movements, as explained at length.

Thirdly, one must say that the tilt of this axis is not fixed, and facing continually toward the same point of the universe, but that it changes its direction from one moment to another. For if the obliquity were always pointed in the same direction, the paths of the spots would never change their appearances;[†] whether they were straight or curved, bent up or down, ascended or descended, they would appear the same at one time as at another. Thus one would have to say that the axis was variable, and found itself sometimes in the plane of the extreme bounding circle of the visible hemisphere (I mean at those times when the passages of the spots appeared to be made in straight lines and were most oblique of all, which occurs twice a year), and then at other times in the meridian plane of the observer, so that one of its poles would fall in the visible solar hemisphere and the other in the hidden one — both being distant from the extreme points (let us call them the poles) of another axis of the sun which would be parallel to the axis of the ecliptic and would necessarily have to be assigned to the sun; as far distant, that is, as the tilt of the axis of revolution of the spots would indicate. Add to this that the pole falling in the visible hemisphere would be in the upper part at one time and in the lower at another. A necessary argument for this is given by what happens to the paths when they are level and at their maximum curvature, once with their convexity toward the lower side and again with it toward the upper part of the solar disc.

And since such states would be continually altering, making the tilting and the curvature now greater and now less, the former being sometimes reduced to complete equilibrium and the latter to perfect straightness, this axis of monthly revolution of the spots would have to be supposed to possess a rotation of its own, by which its poles would describe two circles around the poles of another axis (which would thereby be assigned to the sun), the radius of which circles would correspond to the degree of tilt of this axis. And it would be required that its period should be one year, since that is the time in which all the appearances and diversities in the paths of the spots are repeated. That the rotation of this axis should be made about the poles of another axis parallel to that of the ecliptic, and not around any other points, is clearly indicated by the maximum tilts and the maximum curvatures, which are always of the same magnitude.

Hence finally it will be necessary, in order to keep the earth fixed in the center, to attribute to the sun two movements around its own center, on two different axes, one of which would complete its rotation in a year, and the other in less than a month. To my mind, such an assumption seems very difficult, almost impossible; this arises from having to attribute to the same solar body two other movements about the earth on different axes, tracing out the ecliptic in a year with one of these, and with the other forming spirals or circles parallel to the equinoctial plane, one a day.

And as to that third movement† which must be assigned to the sun itself (I am not speaking of the quasi-monthly one which carries the spots, but of that other one which must convey the axis and the poles of this monthly one), no reason whatever is to be seen why it should complete its motion in a year (as dependent upon the annual motion along the ecliptic) rather than in twenty-four hours (as dependent upon the diurnal motion about the poles of the equinoctial). I know that what I am saying is rather obscure at present, but it will be obvious to you when we come to speak of that third motion (an annual one) assigned by Copernicus to the earth.

Now if these four motions, so incongruous with each other and yet necessarily all attributable to the single body of the sun, could be reduced to a single and very simple one, the sun being assigned one inalterable axis; and if with no innovations in the movements assigned by so many other observations to the terrestrial globe, one could still easily preserve the many peculiar appearances in the movements of the solar spots, then really it seems to me that this decision could not be rejected.

This, Simplicio, is all that occurred to my friend and to myself regarding that which might be adduced in explanation of the appearances in defense of their opinions by the Copernicans and by the Ptolemaics. You may do with it whatever your own judgment persuades you to do.

SIMP. I recognize my own incapacity to take upon myself so important a decision. As to my own ideas, I remain neutral, in the hope that a time will come when the mind will be freed by an illumination from higher contemplations than these of our human reasoning, and all the mists which keep it darkened will be swept away.

SAGR. Simplicio's counsel is excellent and pious, and worthy of being accepted and followed by everyone, since only that which is derived from the highest wisdom and supreme authority may be embraced with complete security. But so far as human reason is allowed to penetrate, confining myself within the bounds of theory and of probable causes, I shall indeed say (with a little more boldness than Simplicio exhibits) that I have not, among all the many profundities that I have ever heard, met with anything which is more wonderful to my intellect or has more decisively captured my mind (outside of pure geometrical and arithmetical proofs) than these two conjectures, one of which is taken from the stoppings and retrograde motions of the five planets, and the other from the peculiarities of movement of the sunspots. And it appears to me that they yield easily and clearly the true cause of such strange phenomena, showing the reason for such phenomena to be a simple motion which is mixed with many others that are also simple but that differ among themselves. Moreover they show this without introducing any difficulties; rather, they remove all those which accompany other viewpoints. So much so that I am rapidly coming to the conclusion that those who remain hostile toward this doctrine must either not have heard it or must not have understood these arguments, which are so numerous and so conclusive.

SALV. I do not give these arguments the status of either conclusiveness or of inconclusiveness, since (as I have said before) my intention has not been to solve anything about this momentous question, but merely to set forth those physical and astronomical reasons which the two sides can give me to set forth. I leave to others the decision, which ultimately should not be ambiguous, since one of the arrangements must be true and the other false. Hence it is not possible within the bounds of human learning that the reasons adopted by the right side should be anything but clearly conclusive, and those opposed to them, vain and ineffective.

SAGR. Then it is now time for us to hear the other side, from that booklet of theses or disquisitions which Simplicio has brought back with him.

SIMP. Here is the book, and here is the place in which the author first briefly describes the system of the world according to the position of Copernicus, saying: *Terram igitur*† *una cum Luna totoque hoc elementari Copernicus* etc. ("Therefore the earth, together with the moon and all this elemental world, Copernicus" etc.)

SALV. Wait a bit, Simplicio; for it seems to me that this author at the very outset declares himself to be very ill-informed about the position he undertakes to refute, when he says that Copernicus makes the earth together with the moon trace out the *orbis magnus* in a year, moving from east to west; a thing which, as it is false and impossible, has accordingly never been uttered by Copernicus. Indeed, he makes it go in the opposite direction (I mean from west to east; that is, in the order of the signs of the zodiac), so that it appears that the annual motion belongs to the sun, which is placed immovably in the center of the zodiac.

You see the excessive boldness of this man's self-confidence, setting himself up to refute another's doctrine while remaining ignorant of the basic foundations upon which the greatest and most important parts of the whole structure are supported. This is a poor beginning for gaining the confidence of the reader, but let us proceed.

SIMP. The system of the universe explained, he begins to propose his objections against the annual movement. The first of these he utters ironically, in derision of Copernicus and his followers, writing that in this fantastic arrangement of the world one must affirm the most sublime inanities: That the sun, Venus, and Mercury are beneath the earth; that heavy material naturally ascends and light stuff descends; that Christ, our Saviour and Redeemer, rose to hell and descended into heaven when He approached the sun. That when Joshua commanded the sun to stand still, the earth stood still — or else the sun moved opposite to the earth; that when the sun is in Cancer, the earth is running through Capricorn, so that the winter signs make the summer and the spring signs the autumn; that the stars do not rise and set for the earth, but the earth for them; and that the east starts in the west while the west begins in the east; in a word, that nearly the whole course of the world is turned inside out.

Objections from a certain booklet, proposed in sarcasm against Copernicus.

SALV. All of this is satisfactory to me except his having mixed passages from the ever venerable and mighty Holy Scriptures among these apish puerilities, and his having tried to utilize sacred things for wounding anybody who might, without either affirming or denying anything, philosophize jokingly and in sport, having made certain assumptions and desiring to argue about them among friends.

SIMP. Truly he scandalized me too, and not a little; especially later, when he adds that if indeed the Copernicans answer these and the like arguments in some distorted way, they still will not be able to answer satisfactorily some things which come later.

SALV. Oh, that is worst of all, for he is pretending to have things which are more effective and convincing than the authority of Holy Writ. But let us, for our part, revere it, and pass on to physical and human arguments. Yet if he does not adduce among his physical arguments matters which make more sense than those set forth up to this point, we may as well abandon him entirely. I am certainly not in favor of wasting words answering such trifling tomfooleries. And as for his saying that the Copernicans do reply to these objections, that is quite false. I cannot believe that any man would put himself to such a pointless waste of time.

Assuming the annual motion to be the earth's, one fixed star must be larger than the whole orbit of the earth.

SIMP. I, too, concur in this decision; let us, then, listen to his other objections, which are more strongly supported. Now here, as you see, he deduces with very precise calculations that if the orbit in which Copernicus makes the earth travel around the sun in a year were scarcely perceptible with respect to the immensity of the stellar sphere, as Copernicus says must be assumed, then one would have to declare and maintain that the fixed stars were at an inconceivable distance from us, and that the smallest of them would be much larger than this whole orbit, while others would be larger than the orbit of Saturn. Yet such bulks are truly too vast, and are incomprehensible and unbelievable.

Tycho's argument founded upon false hypotheses.

Disputants, when in the wrong, fix upon some word accidentally uttered by the other side.

SALV. I have indeed seen something similar argued against Copernicus by Tycho, so this is not the first time that I have revealed the fallacy — or better, the fallacies — of this argument, built as it is upon completely false hypotheses. It is based upon a dictum of Copernicus which is taken by his adversaries with rigorous literalness, as do those quarrelsome people who, being wrong about the principal issue of the case, seize upon some single word accidentally uttered by their opponents and make a great fuss about it without ever letting up.

For your better comprehension, know that Copernicus first explains the remarkable consequences to the various planets deriving from the annual movement of the earth; in particular the forward and retrograde movements of the three outer planets. Then he adds that these apparent mutations which are perceived to be greater in Mars than in Jupiter, from Jupiter's being more distant, and still less in Saturn, from its being farther away than Jupiter, remain imperceptible in the fixed stars because of their immense distance from us in comparison with the distance of Jupiter or of Saturn. Here the adversaries of this opinion rise up, and take what Copernicus has called "imperceptible" as having been assumed by him to be really and absolutely nonexistent. Remarking that even the smallest of the fixed stars is still perceptible, since it strikes our sense of sight, they set themselves to calculating (with the introduction of still more false assumptions), and deduce that in Copernicus's doctrine one must admit that a fixed star is much larger than the orbit of the earth.

Now in order to reveal the folly of their entire method, I shall show that by assuming that a star of the sixth magnitude may be no larger than the sun, one may deduce by means of correct demonstrations that the distance of the fixed stars from us is sufficiently great to make quite imperceptible in them the annual movement of the earth which in turn causes such large and observable variations in the planets. Simultaneously I shall clearly expose to you a gigantic fallacy in the assumptions made by the adversaries of Copernicus.

To begin with, I assume along with Copernicus and in agreement with his opponents that the radius of the earth's orbit, which is the distance from the sun to the earth, contains 1,208 of the earth's radii.† Secondly, I assume with the same concurrence and in accordance with the truth that the apparent diameter of the sun at its average distance is about one-half a degree, or 30 minutes; this is 1,800 seconds, or 108,000 third-order divisions. And since the apparent diameter of a fixed star of the first magnitude is no more than 5 seconds, or 300 thirds, and the diameter of one of the sixth magnitude measures 50 thirds (and here is the greatest error of Copernicus's adversaries), then the diameter of the sun contains the diameter of a fixed star of the sixth magnitude 2,160 times. Therefore if one assumes that a fixed star of the sixth magnitude is really equal to the sun and not larger, this amounts to saying that if the sun moved away until its diameter looked to be 1/2160th of what it now appears to be, its distance would have to be 2,160 times what it is in fact now.

The apparent diversity of motion in the planets remains insensible in the fixed stars.

Assuming that a fixed star of the sixth magnitude is no larger than the sun, the diversity which is great in the planets remains insensible in the fixed stars.

Distance to the sun contains 1,208 radii of the earth.

Diameter of the sun one-half a degree.

Diameter of a fixed star of the first magnitude, and of one of the sixth.

How much the apparent diameter of the sun exceeds that of a fixed star.

Distance of a fixed star of the sixth magnitude, assuming the star equal to the sun.

Diversity of
aspect in fixed
stars caused by
the earth's orbit
is little greater
than that caused
in the sun by the
size of the earth.

This is the same as to say that the distance of a fixed star of the sixth magnitude is 2,160 radii of the earth's orbit. And since the distance from the earth to the sun is commonly granted to contain 1,208 radii of the earth, and the distance of the fixed star is, as we said, 2,160 radii of the orbit, then the radius of the earth in relation to that of its orbit is much greater than (almost double) the radius of that orbit in relation to the stellar sphere. Therefore the difference in aspect of the fixed star caused by the diameter of the earth's orbit would be little more noticeable than that which is observed in the sun due to the radius of the earth.

SAGR. For a first step, this is a bad fall.

The sixth-magni-
tude star as-
sumed by Tycho
and by the
author of the
booklet is ten
million times
what it ought
to be.

SALV. It is indeed wrong, since according to this author a star of the sixth magnitude would have to be as large as the earth's orbit in order to justify the dictum of Copernicus. Yet assuming it to be equal only to the sun, which in turn is rather less than one ten-millionth[†] of that orbit, makes the stellar sphere so large and distant that this alone is sufficient to remove this objection against Copernicus.

SAGR. Please make this computation for me.

Computation of
the size of a fixed
star in terms of
the earth's orbit.

SALV. The calculation is very short and simple. The diameter of the sun is 11 radii of the earth, and the diameter of the earth's orbit contains 2,416 of these radii, as both parties agree. So the diameter of the orbit contains that of the sun approximately 220 times, and since spheres are to each other as the cubes of their diameters, we take the cube of 220 and we have the orbit 10,648,000 times as large as the sun. The author would say that a star of the sixth magnitude would have to be equal to this orbit.

SAGR. Then their error consists in their having been very much deceived in taking the apparent diameter of the fixed stars.

Common delu-
sion of all as-
tronomers about
the sizes of the
stars.

SALV. That is the error, but not the only one. And truly I am quite surprised at the number of astronomers, and famous ones too, who have been quite mistaken in their determinations of the sizes of the fixed as well as the moving stars, only the two great luminaries being excepted. Among these men are al-Fergani, al-Battani, Thabit ben Korah, and more recently Tycho, Clavius, and all the predecessors of our Academician. For they did not take care of the adventitious irradiation which deceptively makes the stars look a hundred or more times as large as they are when seen without haloes. Nor can these men be excused for their carelessness; it was within their power to see the bare stars at their pleasure, for it suffices to look at them when they first appear in the evening, or just before they vanish at dawn. And Venus, if nothing else, should have warned them of their mistake, being frequently seen in daytime so small that it takes sharp eyesight to see it, though in the following night it appears like a great torch. I will not believe that they thought the true disc of a torch was as it appears in profound darkness, rather than as it is when perceived in lighted surroundings; for our lights seen from afar at night look large, but from near at hand their true flames are seen to be small and circumscribed. This alone might have sufficed to make them cautious.

Venus makes inexcusable the error of the astronomers in determining the sizes of the stars.

To speak quite frankly, I thoroughly believe that none of them — not even Tycho himself, accurate as he was in handling astronomical instruments and despite his having built such large and accurate ones without a thought for their enormous expense — ever set himself to determine and measure the apparent diameter of any star except the sun and moon. I think that arbitrarily and, so to speak, by rule of thumb some one among the most ancient astronomers stated that such-and-such was the case, and the later ones without any further experiment adhered to what this first one had declared. For if any of them had applied himself to making any test of the matter, he would doubtless have detected the error.

SAGR. But if they lacked the telescope (for you have already said that our friend came to know the truth of the matter by means of that instrument), they ought to be pardoned, not accused of negligence.

SALV. That would be true if they could not have obtained the result without the telescope. It is true that the telescope, by showing the disc of the star bare and very many times enlarged, renders the operations much easier; but one could carry them on without it, though not with the same accuracy. I have done so, and this is the method I have used. I hung up a light rope in the direction of a star (I made use of Vega, which rises between the north and the northeast) and then by approaching and retreating from this cord placed between me and the star, I found the point where its width just hid the star from me. This done, I found the distance of my eye from the cord, which amounts to the same thing as one of the sides which includes the angle formed at my eye and extending over the breadth of the cord.

Way of measuring the apparent diameter of a star.

This is similar to, or rather equal to, the angle made in the stellar sphere by the diameter of the star. From the ratio of the thickness of the cord to its distance from my eye, using a table of arcs and chords, I immediately found the size of the angle — taking the customary precaution, used in determining such very acute angles, not to put the intersection of the visual rays at the center of my eye, where they would not go if they were not refracted, but beyond the location of the eye where the actual width of the pupil would permit them to converge.

SAGR. I understand this precaution, though I somewhat question it; what bothers me most in this operation is that if it is made in the dark of night, it seems to me that one is measuring the diameter of the irradiated disc and not that of the true and naked star.†

SALV. Not a bit; for the string, by covering the bare body of the star, takes away the halo belonging not to it but to our eyes; of this it is deprived the moment the true disc is hidden. In making the observation you will be astonished to see how thin a rope will cover that great torch which seemed incapable of being hidden except by a much larger obstacle.

Next, in order to determine the thickness of such a cord and to measure it very accurately to see how many thicknesses of such a string comprise the distance to the eye, I do not take a single diameter of it, but join many pieces on a table so that they touch. Then I use a pair of dividers to take the entire space occupied by fifteen or twenty of them, and with this I measure the distance from the cord to the focus of the visual rays, this having been previously marked on another string. By this very precise operation I find that the apparent diameter of a star of the first magnitude (commonly believed to be two minutes, and even put at three by Tycho in his *Astronomical Letters*, p. 167) is no more than five seconds, which is one twenty-fourth or one thirty-sixth of what they thought. Now you see what a serious mistake their doctrine is based upon.

Diameter of a
first-magnitude
fixed star no
more than five
seconds.

SAGR. I see, and I quite understand. But before going further I should like to propose the question which occurred to me about finding the meeting point of the visual rays included within very acute angles. My trouble arises from the impression that this intersection might vary in its location not on account of the greater or smaller size of the object looked at, but because of a certain other respect in which it seems to me that the meeting of the rays might be farther from or closer to the eye when looking at objects of the same size.

SALV. I see already where your perspicacity is leading you, Sagredo. You are a careful observer of nature; I'll wager anything that not more than one out of every thousand people who have observed the extreme contraction and dilation of the pupil in a cat's eye have observed a like effect in the human pupil, depending upon whether it is looking through a well or a poorly lighted medium. In daylight the circlet of the eye is much diminished; when looking at the disc of the sun, it is reduced to a size smaller than a millet seed; but when looking at nonshining objects in a dark medium, it dilates to the size of a pea or larger. In general this enlargement and reduction varies in much more than a tenfold ratio, from which it is obvious that when the pupil is much dilated, the angle of intersection of the rays must be farther away from the eye, as happens when looking at poorly lighted objects. It is Sagredo who has just furnished me with this doctrine, and it warns us that if a very accurate observation of great importance were to be made, we should conduct our investigation of that intersection by performing an experiment concerning this. But in the present case, in order to reveal the error of the astronomers, you do not need such accuracy; for even if we favor them by assuming that the intersection is made right at the pupil itself, it does not much matter, their error being so enormous. I am not sure that this is what you meant, Sagredo.

Pupil of the eye enlarges and contracts.

SAGR. It is, exactly; and I am glad that it was not unreasonable, as I am assured by your being in accord. But I should like to take advantage of this opportunity to hear how the distance to the intersection of the visual rays may be determined.

SALV. The method is very easy, and it is as follows: I take two strips of paper, one black and the other white, making the black strip half the width of the white. I attach the white one to a wall, and fix the other at a distance of some 15 or 20 yards from it on a stick or some other support. Then I move away an equal distance from this in the same direction, and it is obvious enough that at this distance those straight lines intersect which, leaving from the edges of the white paper, would just touch in passing the edges of the strip placed midway between. From this it follows that the eye being placed at this intersection, the black

How to find the distance of convergence of rays from the pupil.

strip in the center would just hide the white one, provided vision took place in a single point. But if we should find that the edges of the white strip could still be seen, it would argue necessarily that the visual rays are not converging at one point alone. And to make the white strip stay hidden by the black one, the eye would have to be brought closer. This done so that the central strip hides the distant one, and the amount of the required approach being noted, this amount will be a safe measure of the distance from the eye of the true intersection of the visual rays in such operations. Moreover, we shall thus have the diameter of the pupil, or rather of the hole upon which the visual rays impinge.† For its proportion to the width of the black paper will be that which is borne to the distance between the two papers by the distance from the intersection of the lines produced along the edges of the papers to the place where the eye was when it first saw the more distant paper hidden by the intermediate one.

Therefore if we wish to measure accurately the apparent diameter of a star, the observations being made in the above manner, it will be necessary to compare the diameter of the cord with the diameter of the pupil. Finding the diameter of the former to be, for example, four times that of the pupil, and the distance from the eye to the cord being 30 yards, we should say the true intersection of the lines produced from the edges of the diameter of the star along the edges of the cord would be found 40 yards from the cord. In this way the ratio between the distance from the cord to the intersection of the said lines and the distances from that intersection to the location of the eye will be in the proper proportion, which must be the same as that which holds between the diameter of the cord and the diameter of the pupil.

SAGR. I understand. Now let us hear what Simplicio has to say in defense of the adversaries of Copernicus.

SIMP. Although Salviati's discourse has greatly lessened that huge and incredible impropriety which these adversaries of Copernicus point out, this does not seem to me to be so completely removed as to have no longer enough force to upset his view. For if I properly understood the last and principal conclusion, then when one assumes the star of the sixth magnitude to be as large as the sun (which seems to me a remarkable assumption), it still remains true that the earth's orbit would necessarily cause changes and variations in the stellar sphere similar to the observable changes produced by the earth's radius in regard to the sun. No such changes, or even smaller ones, being observed among the fixed stars, it appears to me that by this fact the annual movement of the earth is rendered untenable and is overthrown.

SALV. You would do well to conclude so, Simplicio, were there nothing more to be said for Copernicus's side; but a great deal more remains. As to your rejoinder, nothing prevents our supposing that the distance of the fixed stars is still much greater than has been assumed. You yourself, and anyone else there may be who does not want to disparage the propositions accepted by Ptolemy's followers, must find it a very convenient thing to suppose the stellar sphere to be enormously larger than we have said it must be considered thus far. For all astronomers agree that a slower rotation is caused for planets by increasing their orbits, and that it is for this reason that Saturn is slower than Jupiter, and Jupiter than the sun (because the first named must describe a larger circle than the second, and that one than the next, etc.) The orbit of Saturn, for example, is 9 times as far away as that of the sun, and the resulting time of one revolution for Saturn is 30 times as long as that of one circuit of the sun. Now seeing that in Ptolemy's doctrine one revolution of the stellar sphere is completed in 36,000 years, whereas that of Saturn is completed in 30 years and that of the sun in one year, we may reason as follows with such ratios:

If Saturn's orbit, by being 9 times as large as that of the sun, has 30 times as long a period of revolution, then proportionately how large should that orbit be in which the rotation is 36,000 times slower? It comes out that the distance of the stellar sphere must be 10,800 radii of the earth's orbit — which would be just 5 times as large as we calculated a little while ago for it if a star of the sixth magnitude were as large as the sun. Now you see how much less, on this account, should be the variations caused in it by the annual motion of the earth.

And if we wanted to figure out the distance of the stellar sphere from similar relations of Jupiter and Mars, the former would give us 15,000 and the latter 27,000 radii of the earth's orbit; that is, even more (the former by 7 times and the latter by 12) than was derived by supposing the size of a fixed star to be equal to that of the sun.

Astronomers agree that the cause of greater slowness in rotations is the greater size of the orbits.

Under another supposition taken from the astronomers, it is calculated that the distance to the fixed stars must be 10,800 radii of the earth's orbit.

By means of the ratios for Jupiter and Mars, the stellar sphere is also found to be much farther away.

SIMP. It seems to me that to this one might reply that since the time of Ptolemy the motion of the stellar sphere has been observed to be not so slow as he thought it was. I think I have even heard that it was Copernicus himself who observed this.†

SALV. Right you are, but you are not saying anything which is favorable in any way to the cause of the Ptolemaics, who have never rejected the 36,000-year motion of the stellar sphere on account of such slowness making it too vast and too immense. If such immensity is not to be allowed in nature, then they should long ago have denied so slow a rotation, which cannot be adapted with good proportion to any sphere except one of monstrous size.

SAGR. Please, Salviati, let us waste no more time invoking these ratios against people who are ready to accept the most disproportionate things; absolutely nothing is to be gained against them by this route. What more disproportionate ratios can be imagined than those which these people grant and allow to pass without comment? First they write that there cannot be a more suitable way for us to order the celestial spheres than by arranging them according to the variations of their periodic times, putting the slower beyond the faster, and they place the stellar sphere highest, as the slowest of all; then afterward they have put one more, still higher, and thereby still larger, and made it move around in twenty-four hours when the next one beneath it takes 36,000 years! But enough was said yesterday about these monstrosities.

SALV. Simplicio, I wish you could for a moment put aside your affection for the followers of your doctrines and tell me frankly whether you believe that they comprehend in their own minds this magnitude which they subsequently decide cannot be ascribed to the universe because of its immensity. I myself believe that they do not. It seems to me that here the situation is just as it is with the grasp of numbers when one gets up into the thousands of millions, and the imagination becomes confused and can form no concept. The same thing happens in comprehending the magnitudes of immense distances; there comes into our reasoning an effect similar to that which occurs to the senses on a serene night, when I look at the stars and judge by sight that their distance is but a few miles, or that the fixed stars are not a bit farther off than Jupiter, Saturn, or even the moon.

Immense sizes and numbers are incomprehensible to our intellects.

But aside from all this, consider those previous disputes be- tween the astronomers and the Peripatetic philosophers about the reasoning as to the distance of the new stars in Cassiopeia and Sagittarius, the astronomers placing these among the fixed stars and the philosophers believing them to be closer than the moon. How powerless are our senses to distinguish large distances from extremely large ones, even when the latter are in fact many thousands of times the larger!

And finally I ask you, O foolish man:† Does your imagination first comprehend some magnitude for the universe, which you then judge to be too vast? If it does, do you like imagining that your comprehension extends beyond the Divine power? Would you like to imagine to yourself things greater than God can accomplish? And if it does not comprehend this, then why do you pass judgment upon things you do not understand?

SIMP. These arguments are very good, and no one denies that the size of the heavens may exceed our imaginings, since God could have created it even thousands of times larger than it is. But must we not admit that nothing has been created in vain, or is idle, in the universe? Now when we see this beautiful order among the planets, they being arranged around the earth at distances commensurate with their producing upon it their effects for our benefit, to what end would there then be interposed between the highest of their orbits (namely, Saturn's), and the stellar sphere, a vast space without anything in it, superfluous, and vain? For the use and convenience of whom?

SALV. It seems to me that we take too much upon ourselves, Simplicio, when we will have it that merely taking care of us is the adequate work of Divine wisdom and power, and the limit beyond which it creates and disposes of nothing. I should not like to have us tie its hand so. We should be quite content in the knowledge that God and Nature are so occupied with the government of human affairs that they could not apply themselves more to us even if they had no other cares to attend to than those of the human race alone. I believe that I can explain what I mean by a very appropriate and most noble example, derived from the action of the light of the sun. For when the sun draws up some vapors here, or warms a plant there, it draws these and warms this as if it had nothing else to do. Even in ripening a bunch of grapes, or perhaps just a single grape, it applies itself

Nature and God occupy themselves in the care of mankind as if they had no other concerns.

Example of God's care of the human race drawn from the sun.

so effectively that it could not do more even if the goal of all its affairs were just the ripening of this one grape. Now if this grape receives from the sun everything it can receive, and is not deprived of the least thing by the sun simultaneously producing thousands and thousands of other results, then that grape would be guilty of pride or envy if it believed or demanded that the action of the sun's rays should be employed upon itself alone.

I am certain that Divine Providence omits none of the things which look to the government of human affairs, but I cannot bring myself to believe that there may not be other things in the universe dependent upon the infinity of its wisdom, at least so far as my reason informs me; yet if the facts were otherwise, I should not resist believing in reasoning which I had borrowed from a higher understanding. Meanwhile, when I am told that an immense space interposed between the planetary orbits and the starry sphere would be useless and vain, being idle and devoid of stars, and that any immensity going beyond our comprehension would be superfluous for holding the fixed stars, I say that it is brash for our feebleness to attempt to judge the reason for God's actions, and to call everything in the universe vain and superfluous which does not serve us.

It is very brash to call everything in the universe which we do not understand to be created for us.

SAGR. Say rather, and I think you will be speaking more accurately, "which we do not know to serve us." I believe that one of the greatest pieces of arrogance, or rather madness, that can be thought of is to say, "Since I do not know how Jupiter or Saturn is of service to me, they are superfluous, and even do not exist." Because, O deluded man, neither do I know how my arteries are of service to me, nor my cartilages, spleen, or gall; I should not even know that I had gall, or a spleen, or kidneys, if they had not been shown to me in many dissected corpses. Even then I could understand what my spleen does for me only if it were removed. In order to understand how some celestial body acted upon me (since you want all their actions to be directed at me), it would be necessary to remove that body for a while, and say that whatever effect I might then feel to be missing in me depended upon that star.

By depriving the sky of some star, one might get to know how that one acted upon us.

Besides, what does it mean to say that the space between Saturn and the fixed stars, which these men call too vast and useless, is empty of world bodies? That we do not see them, perhaps? Then did the four satellites of Jupiter and the companions of Saturn come into the heavens when we began seeing them, and not before? Were there not innumerable other fixed stars before men began to see them? The nebulae were once only little white patches; have we with our telescopes made them become clusters of many bright and beautiful stars? Oh, the presumptuous, rash ignorance of mankind!

SALV. There is no need, Sagredo, to probe any farther into their fruitless exaggerations. Let us continue our plan, which is to examine the validity of the arguments brought forward by each side without deciding anything, leaving the decision to those who know more about it than we.

Returning to our natural and human reason, I say that these terms "large," "small," "immense," "minute," etc. are not absolute, but relative; the same thing in comparison with various others may be called at one time "immense" and at another "imperceptible," let alone "small." Such being the case, I ask: In relation to what can the stellar sphere of Copernicus be called too vast? So far as I can see, it cannot be compared or said to be too vast except in relation to some other thing of the same kind. Now let us take the smallest thing of the same kind, which will be the orbit of the moon. If the stellar orb must be considered too vast in relation to that of the moon, then every other magnitude which exceeds some other of its kind by a similar or greater ratio ought also to be said to be too vast; and likewise, by the same reasoning, it should be said not to exist in the universe. Then the elephant and the whale will be mere chimeras and poetical fictions, because the former are too vast in comparison with ants (being land animals), and the latter in relation to gudgeons (being fish). And if actually found in nature, they would be immeasurably large; for the elephant and whale certainly exceed the ant and gudgeon in a much greater ratio than the stellar sphere does that of the moon, taking the stellar sphere to be as large as is required by the Copernican system.

Besides, how large is the sphere of Jupiter, and how great is that assigned to Saturn as the receptacle of a single star, though the planet itself is small in comparison with a fixed star! Surely if to each fixed star such a large portion of the space in the universe should be assigned as its container, that orb which contains an innumerable quantity of these would have to be made many thousands of times larger than suffices for the needs of

There may be in the heavens many things invisible to us.

"Great," "small," "immense," etc. are relative terms.

Vanity of the argument of those who judge the stellar sphere would be too vast in the Copernican position.

The space assigned for a fixed star is much less than for a planet.

A star is called
small with re-
spect to the
largeness of the
space surround-
ing it.

From a great
distance, the
whole stellar
sphere might ap-
pear as small as
a single star.

Questions raised
in objection by
the author of
the booklet.

Answers to ques-
tions of the
author of this
booklet.

The author of
the booklet is
confused, and
contradicts him-
self in his
questions.

COPERNICUS. Moreover, do you not call a fixed star very small —
I mean even one of the most conspicuous ones, let alone those
which escape our sight? And we call it so in comparison with the
surrounding space. Now if the whole stellar sphere were one
single blazing body, who is there that does not understand that
in an infinite space there could be assigned a distance so great
that, from there, such a brilliant sphere would appear as small
as or even smaller than a fixed star now appears to us from the
earth? So from such a point we should judge as small the very
things which we now call immeasurably huge.

SAGR. To me, a great ineptitude exists on the part of those who
would have it that God made the universe more in proportion to
the small capacity of their reason than to His immense, His infinite, power.

SIMP. All this that you are saying is good, but what the other
side objects to is having to grant that a fixed star must be not
only equal to, but much greater than, the sun; for both are still
individual bodies located within the stellar orb. And it seems to
me much to the purpose that this author inquires, "To what end
and use are such vast frames? Produced for the earth, perhaps?
That is, for a trifling little dot? And why so remote as to appear
very small and be absolutely unable to act in any way upon the
earth? To what purpose such a disproportionately large abyss
between these and Saturn? All these things are baffling, for they
cannot be maintained by probable reasons."

SALV. From the questions this fellow asks, it seems to me that
one may deduce that if only the sky, the stars, and their distances
were permitted to keep the sizes and magnitudes which he has
believed in up to this point (though he has surely never imagined
for them any comprehensible magnitudes), then he would completely understand and be satisfied about the benefits which
would proceed from them to the earth, which itself would no
longer be such a trifling thing. Nor would these stars any longer
be so remote as to seem quite minute, but large enough to be able
to act upon the earth. And the distance between them and Saturn
would be in good proportion, and he would have very probable
reasons for everything, which I should very much like to have
heard. But seeing how confused and contradictory he is in these
few words leads me to believe that he is very thrifty with or else
hard up for these probable reasons, and that what he calls rea-

sons are more likely fallacies, even shadows of foolish fantasies.

Therefore I ask him whether these celestial bodies really act upon the earth, and whether it was for that purpose that they were made of such-and-such sizes and arranged at such-and-such distances, or whether they have nothing to do with terrestrial affairs? If they have nothing to do with the earth, then it is a great folly for us Terrestrials to want to be arbiters of their sizes and regulators of their local dispositions, we being quite ignorant of all their affairs and interests. But if he says that they do act, and that it is to this end that they are directed, then this amounts to admitting what he denies in another place, and praising what he has just finished condemning when he said that celestial bodies located at such distances from the earth as to appear miniscule could not act upon it in any way. Now, my good man, in the starry sphere, which is already established at whatever distance it is, and which you have just decided is well proportioned for an influence upon terrestrial matters, a multitude of stars do appear quite small, and a hundred times as many are entirely invisible to us — which is to appear smaller than small. Therefore you must now (contradicting yourself) deny their action upon the earth, or else (still contradicting yourself) admit that their appearing small does not detract from their power to act. Or else (and this would be a frank and honest confession) you must grant and freely admit that your judgment about their sizes and distances was folly, not to say presumption or brashness.

SIMP. As a matter of fact, I also saw immediately, upon reading this passage, the obvious contradiction in his saying that the stars of Copernicus, so to speak, could not act upon the earth because they appeared so small, and his not noticing that he had granted action upon the earth to the stars of Ptolemy and his own, these not merely appearing small but being for the most part invisible.

SALV. But now I come to another point. Upon what basis does he say that the stars appear so small? Is it perhaps because that is the way they look to us? Does he not know that this comes about from the instrument which we use in looking at them — that is, our eyes? Or for that matter that by changing instruments we may see them larger and larger, as much as we please? Who knows; perhaps to the earth, which beholds them without eyes, they may appear quite huge and as they really are?

Questions directed to the author of the booklet, by which are shown the ineffectiveness of his own questions.

That distant objects appear small is a defect of the eye, as is demonstrated.

Tycho and his
adherents did
not try to see
whether there
were appear-
ances in the firm-
ament for or
against the
annual motion.

Astronomers
have perhaps not
noticed what ap-
pearances would
follow from the
earth's annual
motion.

Some things not
known by Co-
pernicus for lack
of instruments.

But it is time for us to leave these trifles and get to more important matters. I have already demonstrated two things: first,
at what distance the firmament may be placed so that the diameter of the earth's orbit would make no greater variation in it than
that which the terrestrial diameter makes with respect to the
sun at its distance therefrom, and I then showed that in order
to make a fixed star appear to us as of the size we see, it is not
necessary to assume it to be larger than the sun. Now I should
like to know whether Tycho or any of his disciples has ever tried
to investigate in any way whether any phenomenon is perceived
in the stellar sphere by which one might boldly affirm or deny
the annual motion of the earth.

SAGR. I should answer "no" for them, they having had no need to
do so, since Copernicus himself says that there is no such variation there; and they, arguing *ad hominem*, grant this to him.
Then on this assumption they show the improbability which follows from it; namely, it would be required to make the sphere
so immense that in order for a fixed star to look as large as it
does, it would actually have to be so immense in bulk as to exceed the earth's orbit — a thing which is, as they say, entirely
unbelievable.

SALV. So it seems to me, and I believe that they argue against
the man more in the defense of another man than out of any
great desire to get at the truth. And not only do I believe that
none of them ever applied himself to making such observations,
but I am not even sure that any of them knew what variation
ought to be produced in the fixed stars by the annual movement
of the earth, if the stellar sphere were not at such a distance that
any variation in them would vanish on account of its smallness.
For to stop short of such researches and fall back upon the mere
dictum of Copernicus may suffice to refute the man, but certainly
not to clear up the fact.

Now it might be that there is a variation,† but that it is not
looked for; or that because of its smallness, or through lack of
accurate instruments, it was not known by Copernicus. This
would not be the first thing that he failed to know, either for lack
of instruments or from some other deficiency. Yet, grounded
upon most solid theories, he affirmed what seemed to be contradicted by things he did not understand. For as already said,
without a telescope it cannot be comprehended that Mars does

increase sixty times and Venus forty times in one position as against another, and their differences appeared to be much less than the true ones. Yet since that time it has become certain that such variations are, to a hair, just what the Copernican system required. Hence it would be a good thing to investigate with the greatest possible precision whether one could really observe such a variation as ought to be perceived in the fixed stars, assuming an annual motion of the earth.

This is a thing which I firmly believe has not been done by anyone up to the present. Not only that, but perhaps, as I said, few people have well understood what it is that should be looked for. Nor am I saying this at random, for I have seen a certain manuscript of one of these anti-Copernicans† which says that if this opinion were true, there would necessarily follow a continual rising and falling of the pole every six months, inasmuch as the earth would be going now north and now south during that time over so great a space as the diameter of its orbit; for it also seemed reasonable to him, or even necessary, that we who accompany the earth should have our pole more elevated when we were northerly than when we were southerly. Another very intelligent mathematician fell into this same error although he was a follower of Copernicus, according to what Tycho relates in his *Progymnasmata*, on page 684. This man said that he had observed the polar altitude to vary, differing in summer and winter; and since Tycho denied the merit of the assertion but did not condemn the method (that is, he denied seeing any variation in the altitude of the poles but he did not condemn such an inquiry as inappropriate for the determination of what was sought), this amounts to his saying that he also considered that whether the polar altitude did or did not vary over a six-month period would be a good test for rejecting or accepting the annual motion of the earth.

Tycho and others argue against the annual motion on account of the invariable elevation of the pole.

SIMP. Frankly, Salviati, it seems to me too that this would have to follow. For I do not believe that you will deny to me that if we were to travel only 60 miles to the north, the pole would rise one degree; and likewise, another 60 miles to the north, the pole would be raised for us another degree, etc. Now if approaching or retreating only 60 miles makes such a noticeable change in the polar altitudes, what would be accomplished by transporting the earth, and us along with it, not 60 miles but 60,000 in that direction?

SALV. That ought to make the pole rise a thousand degrees for us, if the same ratio had to be followed. Just see, Simplicio, what can be done by an inveterate impression! Having had it fixed in your mind for so many years that it is the sky which turns around in twenty-four hours, and not the earth, and consequently that the poles of this revolution are in the sky and not in the terrestrial globe, you cannot put off this habit even for an hour and, imagining to yourself that it is the earth alone which moves, disguise yourself as the enemy sufficiently long to conceive what would follow if this masquerade were really the truth. If it is the earth, Simplicio, which moves upon itself every twenty-four hours, then in it are the poles, in it is the axis, in it is the equatorial plane (that is, the great circle passing through the points which are equidistant from the poles), and in it are the infinite other parallels, greater and lesser, which pass through the points on its surface at other distances from the poles. All these things are in the earth, and not in the stellar sphere. That, being immovable, is devoid of all such things, and it is only in imagination that they can be pictured there by prolonging the axis of the earth to where its termination would designate two points placed over our poles, and extending the equatorial plane so that there would appear to be a circle in the sky corresponding to it.

Now if the true axis, the true poles, and the true terrestrial equator do not change on the earth so long as you stay at the same place on the earth, you may take the earth anywhere you please without ever changing your own location with respect to the poles, or to these circles, or to any other terrestrial thing. This is because such a transposition is common to you and to all other terrestrial objects; and motion, where it is in common, is as if it were nonexistent. And as you do not change place with respect to the earth's poles (that is, in such a way as to raise or lower them), likewise you will not change place with respect to the poles imagined in the sky, so long as we mean by "celestial poles" (as previously defined) those two points which would be marked by the terrestrial axis when prolonged to the sky.

It is true that such points in the heavens are changed when the transposition of the earth is carried out in such a way that its axis points to other parts of the immovable celestial sphere, but our situation with respect to them would not be changed so that one would be elevated more than the other. Whoever wants

Motion where it is in common is as if it did not exist.

one of the points in the firmament corresponding to the earth's poles to move upward and the other one downward must travel along the earth toward one and away from the other. Nothing is accomplished by transposing the earth and ourselves along with it, as I have said.

SAGR. Allow me, Salviati, the privilege of explaining this quite clearly by means of an example which, though crude, is nevertheless well suited for the purpose. Simplicio, imagine yourself to be in a ship, standing in the poop, and suppose you have pointed a quadrant or some other instrument at the top of the foremast, as if you wished to take its elevation, which is, say, forty degrees. No doubt if you walk 25 or 30 paces along the deck and again direct the instrument toward the same mast, you will find its elevation to be greater, having increased, for example, ten degrees. But if instead of walking 25 or 30 paces toward the mast, you had remained in the poop and made the whole boat move in that direction, do you believe that because of the 25 or 30 paces it had traveled the elevation of the foremast would appear ten degrees higher to you?

An example suitable for explaining why the altitude of the pole should not vary because of the annual motion of the earth.

SIMP. I understand and believe that the elevation would not increase by so much as a single hair even if the voyage were one of thousands of miles, let alone thirty paces. But all the same, I believe that if upon looking past the top of the foremast one should sight a fixed star in the same direction and then should hold the quadrant fixed, then after the ship had sailed sixty miles toward the star, the quadrant would still strike the top of the mast as before, but no longer the star, which would be one degree higher.

SAGR. But do you think that the sight would not fall upon that point of the stellar sphere which was in the direction of the top of the foremast?

SIMP. No, but this point would be different, and would be lower than the star first observed.

SAGR. That is exactly it. Just as in this example the elevation of the top of the mast corresponds not to the star but to the point of the firmament which lies in the direction of the eye and the top of the mast, so (in the case we are examining) that which corresponds in the firmament to the pole of the earth is not a star, or some other fixed object in the firmament, but it is that point in which the terrestrial axis would terminate if prolonged that

From the annual motion of the earth there might follow changes in some fixed star, but not in the pole.

far. This point is not fixed, but obeys the changes made by the terrestrial pole. Hence Tycho, or whoever brought up this objection, should have said that from such a motion of the earth, if it existed, some variation would be recognized and observed in the elevation or depression not of the pole, but of some fixed star near the place corresponding to our pole.

SIMP. Indeed, I understand their equivocation, but to me this still does not take away the force of the opposing argument, which seems to me considerable if it refers to the variation of the stars and not of the pole. Thus if the movement of the ship a mere sixty miles makes a fixed star rise one degree for me, why shouldn't a similar change, and even a much greater one, happen for me when the ship is transported toward the same star by such a space as the diameter of the earth's orbit, which you say is double the distance from the earth to the sun?

Resolving the equivocation of anyone who believes that the annual motion should make great changes in the elevation of a fixed star.

SAGR. This, Simplicio, is another equivocation on your part, which you know without realizing that you do; I shall try to bring it into your mind. Therefore tell me: If, after having set the quadrant on a fixed star and having found its elevation to be, for example, forty degrees, you should tilt the side of the quadrant (without changing your own place) so that the star would stay elevated above the direction of the quadrant, would you say that on this account the star had acquired a greater altitude?

SIMP. Certainly not, for the change would be made in the instrument and not by the observer having changed position by moving toward the star.

SAGR. But if you sailed or traveled over the surface of the earth, would you say that no change was made in that same quadrant, and that it always kept the same elevation with respect to the sky, so long as you yourself did not tilt it but left it fixed in its original position?

SIMP. Let me think a minute. I should say that undoubtedly it would not keep this same tilt, my voyage being made not over a plane but on the circumference of the terrestrial globe. At every step this changes its inclination with respect to the heavens, and consequently the instrument kept upon it would change.

SAGR. Well said. And you also understand that the larger the circle upon which you move, the longer the voyage would have to be in order to make that star rise one degree for you. And finally, if your motion toward the star were along a straight line, it would be necessary for you to move much farther than along the circumference of any circle, however immense.

SALV. Yes, because ultimately the circumference of an infinite circle and a straight line are the same thing.

SAGR. Oh, that I do not understand, nor do I think Simplicio understands it either. Behind it there must be some deep mystery, because we know that Salviati never speaks at random, or puts in the field any paradox unless it eventuates in some idea not entirely trivial. So at the proper time and place I shall remind you to explain this remark about a straight line being the same as the circumference of an infinite circle; but for now, I do not wish to interrupt the debate we have in hand.

Getting back to the point, I invite Simplicio to consider how the approach and retreat which the earth makes with respect to some fixed star near the pole may be made as if by a straight line, for such is the diameter of the earth's orbit. Hence the attempt to compare the rising and falling of the polestar due to motion along such a diameter with that due to motion over the small circle of the earth strongly indicates a lack of understanding.

SIMP. But we are still in the same difficulty, since not even the small variation which ought to exist is to be found, and if the variation is null, then the annual motion attributed to the earth along its orbit must also be admitted to be null.

SAGR. Now I shall let Salviati resume, who I believe would not shrug off as nonexistent the rising or dropping of the polestar or of some other fixed star. I say this even though such events may not be known to anyone, and were assumed by Copernicus himself to be, I shall not say null, but unobservable because of their smallness.

SALV. I said earlier that I do not believe anyone has set himself the task of observing whether variations which might depend upon an annual movement of the earth are to be perceived in any fixed star at the various seasons of the year, and I added that I doubt whether anyone has very clearly understood just what variations should appear, or among what stars. Therefore it will be good for us to examine this point carefully.

I have indeed found authors writing in general terms that the annual motion of the earth should not be admitted because it is improbable that visible changes would not then be seen in the

A straight line and the circumference of an infinite circle are the same thing.

It is inquired what mutations, and in what stars, ought to be noticed because of the earth's annual motion.

fixed stars. Not having heard anyone go on to say what, in particular, these visible changes ought to be, and in what stars, I
think it quite reasonable to suppose that those who say generally
that the fixed stars remain unchanged have not understood (and
perhaps have not even tried to find out) the nature of these
alterations, or what it is that they mean ought to be seen. In
making this judgment I have been influenced by knowing that
the annual movement attributed to the earth by Copernicus, if
made perceptible in the stellar sphere, would not produce visible
alterations equally among all stars, but would necessarily make
great changes in some, less in others, still less in yet others, and
finally none in some stars, however great the size of the circle
assumed for this annual motion. The alterations which should
be seen, then, are of two sorts; one is an apparent change in size
of these stars, and the other is a variation in their altitudes at
the meridian, which implies as a consequence the varying of
places of rising and setting, of distances from the zenith, etc.

SAGR. I think that what I see coming is like a ball of string so
snarled that without God's help I may never manage to disentangle it; for to confess my deficiencies to Salviati, I have
often thought about this without ever getting hold of the loose
end of it. I say this not so much in reference to things pertaining
to the fixed stars as to an even more terrifying task that you
have brought to my mind by mentioning these meridian altitudes, latitudes of rising, distances from the zenith, etc. The
reeling of my brain has its origin in what I shall now tell you.

Copernicus assumes the stellar sphere to be motionless, with
the sun likewise motionless in the center of it. Therefore all
alterations in the sun or in the fixed stars which may appear to
us must necessarily belong to the earth; that is, be ours. But the
sun rises and sets along a very great arc on our meridian —
almost forty-seven degrees — and its deviations in rising and
setting vary by still greater arcs along the oblique horizons. Now
how can the earth be so remarkably tilted and elevated with respect to the sun, and not at all so with regard to the fixed stars —
or so little as to be imperceptible? This is the knot which has
never passed through my comb, and if you untie it for me I shall
consider you greater than an Alexander.

SALV. These difficulties do credit to Sagredo's ingenuity; the
question is one which Copernicus himself despaired of explain-

ing in such a way as to make it intelligible, as will be seen both from his own admission of its obscurity and from his setting out twice to explain it, in two different ways. And without affectation I admit not having understood his explanation myself, until I had made it intelligible in still another way which is quite plain and clear, and this only after a long and laborious application of my mind.

SIMP. Aristotle saw the same objection, and made use of it to disprove some of the ancients who would have had it that the earth was a planet. Against them he reasoned that if it were, it would be necessary for it, like the other planets, to have more than one movement, producing these variations in the risings and settings of the fixed stars as well as in their meridian altitudes. And since he raised the difficulty without solving it, it must necessarily be very difficult of solution, if not entirely impossible.

Aristotle's argument against the ancients who would have had the earth a planet.

SALV. The strength and force of the knotting make the untying the more beautiful and admirable, but this I do not promise you today; you must excuse me until tomorrow. For the present, let us go on considering and explaining these alterations and differences which ought to be perceived in the fixed stars on account of the annual movement, as we were just saying. In the explanation of this, certain points suggest themselves as preparation for the solution of the chief difficulty.

Now going back once more to the two movements attributed to the earth (I say two, because the third is not unquestionably a motion, as I shall explain in the proper place), the annual and the diurnal, the former must be understood to be made by the center of the earth in the circumference of its orbit, which is a large circle described in the plane of the ecliptic, and is fixed and immutable. The other (that is, the diurnal) is made by the earth's globe upon itself around its own center and axis, and not vertical to the plane of the ecliptic, but inclined to that with a tilt of about twenty-three and one-half degrees, which inclination is maintained throughout the year. And what must be especially noted is that it keeps this tilt always toward the same part of the sky, so that the axis of diurnal motion is maintained always parallel to itself. Hence if we imagine this axis prolonged all the way to the fixed stars, then while the earth is going around the whole ecliptic in a year this axis describes an oblique cylindrical

Annual motion is made by the earth's center along the ecliptic, and the diurnal motion by the earth about its own center.

The earth's axis keeps always parallel to itself, and describes a cylindrical surface inclined to its orbit.

The earth's globe
never tilts, but
immutably
maintains itself.

surface which has for one of its bases the said annual circle, and for the other a similar imaginary circle traced by its extremity — or let us say its pole — among the fixed stars. This cylinder is oblique to the plane of the ecliptic according to the inclination of the axis which describes it, and this we have said to be twenty-three and one-half degrees. This remains perpetually the same, except for some small variation in many thousands of years which is not significant in the present connection. Thus the terrestrial globe neither tilts further nor straightens up, but is kept immutable. From this it follows that with regard to alterations observed in the fixed stars and depending only upon the annual movement, these will occur in the same way for any point upon the earth's surface as they would for the very center of the earth. Hence in the present explanations we shall make use of the center as if it were any point upon the surface.

For a clearer understanding of the whole matter, let us draw a diagram. First we shall designate in the plane of the ecliptic the circle ANBO; let us suppose the points A and B to be the extremities toward the north and south — that is, the beginning of Cancer and of Capricorn — and extend the diameter AB without limit through D and C toward the stellar sphere.

Fixed stars
placed on the
ecliptic never go
up or down on
account of the
annual motion of
the earth, al-
though they do
approach and
retreat.

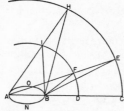

Fig. 24

Now I say, first, that none of the fixed stars in the ecliptic will ever vary in elevation no matter what motions the earth makes in the plane of the ecliptic, but will always be perceived in the same plane, though they will approach and recede from the earth by as great a space as the diameter of the earth's orbit. This is easily seen in the figure, for whether the earth is at the point A or at B, the star C is always seen along the same line ABC, although the distance BC is smaller than CA by the entire diameter BA. Therefore what might be discovered in the star C, or in any other star placed in the ecliptic, is a growth or diminution in apparent size due to the approach or retreat of the earth.

SAGR. Wait a moment, please, because I am somewhat ill at ease about this. That the star C is seen along the same line ABC when the earth is at A and at B, I understand perfectly. And I would also understand that the same would hold for all points in the line AB, if the earth passed from A to B along that line. But since it passes according to our assumptions along the arc ANB, it is obvious that when it is at the point N (or any other point outside of A and B), the star will no longer be seen along the line AB but along one of many others. Now if being seen along different lines should cause visible changes, such variations ought to be perceived.

And I shall say further, with that philosophical freedom which should be permitted among philosophical friends, that it seems to me you are contradicting yourself and denying now something that this very day you have explained, to our astonishment, as being a remarkable and perfectly true thing. I mean that which occurs among the planets, and especially the three outer ones; these, being continually in the ecliptic or very close to it, not only look close to us at one time and very distant at another, but are so variable in the rules of their movements that they appear sometimes stationary, and at other times retrograde in differing degrees — and all for no other cause than the annual movement of the earth.

SALV. Although I have made sure of Sagredo's perspicacity a thousand times, yet I wanted with this new trial to assure myself further as to how much I might expect from his ingenuity. This is for my own purposes, since if my propositions can stand fast against the hammer and furnace of his judgment, I may be certain that they are of good metal and can compare with any. Hence I say that I have deliberately pretended to overlook this objection, but not in order to deceive you or to persuade you of anything false, as might happen if an objection had been ignored by me and overlooked by you which was in fact what this one seems to be; namely, truly strong and conclusive. But it is not so; rather, I now wonder whether you are pretending not to recognize its emptiness just to test me. Well, on this particular I want to be more sly than you are, by forcibly drawing from your own mouth what you are craftily concealing within it. So tell me how it is that you are aware that the stoppings and retrograde movements of the planets are due to the annual motion, and how you know it is large enough so that at least some traces of a similar effect ought to be recognized among the stars in the ecliptic.

SAGR. This demand of yours includes two questions to which I must reply; the first concerns the imputation which you put upon me, of being a hypocrite; and the other bears on what may appear in the stars, etc. As to the first, permit me to say that it is not true that I was merely pretending not to know the invalidity of this objection. And to reassure you about this, I tell you right now that I understand its emptiness quite well.

SALV. Well, I certainly do not understand how it can be that you were not speaking hypocritically when you claimed not to understand as a fallacy that which you now admit you understand very well to be one.

SAGR. The very confession of understanding may assure you that I was not simulating when I said that I did not understand; for if I had wished to simulate and had done so, who would there be to stop me from continuing the sham by still denying that I see the fallacy? I say, then, that I did not understand it at the time, but that I see it clearly now, thanks to your having awakened my mind, first by telling me positively that a fallacy existed, and next by commencing to interrogate me in general about the means of my recognizing the stoppings and retrograde motions

Stopping, advancing, and retrogressing of the planets are known in relation to the fixed stars.

of the planets. Now, this is known by comparing the planets with the fixed stars, in relation to which they are seen to vary their movements now westward, now eastward, and sometimes to remain practically motionless. But beyond the stellar sphere there is not another sphere, immensely more remote and visible to us, with which we might compare the fixed stars. Hence not a trace could we discover in them of anything corresponding to what appears among the planets. I believe that this is what you were so anxious to draw from my mouth.

SALV. And there it is, with the addition of your most subtle insight to boot. And if I, with my little joke, opened your mind, you with yours have reminded me that it is not entirely impos-

An indication in the fixed stars, similar to what is seen in the planets, as an argument of the annual motion of the earth.

sible for something some time to become observable among the fixed stars by which it might be discovered what the annual motion does reside in. Then they, too, no less than the planets and the sun itself, would appear in court to give witness to such motion in favor of the earth. For I do not believe that the stars are spread over a spherical surface at equal distances from one center; I suppose their distances from us to vary so much that some are two or three times as remote as others. Thus if some tiny star were found by the telescope quite close to some of the larger ones, and if that one were therefore very very remote, it might happen that some sensible alterations would take place among them corresponding to those of the outer planets.

So much for the moment with regard to the special case of stars placed in the ecliptic. Let us now go to the fixed stars outside the ecliptic, and assume a great circle vertical to its plane, for example a circle that would correspond in the stellar sphere to the solstitial colure. This we shall mark CEH, and it will be a meridian at the same time. Let us take in it a star outside the ecliptic, which can be E here. Now this will indeed vary its elevation with the movement of the earth, because from the earth at A it will be seen along the ray AE, with the elevation of the angle EAC, but from the earth at B it will be seen along the ray BE, with an angle of elevation EBC. This is greater than EAC, on account of its being an exterior angle of the triangle EAB, while the other is the opposite interior angle. Hence the distance of the star E from the ecliptic would be seen to be changed, and also its meridian altitude would be greater in position B than in the place A, in proportion as the angle EBC exceeds EAC; that is, by the angle AEB. For the side AB of the triangle EAB being produced to C, the exterior angle EBC (being equal to the two opposite interior angles E and A) exceeds A by the size of the angle E. And if we take another star in the same meridian farther from the ecliptic — let this be the star H — then this will be even greater in variation when seen from the two positions A and B, according as the angle AHB becomes greater than the angle E. This angle will continue to increase in proportion as the star observed gets farther from the ecliptic, until finally the maximum alteration will appear in that star which is placed at the very pole of the ecliptic. For a complete understanding, this may be demonstrated as follows:

Let the diameter of the earth's orbit be AB, whose center is G, and assume it to be extended out to the stellar sphere in the points D and C. From the center G, let the axis GF of the ecliptic be erected as far as the same sphere, in which a meridian DFC vertical to the plane of the ecliptic is assumed to be described. Taking, in the arc FC, any points H and E as places of fixed stars, add the lines FA, FB, AH, HG, HB, AE, GE, and BE. Then AFB is the angle of difference (or we may say the parallax)

Fixed stars outside the ecliptic go up and down more and less, according to their distances from the ecliptic.

of the star placed at the pole F; that of the star at H is the angle AHB, and for the star at E it is the angle AEB. I say that the angle of difference of the polestar F is the maximum; of the others, those closest to this maximum are larger than those more distant from it. That is, the angle F is greater than the angle H, and this is greater than the angle E.

Suppose a circle described about the triangle FAB. Since the angle F is acute, its base AB being less than the diameter DC of the semicircle DFC, it will fall in

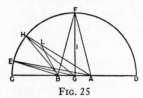

Fig. 25

the larger portion of the circumscribed circle cut by the base AB. And since AB is divided in the center and at right angles to FG, the center of the circumscribed circle will be in the line FG. Let this be the point I. Now of all the lines drawn to the circumference of the circumscribed circle from the point G, which is not its center, the greatest is that which passes through the center. Hence FG will be greater than any other line drawn through G to the circumference of the same circle, and therefore this circumference will cut the line GH, which is equal to the line GF, and cutting GH it will also cut AH. Let it cut that in L, and add the line LB. Then the two angles AFB and ALB will be equal, being included in the same portion of the circumscribed circle. But ALB, an exterior angle, is greater than the interior angle H; therefore angle F is greater than angle H.

By the same method we may show that the angle H is greater than the angle E, because the center of the circle described about the triangle AHB is on the perpendicular GF, to which the line GH is closer than the line GE; hence its circumference cuts GE and also AE, from which the proposition is obvious.

From this we conclude that the alteration of appearance (which, using the proper technical term, we may call the parallax of the fixed stars) is greater or less according as the stars observed are more or less close to the pole of the ecliptic, and that finally for stars on the ecliptic itself the alteration is reduced to nothing. Next, as to the earth approaching and retreating from the stars by its motion, those stars which are on the ecliptic are made nearer or farther by the entire diameter of the earth's

The earth approaches and retreats from the fixed stars on the ecliptic by as much as the diameter of its orbit.

orbit, as we have already seen. For those which lie near the pole of the ecliptic, this approach and retreat is almost nothing, while for others the alteration is made greater as the stars become closer to the ecliptic.

In the third place we may see that this alteration of appearance is greater or less according as the observed star is closer to or more remote from us. For if we draw another meridian less distant from the earth (which shall be DFI here), a star placed at F and seen along the same ray AFE with the earth at A, when it is later observed from the earth at B will be seen along the ray BF, and will make the angle of difference BFA greater than the first one, AEB, being exterior to the triangle BFE.

Greater variation produced in closer stars than those more remote.

SAGR. I have listened to your discourse with great pleasure, and with profit too; now, to make sure that I have understood everything, I shall state briefly the heart of your conclusions. It seems to me that you have explained to us two sorts of differing appearances as being those which because of the annual motion of the earth we might observe in the fixed stars. One is their variation in apparent size as we, carried by the earth, approach them or recede from them; the other (which likewise depends upon this same approach and retreat) is their appearing to us to be now more elevated and now less so on the same meridian. Besides this you tell us (and I thoroughly understand) that these two alterations do not occur equally in all stars, but to a greater extent in some, to a lesser in others, and not at all in still others. The approach and retreat by which the same star ought to appear larger at one time and smaller at another is imperceptible and practically nonexistent for stars which are close to the pole of the ecliptic, but it is great for the stars placed in the ecliptic itself, being intermediate for those in between. The reverse is true of the other alteration; that is, the elevation or lowering is nil for stars along the ecliptic and large for those encircling the pole of the ecliptic, being intermediate for those in the middle.

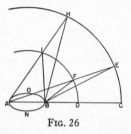

FIG. 26

Summary of the appearances of the fixed stars caused by the earth's annual motion.

Furthermore, both these alterations are more perceptible in the closest stars, less sensible in those more distant, and would ultimately vanish for those extremely remote.

So much for my part. The next thing, so far as I can see, is to convince Simplicio. I think he will not easily be reconciled to admitting such alterations as these to be imperceptible, stemming as they do from such a vast movement of the earth and from a change that carries the earth to places twice as far apart as our distance from the sun.

SIMP. Really, to be quite frank, I do feel a great repugnance against having to concede the distance of the fixed stars to be so great that the alterations just explained would have to remain entirely imperceptible in them.

SALV. Do not completely despair, Simplicio; perhaps there is yet some way of tempering your difficulties. First of all, that the apparent size of the stars is not seen to alter visibly need not appear entirely improbable to you when you see that men's estimates in such a matter may be so grossly in error, particularly when looking at brilliant objects. Looking, for example, at a burning torch from a distance of two hundred paces, and then coming closer by three or four yards, do you believe that you yourself would perceive it as larger? For my part, I should certainly not discover this even if I approached by twenty or thirty paces; sometimes I have even happened to see such a light at a distance, and been unable to decide whether it was coming toward me or going away, when in fact it was approaching. Now what of this? If the same approach and retreat of Saturn (I mean double the distance from the sun to us) is almost entirely imperceptible, and if it is scarcely noticeable in Jupiter, what could it amount to in the fixed stars, which I believe you would not hesitate to place twice as far away as Saturn? In Mars, which while approaching us . . .

> In very distant and very bright objects, a small approach or retreat is imperceptible.

SIMP. Please do not labor this point, for I am indeed convinced that what you have said about the unaltered appearance of the apparent sizes of the fixed stars may very well be the case. But what shall we say to that other difficulty which arises from no variation at all being seen in their changing aspects?

SALV. Let us say something which will perhaps satisfy you also on this point. Briefly, would you be content if those alterations really were perceived in the stars which seem to you so necessary if the annual motion belongs to the earth?

SIMP. I should indeed be, so far as this particular is concerned.

SALV. I wish you had said that if such a variation were perceived, nothing would remain that could cast doubt upon the earth's mobility, since no counter could be found to such an event. But even though this may not make itself visible to us, the earth's mobility is not thereby excluded, nor its immobility necessarily proved. It is possible, Copernicus declares, that the immense distance of the starry sphere makes such small phenomena unobservable. And as has already been remarked, it may be that up to the present they have not even been looked for, or, if looked for, not sought out in such a way as they need to be; that is, with all necessary precision and minute accuracy. It is hard to achieve this precision, both on account of the imperfection of astronomical instruments, which are subject to much variation, and because of the shortcomings of those who handle them with less care than is required. A cogent reason for putting little faith in such observations is the disagreement we find among astronomers in assigning the places, I shall say not merely of novas and of comets, but of the fixed stars themselves, and even of polar altitudes, about which they disagree most of the time by many minutes.

As a matter of fact, how would you expect anyone to be sure, with a quadrant or sextant that customarily has an arm three or four yards long, that he is not out by two or three minutes in the setting of the perpendicular or the alignment of the alidade? For on such a circumference this will be no more than the thickness of a millet seed. Besides which, it is almost impossible for the instrument to be constructed absolutely accurate and then maintained so. Ptolemy distrusted an armillary instrument constructed by Archimedes himself for determining the entry of the sun into the equinox.

SIMP. But if the instruments are thus suspect, and the observations are so dubious, how can we ever safely accept them and free them from error? I have heard great vauntings of Tycho's instruments, which were made at enormous expense, and of his remarkable skill in making observations.

SALV. I grant you all this, but neither the one fact nor the other suffices to make us certain in affairs of such importance. I want to have us use instruments far larger than those of Tycho's; quite precise ones, and made at minimum cost, whose sides will be four, six, twenty, thirty, or fifty miles, so that a degree is a mile wide, a minute is fifty yards, and a second is little less than

If some annual mutation were perceived in the fixed stars, the motion of the earth would brook no contradiction.

Proving how little faith in astronomical instruments is justified for minute observations.

Ptolemy did not trust an instrument made by Archimedes.

Tycho's instruments made at great expense.

What kinds of instruments would be suitable for the most precise observations.

Exact observa-
tions of the
arrival and de-
parture of the
sun from the
summer solstice.

a yard. In a word, we may have them as large as we please, without their costing us a thing.

Being at a villa of mine near Florence, I plainly observed the
arrival of the sun at the summer solstice and its subsequent departure. For one evening at its setting it hid itself behind a cliff
in the Pietrapana Mountains, about sixty miles away, leaving
only a small shred of itself revealed to the north, the breadth of
which was not the hundredth part of its diameter. But the following evening, at the same position of setting, it left a like part
of itself showing which was noticeably thinner. This is a conclusive proof that it had commenced to move away from the
tropic; yet the sun's return between the first and second observations surely did not amount to one second of arc along the
horizon. Making the observation later with a fine telescope
which would multiply the disc of the sun more than a thousand-
fold turned out to be pleasant and easy.

Now my idea is for us to make our observations of the fixed
stars with similar instruments, utilizing some star in which the
changes would be conspicuous. These are, as I have already explained, the ones which are farthest from the ecliptic. Among
them Vega,[†] a very large star close to the pole of the ecliptic,
would be the most convenient when operating in the manner I
am about to describe to you, so far as the more northern countries are concerned, though I am going to make use of another
star. I have already been looking by myself for a place well

A suitable place
for the observa-
tion of fixed stars
as related to the
annual motion
of the earth.

adapted for such observations. The place is an open plain, above
which there rises to the north a very prominent mountain, at the
summit of which is built a little chapel facing west and east, so
that the ridgepole of its roof may cut at right angles the meridian
over some house situated in the plain. I wish to affix a beam
parallel to that ridgepole and about a yard above it. This done,
I shall seek in the plain that place from which one of the stars
of the Big Dipper is hidden by this beam which I have placed,
just when the star crosses the meridian. Or else, if the beam is
not large enough to hide the star, I shall find the place from
which the disc of the star is seen to be cut in half by the beam —
an effect which can be discerned perfectly by means of a fine
telescope. It will be very convenient if there happens to be some
house at the place from which this event can be perceived, but
if not, then I shall drive a stick firmly into the ground and affix

a mark to indicate where the eye is to be placed whenever the observation is to be repeated. I shall make the first of these observations at the summer solstice, in order to continue them from month to month, or whenever I please, until the other solstice.

By means of such observations, the star's rising or lowering can be perceived no matter how small it may be. And if in the course of these operations any such variation shall happen to become known, how great an achievement will be made in astronomy! For by this means, besides ascertaining the annual motion, we shall be able to gain a knowledge of the size and distance of that same star.

SAGR. I thoroughly understand the whole procedure, and the operations seem to me to be so easy and so well adapted to what is wanted, that it may very reasonably be believed that Copernicus himself, or some other astronomer, has actually performed them.

SALV. It seems the other way around to me, for it is improbable that if anyone had tried this he would not have mentioned the result, whichever opinion it turned out to favor. But no one is known to have availed himself of this method, for the above or for any other purpose; and without a fine telescope it could not very well be put into effect.

SAGR. What you say completely satisfies me.

Now, since quite a while remains until the night, if you want me to find any rest then, I hope it will not be too much trouble for you to explain to us those problems which a little while ago you asked us to put off until tomorrow. Please give us back the reprieve which we extended to you, and abandoning all other arguments explain to us how (assuming the motions which Copernicus attributes to the earth, and keeping immovable the sun and the fixed stars) such events may follow as pertain to the elevation and lowering of the sun, the changing of the seasons, and the inequalities of nights and days, in just the way that is so easily understood to take place in the Ptolemaic system.

SALV. I must not and cannot refuse anything which Sagredo pleads for. The delay that I requested was only to give me time to rearrange in my mind the premises which are useful for a clear and comprehensive explanation of the manner in which these events take place in the Copernican as well as in the Ptolemaic system. Indeed, more easily and simply in the former than

Copernican system difficult to understand and easy in its operation.

in the latter, so that it may be clearly seen that the former hypothesis is as easy for nature to put into effect as it is hard for the intellect to comprehend. Nevertheless I hope, by utilizing explanations other than those resorted to by Copernicus, to make even the learning of it very much less obscure. In order to do this, I shall set forth some assumptions as known and self-evident, as follows:

First. I assume that the earth is a spherical body which rotates about its own axis and poles, and that every point on its surface traces out the circumference of a circle, greater or lesser according as the designated point is more or less distant from the poles. Of these circles, that one is greatest which is traced out by a point equidistant from the poles. All these circles are parallel to one another, and we shall refer to them as *parallels*.

Second. The earth being spherical in shape and its material being opaque, half its surface is continually lighted and the rest is dark. The boundary which separates the lighted part from the dark being a great circle, we shall call this the *boundary circle of light*.

Third. When the boundary circle of light passes through the earth's poles it will cut all the parallels into equal sections, it being a great circle; but, not passing through the poles, it will cut them all into unequal parts except the central circle; this, being also a great circle, will be cut into equal parts in any case.

Fourth. Since the earth turns about its own poles, the length of day and night is determined by the arcs of the parallels cut by the boundary circle of light. The arc which remains in the illuminated hemisphere determines the length of the day, and the remainder that of the night.

These things being set forth, we may wish to draw a diagram for a clearer understanding of what comes next. First let us indicate the circumference of a circle, to represent for us the orbit of the earth, described in the plane of the ecliptic. This we may divide by two diameters into four equal parts; Capricorn, Cancer, Libra, and Aries, which shall here represent at the same time the four cardinal points; that is, the two solstices and the two equinoxes. And in the center of this circle, let us denote the sun, O, fixed and immovable.

Now with the four points Capricorn, Cancer, Libra, and Aries as centers, we shall draw four equal circles which to us will rep-

Propositions
necessary for
understanding
well the conse-
quences of the
earth's move-
ments.

A simple sketch
to represent the
Copernican ar-
rangement and
its consequences.

resent the earth at these four different seasons. The center of
the earth travels in the space of a year around the whole circumference Capricorn–Aries–Cancer–Libra, moving from west
to east in the order of the signs of the zodiac. It is already
evident that when the earth is in Capricorn the sun will appear

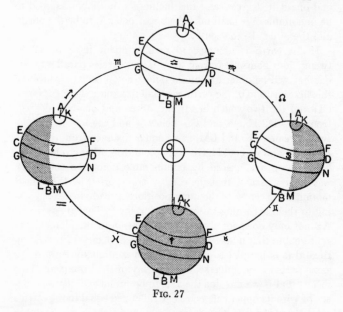

FIG. 27

in Cancer; the earth moving along the arc from Capricorn to
Aries, the sun will appear to be moving along the arc from
Cancer to Libra. In a word, it will run through the signs of the
zodiac in their order during the space of a year. So with this
first assumption, the apparent annual motion of the sun around
the ecliptic is satisfied beyond any argument.

Coming now to the other movement — that is, the diurnal
motion of the earth about itself — its poles and axis must be
established. These must be understood to be not perpendicularly
erect to the plane of the ecliptic; that is, not parallel to the axis
of the earth's orbit, but inclined from right angles about twenty-
three and one-half degrees, with the North Pole toward the axis
of the earth's orbit when the center of the earth is at the solstitial

Annual motion
of the sun as it
follows from
Copernicus's
method.

point in Capricorn. Assuming, then, that the center of the terrestrial globe is at that point, let us indicate the poles and the
axis AB, tilted twenty-three and one-half degrees from the per-
pendicular† on the Capricorn–Cancer diameter, so that the angle
A–Capricorn–Cancer amounts to the complement, or sixty-six
and one-half degrees, and this inclination must be assumed to
be immutable. We shall take the upper pole, A, to be the north,
and the other, B, the south.

If the earth is assumed to revolve about its axis AB in
twenty-four hours, also from west to east, circles parallel to one
another will be described by all points noted on its surface. In
this first position of the earth, we shall designate the great circle
CD and the two which are twenty-three and one-half degrees
from it — EF above, and GN below — and these others at the
two extremes, IK and LM, at a similar distance from the poles
A and B; and we could have drawn countless other circles parallel to these five, traced by innumerable points on the earth.
Let us now assume that the earth is transported by the annual
motion of its center to the other positions already marked, passing to them according to the following laws: That its own axis
AB not only does not change its inclination to the plane of the
ecliptic, but that it does not vary its direction, either; remaining
thus always parallel to itself, it points continually toward the
same parts of the universe, or let us say of the firmament. This
means that if we imagine the axis to be prolonged, it would describe with its upper end a circle parallel and equal to the earth's
orbit through Libra, Capricorn, Aries, and Cancer, as the upper
base of a cylinder described by itself in its annual motion upon
the lower base, Libra–Capricorn–Aries–Cancer. Hence, because
of this unchanging tilt, let us draw these other three figures
around the centers of Aries, Cancer, and Libra, exactly similar
to the one drawn around the center of Capricorn.

Next let us consider the first diagram of the earth. Because
of the axis AB being inclined at twenty-three and one-half degrees toward the sun, and since the arc AI is also twenty-three
and one-half degrees, the light of the sun illumines the hemisphere of the terrestrial globe exposed to the sun (of which only
half is seen here), divided from the dark part by the boundary
of light, IM. The parallel CD, being a great circle, will be divided
into equal parts by this, but all others will be cut into unequal

parts because the boundary of light IM does not pass through the poles A and B. The parallel IK, together with all others described between it and the pole A, will be entirely within the illuminated part, just as on the other hand the opposite ones toward the pole B and contained within the parallel LM will remain in the dark.

Besides this, since the arc AI is equal to the arc FD, and the arc AF is common to IKF and AFD, the latter two are equal, each being one quadrant; and since the whole arc IFM is a semicircle, the arc MF will also be a quadrant and equal to FKI. Hence the sun, O, in this position of the earth, will be vertical to anyone at the point F. But through the diurnal revolution around the fixed axis AB, all points on the parallel EF pass through this same point F, and therefore on such a day the sun at midday will be overhead to all inhabitants of the parallel EF; and to them it will seem to describe by its motion that circle which we call the tropic of Cancer.

But to the inhabitants of all parallels above the parallel EF toward the North Pole, A, the sun is below their zenith toward the south. On the other hand, to all inhabitants of the parallels below EF toward the equator CD and the South Pole B, the midday sun is elevated above the zenith toward the North Pole, A.

Next you may see how of all parallels, only the great circle CD is cut into equal parts by the boundary of light IM, the others above and below this all being cut into unequal parts. Of the upper ones, the semidiurnal arcs (which are those in the part of the earth lighted by the sun) are greater than the seminocturnal ones, which remain in the dark. The contrary happens for the remainder which are beneath the great circle CD toward the pole B; of these, the semidiurnal arcs are smaller than the seminocturnal. Also you may see quite plainly that the differences of these arcs go on increasing as the parallels become closer to the poles, until the parallel IK stays entirely in the lighted part, and its inhabitants have a twenty-four-hour day without night. In contrast to this the parallel LM, remaining all in the dark, has a night of twenty-four hours without day.

Next let us proceed to the third diagram of the earth, here placed with its center at the Cancer point, from which the sun would appear to be at the first point of Capricorn. It is indeed easy to see that as the axis AB has not changed its tilt, but has remained parallel to itself, the appearance and situation of the earth are precisely the same as in the first diagram, except that the hemisphere which in the first was lighted by the sun remains in shadow here, and the one which was previously dark now becomes illuminated. Hence what occurred in the first diagram is now reversed with respect to the differences of days and nights and their relative length or shortness.

The first thing noticed is that where in the first figure, the circle IK was entirely in the light, it is now all in the dark; and LM, which is opposite, is now entirely in the light, where it was previously completely in shadow. Of the parallels between the great circle CD and the pole A, the semidiurnal arcs are now smaller than the seminocturnal, which is the opposite of the first; and of the others toward the pole B, the semidiurnal arcs are now longer than the seminocturnal, likewise the opposite of what took place in the other position of the earth. You may now see the sun made vertical to the inhabitants of the tropic GN, and for those of the parallel EF it is depressed southward through the entire arc ECG; that is, forty-seven degrees. It has, in short, gone from one tropic to the other, passing through the equator, being raised and then dropped along the meridian through the said interval of forty-seven degrees. This entire change has its origin not in any dropping or rising of the earth; on the contrary, in its never dropping nor rising, but in generally keeping itself always in the same location with respect to the universe and merely going around the sun, which is situated at the center of this same plane in which the earth moves around it in the annual motion.

Marvelous phe-
nomenon de-
pending upon
the axis of the
earth not vary-
ing in tilt.

Here a remarkable phenomenon must be noticed, which is that just as the preservation of the axis of the earth in the same direction with respect to the universe (or let us say toward the highest fixed stars) makes the sun appear to us to rise and fall by as much as forty-seven degrees without any rise or drop in the fixed stars at all, so if on the contrary the earth's axis were continually kept at a given inclination toward the sun (or we might say toward the axis of the zodiac), no alteration of ascent or descent would appear to be made by the sun. Thus the inhabitants of a given place would always have the same periods of night and day, and the same kind of season; that is, some people would always have winter, some always summer, some spring, etc. But on the other hand, the changes in the fixed stars with regard to rising and falling would then appear enormous to us, amounting to this same forty-seven degrees. For an understanding of this let us go back to a consideration of the position of the earth in the first diagram, where the axis AB is seen with its upper pole A tilted toward the sun. In the third figure the same axis has kept the same direction toward the highest sphere by remaining parallel to itself, so the upper pole A no longer tilts toward the sun but tilts away from it, and lies forty-seven degrees from its first position. Thus, in order to reproduce the same inclination of the pole A toward the sun, it would be required (by turning the globe along its circumference ACBD) to take it forty-seven degrees toward E; and any fixed star observed on the meridian would be raised or lowered by that many degrees.

Now let us proceed with an explanation of the rest, and consider the earth placed in the fourth diagram with its center at the first point of Libra, the sun appearing in the beginning of Aries. Thus the earth's axis, which in the first diagram was assumed to be inclined to the Capricorn–Cancer diameter and hence to be in the same plane as that which cuts the earth's orbit perpendicularly in the Capricorn–Cancer line, when transferred to the fourth figure (being kept always parallel to itself, as we have said), comes to be in a plane which is likewise vertical to the plane of the earth's orbit, and parallel to the one which cuts the latter at right angles along the Capricorn–Cancer diameter. Hence the line from the center of the sun to the center of the earth (from O to Libra) will be perpendicular to the axis BA. But this same line from the center of the sun to the center of the earth is always perpendicular also to the boundary circle of light; therefore this same circle will pass through the poles A and B in the fourth figure, and the axis AB will lie in its plane. But the great circle, passing through the poles of the parallels, will divide them all into equal parts; therefore the arcs IK, EF, CD, GN, and LM will all be semicircles, and the lighted hemisphere will be this one which faces us and the sun, and the boundary circle of light will be this very circumference ACBD. And when the earth is at this place, the equinox will occur for all its inhabitants.

The same will happen in the second diagram, where the earth having its lighted hemisphere toward the sun shows to us its dark side with the nocturnal arcs. These are also all semicircles, and consequently also make an equinox. Finally, since the line produced from the center of the sun to the center of the earth is perpendicular to the axis AB, to which likewise the great circle CD among the parallels is perpendicular, the same line O–Libra necessarily passes through the same plane as the parallel CD, cutting its circumference in the center of the daytime arc CD; therefore the sun will be vertical to anyone located in that cut. But all inhabitants of that parallel pass by there, carried by the earth's rotation, and have the midday sun directly overhead; therefore the sun will appear to all inhabitants of the earth to be tracing out the greatest parallel, called the equatorial circle.

Moreover, the earth being at either of the solstitial points, one of the polar circles IK or LM is entirely in the light and the other in the shadow; but when the earth is at the equinoctial points, half of each of these polar circles is in the light and the balance in the dark. It should not be hard to see how the earth in passing, for example, from Cancer (where the parallel IK is entirely dark) to Leo, a part of the parallel IK toward the point I will commence to enter the light, and the boundary of light IM will begin to retreat toward the poles A and B, cutting the circle ACBD no longer at I and M, but in two other points falling between the endpoints I, A, M, and B, of the arcs IA and MB. Thus the inhabitants of the circle IK begin to enjoy the light, and those of the circle LM to experience the darkness.

See, then, how two simple noncontradictory motions assigned to the earth, performed in periods well suited to their sizes, and also conducted from west to east as in the case of all movable world bodies, supply adequate causes for all the visible phenomena. These phenomena can be reconciled with a fixed earth only by renouncing all the symmetry that is seen among the speeds and sizes of moving bodies, and attributing an inconceivable velocity to an enormous sphere beyond all the others, while lesser spheres move very slowly. Besides, one must make the motion of the former contrary to that of the latter, and to increase the improbability, must have the highest sphere transport all the lower ones opposite to their own inclination. I leave it to your judgment which has the more likelihood in it.

SAGR. For my part, so far as my senses are concerned, there is a great difference between the simplicity and ease of effecting results by the means given in this new arrangement and the multiplicity, confusion, and difficulty found in the ancient and generally accepted one. For if the universe were ordered according to such a multiplicity, one would have to remove from philosophy many axioms commonly adopted by all philosophers. Thus it is said that Nature does not multiply things unnecessarily; that she makes use of the easiest and simplest means for producing her effects; that she does nothing in vain, and the like.

I must confess that I have not heard anything more admirable than this, nor can I believe that the human mind has ever penetrated into subtler speculations. I do not know how it looks to Simplicio.

SIMP. If I must tell you frankly how it looks to me, these appear to me to be some of those geometrical subtleties which Aristotle reprehended in Plato when he accused him of departing from sound philosophy by too much study of geometry. I have known some very great Peripatetic philosophers, and heard them advise their pupils against the study of mathematics as something which makes the intellect sophistical and inept for true philosophizing; a doctrine diametrically opposed to that of Plato, who would admit no one into philosophy who had not first mastered geometry.

SALV. I endorse the policy of these Peripatetics of yours in dissuading their disciples from the study of geometry, since there is no art better suited for the disclosure of their fallacies. You see how different they are from the mathematical philosophers, who much prefer dealing with those who are well informed about the general Peripatetic philosophy than with those who lack such information and because of that deficiency are unable to make comparisons between one doctrine and the other.

But setting all this aside, please tell me what absurdities or excessive subtleties make this Copernican arrangement the less plausible so far as you are concerned.

SIMP. As a matter of fact, I did not completely understand it, perhaps because I am not very well versed either in the way the same effects are produced by Ptolemy — I mean these planetary stoppings, retrograde movements, approaches and retreats, lengthenings and shortenings of the day, alterations of the seasons, etc. But passing over the consequences which stem from the basic assumptions, I feel no small difficulties to exist in these assumptions themselves, and if the assumptions fall to the ground then they bring the whole structure into ruin. Now since the whole framework of Copernicus seems to me to be built upon a weak foundation (being supported upon the mobility of the earth), then if this were removed, there would be no room for further argument. And to remove it, Aristotle's axiom that to a simple body only one simple motion can be natural appears to be sufficient. Here three movements, if not four, are assigned to the earth, a simple body; and all of them are quite different from one another. For besides the straight motion toward the center, which cannot be denied to it as a heavy body, there are ascribed to it a circular motion in a great circle around the sun in one year, and a whirling upon itself every twenty-four hours, and (what is most extreme, and possibly for that reason you have remained silent about this) another whirling about its own center, completed in a year, and opposite to the previously mentioned twenty-four-hour motion. My mind feels a great repugnance to this.

SALV. As to the motion downward, that has already been proved not to belong to the terrestrial globe at all, which never has moved with any such movement and never will. This belongs to its parts, and to them only in order to rejoin them with their whole.

Then as to the annual and diurnal movements, these, being made in the same direction, are quite compatible, in the same way that if we were to let a ball run down a steep surface, it would, in descending spontaneously along that, turn upon itself.

Concerning this third motion about itself in one year, attributed to the earth by Copernicus merely to keep its axis tilted and pointed toward the same part of the firmament, I am going to tell you something which deserves your most careful consideration. Far from there being any repugnance or difficulty in it (though it is opposite to the other annual motion), it is naturally suited to any suspended and balanced body you please, and without requiring any cause of motion. Such a body, if carried around along the circumference of a circle, immediately acquires by itself a rotation about its own center opposite to that which carries it around; and the speed of this is such that both motions will finish one revolution in precisely the same time. You may see this wonderful effect, which suits our present purposes so well, by putting into a basin of water a floating ball and holding the bowl in your hand. If you turn around on your toe, the ball will promptly commence to revolve upon itself with a motion opposite to that of the bowl, and will complete its rotation when that of the bowl is completed.

Now what else is the earth but a globe, suspended and balanced in thin and yielding air, which, carried around the circumference of a great circle in one year, must indeed acquire — with no other mover — an annual spin around its own center opposite to that annual motion? You will see the effect, but if you proceed to reflect correctly about it you will discover that it is not a real thing, but a mere appearance, and that what looks to you like a revolving about itself is a motionlessness and a conservation of the whole unchanged with respect to everything which remains stationary outside of yourself and the bowl. For if you make some mark upon the ball and consider in what direction this points (toward what part of the wall of the room you are in, or the field, or the sky), you will see that the mark always points the same way during the revolution of the bowl and yourself. But comparing it with the bowl and with yourself (these being moving), it will indeed appear to keep on changing direction and to box the compass in its rotation, with a motion contrary to that of the bowl and yourself. Thus it may be more correctly said that you and the bowl are rotating around the motionless ball than that the latter is turning around in the bowl. In such a manner is the earth suspended and balanced in the circumference of its orbit, and so located that one of its markings (which could be, for example, the North Pole) points toward such-and-such a star, or other part of the firmament, and is kept always directed toward this, despite its being carried around the circumference of its orbit in the annual motion.

This alone is enough to put an end to your surprise and to remove every difficulty. But what will Simplicio say if, to this independence of any coöperating cause, we add a remarkable force inhering in the terrestrial globe and making it point with definite parts of itself toward definite parts of the firmament? I am speaking of magnetic force, in which every piece of lodestone constantly participates. And if every tiny particle of such stone has in it such a force, who can doubt that the same force resides to a still higher degree within the whole of this terrene globe, which abounds in this material? Or that perhaps the globe itself

Experiment which shows sensibly that two contrary motions agree naturally in the same movable body.

Third motion attributed to the earth is rather a kind of steadiness.

Wonderful power intrinsic to the terrestrial globe, of pointing always to the same part of the sky.

Terrestrial globe made of lodestone.

Magnetic phi-
losophy of
William Gilbert.

Cowardice of the
popular mind.

is, as to its internal and primary substance, nothing but an immense mass of lodestone?

SIMP. Then you are one of those people who adhere to the magnetic philosophy of William Gilbert?

SALV. Certainly I am, and I believe that I have for company
every man who has attentively read his book and carried out his
experiments. Nor am I without hope that what has happened to
me in this regard may happen to you also, whenever a curiosity
similar to mine, and a realization that numberless things in nature remain unknown to the human intellect, frees you from
slavery to one particular writer or another on the subject of
natural phenomena, thereby slackening the reins on your reasoning and softening your stubborn defiance of your senses, so that
some day you will not deny them by giving ear to voices which
are heard no more.

Now, the cowardice (if we may be permitted to use this term)
of ordinary minds has gone to such lengths that not only do they
blindly make a gift — nay, a tribute — of their own assent to
everything they find written by those authors who were lauded
by their teachers in the first infancy of their studies, but they
refuse even to listen to, let alone examine, any new proposition
or problem, even when it not only has not been refuted by their
authorities, but not so much as examined or considered. One of
these problems is the investigation of what is the true, proper,
basic, internal, and general matter and substance of this terrestrial globe of ours. Even though neither Aristotle nor anybody
else before Gilbert ever took it into his head to consider whether
this substance might be lodestone (let alone Aristotle or anybody
else having disproved such an opinion), I have met many who
have started back at the first hint of this like a horse at his
shadow, and avoided discussing such an idea, making it out to
be a vain hallucination, or rather a mighty madness. And perhaps Gilbert's book would never have come into my hands if a
famous Peripatetic philosopher had not made me a present of it,
I think in order to protect his library from its contagion.

SIMP. I frankly confess myself to have been one of these ordinary
minds, and it is only since I have been allowed during the past
few days to take part in these conferences of yours that I am
aware of having wandered somewhat from the trite and popular
path. But I do not yet feel so much awakened that the roughness

of this new and curious opinion does not make it seem to me very laborious and difficult to master.

SALV. If what Gilbert writes is true, it is not an opinion; it is a scientific subject; it is not a new thing, but as ancient as the earth itself; and if true, it cannot be rough or difficult, but must be smooth and very easy. If you like, I can make it evident to you that you are creating the darkness for yourself, and feeling a horror of things which are not in themselves dreadful — like a little boy who is afraid of bugaboos without knowing anything about them except their name, since nothing else exists beyond the name.

SIMP. I should enjoy being enlightened and removed from error.

SALV. Then answer the questions I am about to ask you. First, tell me whether you believe that this globe of ours, which we inhabit and call "earth," consists of a single and simple material, or an aggregate of different materials.

SIMP. I can see that it is composed of very diverse substances and bodies. In the first place, I see water and earth as its major components, which are quite different from each other.

Terrestrial globe composed of divers materials

SALV. For the present let us leave out the oceans and other waters, and consider just the solid parts. Tell me whether these seem to you to be all one thing, or various things.

SIMP. As to appearances, I see them various, finding great fields of sterile sand, and others of fertile and fruitful soil; innumerable barren and rugged mountains are to be seen, full of hard rocks and stones of the most various kinds, such as porphyry, alabaster, jasper, and countless sorts of marble; there are vast mines of many species of metal, and, in a word, such a diversity of materials that a whole day would not suffice to enumerate these alone.

SALV. Now of all these different materials, do you believe that in the composition of this great mass they occur in equal proportions? Or rather that among them all there is one part which far exceeds the others and is in effect the principal matter and substance of this huge bulk?

SIMP. I believe that the stones, the marbles, the metals, the gems, and other materials so diverse are exactly like jewels and ornaments, external and superficial to the original globe, which I think immeasurably exceeds in bulk all these other things.

SALV. Now this vast principal bulk, of which the things you have named resemble excrescences and ornaments: Of what do you believe this to be made?

SIMP. I think it is the simple, less impure, element of earth.

SALV. But what is it that you understand by "earth"? Is it perhaps that which is spread over fields, which is broken with spades and plows, in which grain and fruit are sown and great forests spring up spontaneously? Which, in a word, is the habitat of all animals and the womb of all vegetation?

SIMP. This, I should say, is the primary substance of our globe.

SALV. Well, that does not seem to me to be a very good thing to say. For this earth that is broken, sown, planted, and that bears fruit is one part of the surface of the globe, and quite a shallow part. It does not go very deep in relation to the distance to the center, and experience shows that by digging not far down materials are to be found very different from the external crust; harder, and not any good for producing vegetation. Besides, the more central parts may be supposed, from being compressed by the very heavy weights which rest upon them, to be compacted together and to be as hard as the most solid rock. Add to this that it would be vain to endow with fertility material never destined to produce crops, but merely to remain buried forever in the deep dark abysses of the earth.

SIMP. Who is to say that the interior parts, close to the center, are sterile? Perhaps they also have their produce of things unknown to us.

SALV. Why, you, of all people, since you understand so well that all the integral parts of the universe are produced for man's benefit alone — you ought to be most certain that this above all should be destined for the sole convenience of us inhabitants of it. And what good could we get out of materials so hidden from us and so remote that we can never make them available? The interior substance of this globe of ours, then, cannot be material which can be broken or dissipated, or is loose like this topsoil which we call "earth," but must be a very dense and solid body; in a word, very hard rock. And if it must be such, what reason have you for being more reluctant to believe that it is lodestone than that it is porphyry, jasper, or some other hard stone? If Gilbert had written that the inside of this globe is made of sandstone, or chalcedony, perhaps the paradox would seem less strange to you?

Internal parts of the terrestrial globe must be most solid.

SIMP. I grant that the most central parts of this globe are much compressed, and therefore compacted together and solid, more and more so as they go deeper; Aristotle also concedes this. But I am not aware of any reasons which oblige me to believe that they degenerate and become other than earth of the same sort as this on the surface.

SALV. I did not interject this argument for the purpose of proving conclusively to you that the primary and real substance of this globe of ours is lodestone, but merely to show you that there is no reason for people to be more reluctant to grant that it is lodestone than any other material. And if you think it over, you will find that it is not improbable that merely a single and arbitrary name motivated men to believe that this substance is earth, from the name "earth" being commonly used to signify that material which we plow and sow, as well as to name this globe of ours. But if the name for the latter had been taken from stone (as it might just as well have been as from earth) then saying that its primary substance was stone would surely not have met resistance or contradiction from anybody. Indeed, this is much more probable; I think it certain that if one could husk this great globe, taking off only a bulk of one or two thousand yards, and then separate the stones from the earth, the pile of rocks would be much, much larger than that of fertile earth.

Our globe would be called "stone" instead of "earth," if that name had been given to it from the beginning.

Now I have not adduced for you any of the reasons which conclusively prove *de facto* that our globe is made of lodestone, nor is this the time to go into those, the more so as you may look them up in Gilbert at your leisure. I am merely going to explain, with a certain likeness to my own, his method of procedure in philosophizing, in order that I may stimulate you to read it. I know that you understand quite well how much a knowledge of events contributes to an investigation of the substance and essence of things; therefore I wish you to take care to inform yourself thoroughly about many events and properties that are found uniquely in lodestone. Examples of this are its attraction of iron, and its conferring this same power upon iron merely by its presence; likewise its communicating to iron the property of pointing toward the poles, just as it retains this power in itself. Moreover, I want you to make a visual test of how there resides in it a power of conferring upon the compass needle not only the property of pointing toward the poles with a horizontal motion

Gilbert's method in philosophizing.

Multiple properties of lodestone.

under the meridian — a property long since known — but also a newly observed faculty of vertical dip when it is balanced upon a small sphere of lodestone on which this meridian has been previously marked. I mean that the needle declines from a given mark, a greater or less amount according as the needle is taken closer to or farther from the pole, until at the pole itself it stands erect and perpendicular, while in the equatorial regions it remains parallel to the axis.

Next, make a test of the power of attraction being more active in every piece of lodestone, nearer the poles than at the middle, and noticeably stronger at one pole than at the other, the stronger pole being the one which points toward the south. Note that in a small lodestone this stronger south pole becomes weaker whenever it is required to support some iron in the presence of the north pole of a much larger lodestone. To make a long story short, you may ascertain by experiment these and many other properties described by Gilbert, all of which belong to lodestone and none to any other material.

Conclusive argument for the terrestrial globe being a lodestone.

Now, Simplicio, suppose that a thousand pieces of different materials were set before you, each one covered and enclosed in cloth under which it was hidden, and that you were asked to find out from external indications the material of each one without uncovering it. If, in attempting to do this, you should hit upon one which plainly showed itself to have all the properties which you had already recognized as residing only in lodestone and not in any other material, what would you judge to be the essence of that material? Would you say that it might be a piece of ebony, or alabaster, or tin?

SIMP. There is no question at all that I should say it was a piece of lodestone.

SALV. In that case, declare boldly that under this covering or wrapper of earth, stone, metal, water, etc. there is concealed a huge lodestone. For in regard to this there are recognized, by anyone who observes carefully, all the same events which are perceived to belong to a true and unconcealed sphere of lodestone. If nothing more were to be observed than the dipping of the needle, which, carried around the earth, tilts more upon its approach to the pole and less as it goes toward the equator, where it finally becomes balanced, this alone ought to persuade the most stubborn judgment. I say nothing of another remarkable effect which is plainly seen in all pieces of lodestone and causes the south pole of a lodestone to be stronger than the other† for us inhabitants of the Northern Hemisphere. This difference is found to be the greater, the more one departs from the equator; at the equator, both sides are of equal strength, though noticeably weaker. But in the southern regions, far from the equator, it changes its nature and the side which is the weaker for us acquires power over the other. All this conforms with what we see done by a little piece of lodestone in the presence of a big one whose force prevails over the smaller and makes it subservient, so that according as it is held near to or far from the equator of the large one, it makes just such variations as I have told you are made by every lodestone carried near to or far from the earth's equator.

SAGR. I was convinced at my first perusal of Gilbert's book, and, having found an excellent piece of lodestone, I made many observations over a long period, all of which merited the greatest wonder. But what seemed most astonishing of all to me was the great increase in its power of sustaining iron when provided with an armature† in the manner taught by this same author. By thus equipping my piece I multiplied its strength by eight, and where previously it would scarcely hold up nine ounces of iron, with the armature it would sustain more than six pounds. Perhaps you have seen this very piece, sustaining two little iron anchors, in the gallery of your Most Serene Grand Duke, on whose behalf I parted with it.†

SALV. I used to look at it frequently with great amazement, until a still greater admiration seized me because of a little specimen in the possession of our Academician. This, being not over six ounces in weight and sustaining no more than two ounces unarmatured, supports one hundred sixty ounces when so equipped. Thus it bears eighty times as much with an armature as without, and holds up twenty-six times its own weight. This is a greater marvel than Gilbert was able to behold, since he writes that he was never able to get a lodestone which succeeded in sustaining four times its own weight.

SAGR. It seems to me that this stone opens to the human mind a large field for philosophizing, and I have often speculated to myself on how it imparts to the iron which arms it a force so greatly superior to its own. But I was unable ever to find any satisfactory solution, nor did I find anything to much advantage in what Gilbert has to say on this particular. I wonder whether the same is true of you.

SALV. I have the highest praise, admiration, and envy for this author, who framed such a stupendous concept regarding an object which innumerable men of splendid intellect had handled without paying any attention to it. He seems to me worthy of great acclaim also for the many new and sound observations which he made, to the shame of the many foolish and mendacious authors who write not just what they know, but also all the vulgar foolishness they hear, without trying to verify it by experiment; perhaps they do this in order not to diminish the size of their books. What I might have wished for in Gilbert would be a little more of the mathematician, and especially a thorough grounding in geometry, a discipline which would have rendered him less rash about accepting as rigorous proofs those reasons which he puts forward as *verae causae* for the correct conclusions he himself had observed. His reasons, candidly speaking, are not rigorous, and lack that force which must unquestionably be present in those adduced as necessary and eternal scientific conclusions.

I do not doubt that in the course of time this new science will be improved with still further observations, and even more by true and conclusive demonstrations. But this need not diminish the glory of the first observer. I do not have a lesser regard for the original inventor of the harp because of the certainty that his instrument was very crudely constructed and more crudely played; rather, I admire him much more than a hundred artists who in ensuing centuries have brought this profession to the highest perfection. And it seems to me most reasonable for the ancients to have counted among the gods those first inventors of the fine arts, since we see that the ordinary human mind has so little curiosity and cares so little for rare and gentle things that no desire to learn is stirred within it by seeing and hearing these practiced exquisitely by experts. Now consider for yourself whether minds of that sort would ever have been applied to the construction of a lyre or to the invention of music, charmed by the mere whistling of dry tortoise tendons, or the striking of four hammers!† To apply oneself to great inventions, starting from the smallest beginnings, and to judge that wonderful arts

The earliest observers and inventors deserve admiration.

lie hidden behind trivial and childish things is not for ordinary minds; these are concepts and ideas for superhuman souls.

Now, in answer to your question, I say that I also thought for a long time to find the cause for this tenacious and powerful connection that we see between the iron armature of a lodestone and the other iron which joins itself to it. In the first place, I am certain that the power and force of the stone is not increased at all by its having an armature, for it does not attract through a longer distance. Nor does it attract a piece of iron as strongly if a thin slip of paper is introduced between this and the armature; even if a piece of gold leaf is interposed, the bare lodestone will sustain more iron than the armature. Hence there is no change here in the force, but merely something new in its effect.

True cause of the great multiplication of force in a lodestone by means of an armature.

And since for a new effect there must be a new cause, we seek what is newly introduced by the act of supporting the iron via the armature, and no other change is to be found than a difference in contact. For where iron originally touched lodestone, now iron touches iron, and it is necessary to conclude that the difference in these contacts causes the difference in the results. Next, the difference between the contacts must come, so far as I can see, from the substance of the iron being finer, purer, and denser in its particles than is that of the lodestone, whose parts are coarser, less pure, and less dense. From this it follows that the surfaces of the two pieces of iron which are to touch, when perfectly smoothed, polished, and burnished, fit together so exactly that all the infinity of points on one touch the infinity of points on the other. Thus the threads which unite the pieces of iron are, so to speak, more numerous than those which join lodestone to iron, on account of the substance of lodestone being more porous and less integrated, so that not all the points and threads on the surface of the iron find counterparts to unite with on the surface of the lodestone.

For a new effect, the cause must be a new one.

Now we may see that the substance of iron (especially when much refined, as is the finest steel) is much more dense, fine, and pure in its particles than is the material of lodestone, from the possibility of bringing the former to an extremely thin edge, such as a razor edge, which can never be done to a piece of lodestone with any success. The impurity of the lodestone and its adulteration with other kinds of stone can next be sensibly observed; in the first place by the color of some little spots, gray

Showing that iron is made of finer, purer, and denser parts than is lodestone.

Showing to the senses the impurity of lodestone.

for the most part, and secondly by bringing it near a needle suspended on a thread. The needle cannot come to rest at these little stony places; it is attracted by the surrounding portions, and appears to leap toward these and flee from the former spots. And since some of these heterogeneous spots are large enough to be easily visible, we may believe that others are scattered in great quantity throughout the mass but are not noticeable because of their small size.

What I am telling you (that is, that the great abundance of contacts made between iron and iron is the cause of so solid an attachment) is confirmed by an experiment. If we present the sharp point of a needle to the armature of a lodestone, it attaches itself no more strongly than it would to the bare lodestone; this can result only from the two contacts being equal, both being made at a single point. But now see what follows. A needle is placed upon the lodestone so that one of its ends sticks out somewhat beyond, and a nail is brought up to this. Instantly the needle will attach itself to it so firmly that upon the nail being drawn back, the needle can be suspended with one end attached to the lodestone and the other to the nail. Withdrawing the nail still farther, the needle will come loose from the lodestone if the needle's eye is attached to the nail and its point to the lodestone; but if the eye is toward the lodestone, the needle will remain attached to the lodestone upon withdrawing the nail. In my judgment, this is for no other reason than that the needle, being larger at the eye, makes contact in more places than it does at its very sharp point.

Sagr. The entire argument looks convincing to me, and I rank these experiments with the needle very little lower than mathematical proof. I frankly admit that in the entire magnetic science I have not heard or read anything which gives so cogently the reasons for any of its other remarkable phenomena. If their causes were to be explained to us this clearly, I can think of nothing pleasanter that our intellects could wish for.

Salv. In investigating the unknown causes of our conclusions, one must be lucky enough right from the start to direct one's reasoning along the road of truth. When traveling along that road, it may easily happen that other propositions will be encountered which are recognized as true either through reason or experience. And from the certainty of these, the truth of our own will acquire strength and evidence. This is exactly what happened for me in the present instance. Wishing to assure myself by some other observation that the cause I had turned up was correct (that is, that the substance of the lodestone really was much less continuous than that of iron or steel), I had the artisans who work in the museum of my lord the Grand Duke smooth for me one face of that same piece of lodestone which was formerly yours, and then polish and burnish it as much as possible. To my great satisfaction, this enabled me to experience directly just what I sought. For there I found many spots of different color from the rest, bright and shiny as any very dense, hard stone; the rest of the field was polished only to the touch, being not the least bit shiny, but rather as if covered with mist. This was the substance of the lodestone, and the shiny parts were of other stones mixed with it, as was sensibly recognized by bringing the smooth face toward some iron filings, which leaped in large quantities to the lodestone. But not a single grain went to the spots mentioned, of which there were many, some as large as a quarter of a fingernail, some rather smaller, and many quite small; those which were scarcely visible were almost innumerable.

Thus was I assured that my idea had been quite correct when I first judged that the substance of the lodestone must be not continuous and compact, but porous. Better yet, spongy; though with this difference: where the cavities and cells of a sponge contain air or water, those of the lodestone are filled with hard and heavy stone, as shown by the high lustre that they take on. Whence, as I said at the outset, upon applying the surface of iron to the surface of a lodestone, the minute particles of iron — though continuous in perhaps a greater degree than those of any other material, as shown by their shining more than any other material — do not all meet solid lodestone, but only a few of them; and the contacts being few, the attachment is weak. But the armature of a lodestone, in addition to touching a large part of its surface, is also vested with the force of the closer parts even though not touching them; and being quite flat on the side applied to the suspended iron (this also being well smoothed), contact is made by innumerable tiny particles if not by the infinity of points on both surfaces, which yields a very strong attachment.

This experiment of smoothing the surfaces of the pieces of iron which are to touch was not performed by Gilbert; instead, he makes the irons convex, so that their contact is small, from which it comes about that the tenacity with which those irons stick together is very much less.

SAGR. The reason you give, as I have just remarked, satisfies me little less than it would if it were a pure geometrical proof. And since it concerns a physical problem, I suppose Simplicio is also convinced as fully as is permitted by natural science, in which he is aware that geometrical evidence cannot be demanded.

SIMP. Truly, I think that Salviati's eloquence has so clearly explained the cause of this effect that the most mediocre mind, however unscientific, would be persuaded. But we who restrict ourselves to philosophical terminology reduce the cause of this and other similar effects to *sympathy*, which is a certain agreement and mutual desire that arise between things which are similar in quality among themselves, just as on the other hand that hatred and enmity through which other things naturally fly apart and abhor each other is called by us *antipathy*.

Sympathy and *antipathy* are the terms used by philosophers for rendering reasons easily for many physical effects.

SAGR. And thus, by means of two words, causes are given for a large number of events and effects which we behold with amazement when they occur in nature. Now this method of philosophizing seems to me to have great sympathy with a certain manner of painting used by a friend of mine. He would write on the canvas with chalk: "This is where I'll have the fountain, with Diana and her nymphs; here, some greyhounds; there, a hunter with a stag's head. The rest is a field, a forest, and hillocks." He left everything else to be filled in with color by a painter, and with this he was satisfied that he himself had painted the story of Acteon — not having contributed anything of his own except the title.

Amusing example explaining the small efficacy of some philosophical arguments.

But whither are we wandering with so long a digression, contrary to our established arrangements? I have almost forgotten what we were talking about when we veered into this discourse on magnetism. Still, I had something in mind to say on the subject, whatever it was.

SALV. We were proving that the third motion attributed to the earth by Copernicus was not a movement at all, but a state of rest and an immutable keeping of definite parts pointed toward the same definite points in the universe; that is, a perpetual maintenance of the axis of diurnal rotation parallel to itself and pointing at certain fixed stars. This perfectly constant position, we were saying, belongs naturally to every body which is balanced and suspended in a fluid and yielding medium; for, though turned around, it does not change direction with respect to external things, but merely seems to turn upon itself with respect to the person carrying it and to the bowl in which it is carried.

Let us add next, to this simple and natural event, the magnetic force by which the terrestrial globe may be kept so much the more solidly immutable, etc.

SAGR. Now I remember the whole thing. What was passing through my mind at that time, and what I wanted to bring out, was a certain consideration regarding the difficulty and objection raised by Simplicio against the earth's mobility. This was based upon the impossibility of attributing a multiplicity of motions to a simple body, for which in Aristotle's doctrine only one single simple motion can be natural.

What I wanted to bring up for consideration was precisely the lodestone, to which three movements are sensibly seen to belong naturally: One toward the center of the earth as a heavy object; a second is the horizontal circular motion by which it restores and conserves its axis in the direction of certain parts of the universe; and third is this one discovered by Gilbert,† of dipping its axis in the meridian plane toward the surface of the earth, in greater or less degree proportionate to its distance from the equator (where it remains parallel to the axis of the earth). Besides these three, it is perhaps not improbable that it may have a fourth motion of turning about its own axis, whenever it is balanced and suspended in air or some other fluid and yielding medium and all external and accidental impediments are taken away; Gilbert himself also shows his approval of this idea. So you see, Simplicio, how shaky Aristotle's axiom is.

Three diverse natural motions of lodestone.

SIMP. This not only does not hit his maxim, but is not even aimed at it, since he was talking about a simple body and what can be naturally adapted to that, while you oppose him with what is done by a compound body. Nor are you saying anything that is new to Aristotle's doctrine, for he also grants to compound bodies compound motions, etc.

Aristotle concedes mixed motions to compound bodies.

SAGR. Wait a moment, Simplicio. Answer the questions I am going to ask you. You say that the lodestone is not a simple body, but a compound one; now I ask you what are the simple bodies which are mixed in the compounding of lodestone?

SIMP. I cannot tell you the ingredients or the exact proportions, but it is sufficient that they are elementary bodies.

SAGR. That is enough for me, too. And what are the natural motions of these elemental bodies?

SIMP. They are the two simple straight motions, *sursum* and *deorsum*.

SAGR. Next, tell me this: Do you believe that the motion which is natural to such a compound body must be one which could result from the combination of the two simple natural motions of the component simple bodies? Or might it be still another motion, one not possible to compound from those?

SIMP. I believe that it will move with that motion resultant from the composition of the motions of the component simple bodies, and that it could not move with any motion impossible to compound from these.

SAGR. But Simplicio, you can never compound one circular motion from the two simple straight motions, and the lodestone has two or three different circular motions. So you see the trouble into which badly founded principles lead — or, rather, badly drawn consequences from good principles. For next you will be forced to say that the lodestone is a compound composed of elemental and celestial substances, if you wish to maintain that straight motion belongs only to the elements, and circular to the heavenly bodies. Therefore if you want to philosophize with assurance, say that the integral bodies of the universe which are naturally movable all move circularly, and that consequently lodestone, as a part of the true primary and integral substances of our globe, partakes of this same nature.

And please note that by your fallacious reasoning you are calling lodestone a compound body, and the terrestrial globe a simple body; yet the latter may be seen to be a hundred thousand times more compounded, since besides containing thousands and thousands of materials quite different from each other, it contains a great abundance of the very thing you call compound; I mean lodestone. This seems to me the same as if someone were to call bread a compound body, and hash a simple body, though into hash there enters no small quantity of bread, besides a hundred different foods which are eaten with bread.

Motion of compounds must be such as can result from the composition of the motions of their simple component bodies.

With two straight motions, one cannot compound circular motions.

Philosophers are obliged to confess that lodestone is compounded of celestial and elemental substances.

Fallacy of those who call lodestone a mixed body and the terrestrial globe a simple body.

It really seems to me a remarkable thing (among others) that the Peripatetics concede — as indeed they cannot deny — that our terrestrial globe is *de facto* a compound of infinitely diverse materials; that they next concede that the motions of compound bodies must be compound; that the motions which can be compounded are the straight and the circular, since the two straight motions are incompatible on account of being contrary to one another; they affirm that the pure element of earth is not to be found; and they grant that the earth is never moved with any local motion. Finally they want to place in nature this body which is nowhere to be found, and make it movable with a motion which it has never employed and never will employ; but to this actual body which does exist and always has existed they deny that very motion which they originally conceded must be naturally suited to it!

Peripatetic reasoning full of fallacies and contradictions.

SALV. Sagredo, please let us weary ourselves no longer with these particulars, especially since you know that our goal is not to judge rashly or accept as true either one opinion or the other, but merely to set forth for our own pleasure those arguments and counterarguments which can be adduced for one side and for the other. Simplicio answers thus in order to rescue his Peripatetics; therefore we shall suspend judgment, and leave this in the hands of whoever knows more about it than we do.

And since it seems to me that in these three days the system of the universe has been discussed at great length, it is now time for us to take up that principal event from which our discussions took their rise;† I mean the ebb and flow of the oceans, whose cause may be assigned very probably to the movement of the earth. But this we shall postpone until tomorrow, if that is satisfactory to you.

Meanwhile, lest I forget, I want to tell you about one particular to which I wish Gilbert had not lent his ear. This is his concession that if a small sphere of lodestone were exactly balanced, it would revolve upon itself; for this no cause whatever exists. For if the entire terrestrial globe has by its nature a rotation about its own center every twenty-four hours, and all its parts must also rotate together with the whole around its center in twenty-four hours, then by being on the earth they already actually have this motion, turning together with the earth, and to assign to them a motion around their own centers would be to

An improbable effect which Gilbert grants to lodestone.

attribute to them a second movement quite different from the first. Thus they would have two motions; that is, a rotation in twenty-four hours about the center of the whole, and a revolution upon their own centers. Now this second motion is arbitrary, and there is no reason whatever for introducing it. If, upon becoming detached from the whole natural mass, a piece of lodestone were deprived of the property of following that mass as it did while they were joined together (so that it would be deprived of the revolution about the universal center of the terrestrial globe), there might perhaps be a greater probability for believing that it would take upon itself a new whirling about its own particular center. But if it always continues its original natural and perpetual course whether separated or attached, then to what purpose would another new one be added?

SAGR. I see what you mean, and it puts me in mind of an argument very similar to this in its inanity; it is set forth by certain writers on spherical astronomy, Sacrobosco among others, if my memory serves me correctly. In order to prove that the element of water is shaped into a spherical surface together with the land, the two of them forming this globe of ours, he writes that a conclusive proof of this is the seeing of minute particles of water shaping themselves into a rounded form, as in the dewdrops seen daily on the leaves of many plants. And therefore, according to the commonplace axiom, "The same applies to the whole which applies to the parts," since the parts assume this shape, the entire element does. Now it seems to me very muddle-headed on the part of such authors not to see their obvious trifling here, and not to consider that if their argument were correct, then not only the minute drops, but any larger quantity of water you please would reduce itself to a ball when separated from the whole of its element. Such is not the case at all; indeed, one may sensibly see and intellectually understand that since the element of water likes to form into a spherical shape about the common center of gravity toward which all heavy bodies tend (which is the center of the terrestrial globe), all its parts follow it in this, in accordance with the axiom, so that the surfaces of all seas, lakes, pools, and in short all portions of water contained in vessels, do extend themselves in a spherical shape. But this sphere has for its center the center of the terrestrial globe, and bodies of water do not form individual spheres of their own.

Foolish argument of some for proving the element of water to be of spherical surface.

SALV. The error is indeed puerile, and if it were only Sacrobosco who had made it I should freely excuse this in him. But I cannot pardon it likewise in his commentators and in other famous men, and in even Ptolemy himself, without blushing for their reputations.

But now it is time to take leave, for it is getting late; and tomorrow we shall meet at the usual time, for the end and goal of all our previous discussions.

End of the Third Day

THE FOURTH DAY

SAGREDO. I do not know whether you are really arriving later than usual for our accustomed discussion or whether it just seems so to me because of my desire to hear Salviati's thoughts on such an interesting matter. I have been watching through the window for a long time, hoping from one moment to the next to see the gondola come into view which I sent to fetch you.

SALV. I believe it is only your imagination that has made the time drag, rather than any tardiness on our part. But in order not to stretch it still further it will be good for us to get to the matter in hand without wasting any more words.

Nature's whim to make the flow and ebb of the seas endorse the earth's mobility.

Let us see, then, how nature has allowed (whether the facts are actually such, or whether at a whim and as if to play upon our fancies) — has allowed, I say, the movements that have long been attributed to the earth for every reason except as an explanation of the ocean tides to be found now to serve that purpose too, with equal precision; and how, reciprocally, this ebb and flow itself coöperates in confirming the earth's mobility.† Up to this point the indications of that mobility have been taken

The tides and the earth's mobility reciprocally confirm one another.

from celestial phenomena, seeing that nothing which takes place on the earth has been powerful enough to establish the one position any more than the other. This we have already examined at length by showing that all terrestrial events from which it is ordinarily held that the earth stands still and the sun and the fixed stars are moving would necessarily appear just the same to us if the earth moved and the others stood still. Among all sublunary things it is only in the element of water (as something

All terrestrial events except the ocean tides are impartial as to the earth's motion or rest.

which is very vast and is not joined and linked with the terrestrial
globe as are all its solid parts, but is rather, because of its fluidity,
free and separate and a law unto itself) that we may recognize
some trace or indication of the earth's behavior in regard to
motion and rest. After having many times examined for myself
the effects and events, partly seen and partly heard from other
people, which are observed in the movements of the water; after,
moreover, having read and listened to the great follies which
many people have put forth as causes for these events, I have
arrived at two conclusions which were not lightly to be drawn
and granted. Certain necessary assumptions having been made, First general
these are that if the terrestrial globe were immovable, the ebb conclusion: No
ebb and flow if
and flow of the oceans could not occur naturally; and that when the terrestrial
we confer upon the globe the movements just assigned to it, the globe were
immovable.
seas are necessarily subjected to an ebb and flow agreeing in all
respects with what is to be observed in them.

SAGR. The proposition is crucial, both in itself and in what
follows as a consequence; therefore I shall be so much the more
attentive in listening to its explanation and verification.

SALV. In questions of natural science like this one at hand, a
knowledge of the effects is what leads to an investigation and Knowledge of
effects leads to
discovery of the causes. Without this, ours would be a blind investigation
journey, or one even more uncertain than that; for we should of causes.
not know where we wanted to come out, whereas the blind at
least know where they wish to arrive. Hence before all else it is
necessary to have a knowledge of the effects whose causes we
are seeking. Of those effects you, Sagredo, must be more fully
and surely informed than I am, since besides being born in
Venice and having long resided here where the tides are famous
for their size, you have also sailed to Syria, and, having a clever
and curious mind, you must have made many observations. But
I, who have only been able to observe for rather a short time
what happens here at this end of the Adriatic Gulf, and in our
lower sea on the shores of the Tyrrhenian, must often depend
upon what others tell me — which, being for the most part not
in good agreement and accordingly rather unreliable, may con-
tribute confusion rather than confirmation to our reflections.

Still, from those accounts which we are sure of, and which
happen to cover the principal events, it seems to me possible to
arrive at the true and primary causes. I do not presume to be

able to adduce all the proper and sufficient causes of those effects which are new to me and which consequently I have had no chance to think about; what I am about to say, I propose merely as a key to open portals to a road never before trodden by anyone, in a firm hope that minds more acute than mine will broaden this road and penetrate further along it than I have done in my first revealing of it. And though in other seas remote from us events may take place which do not occur in our Mediterranean, nevertheless the reason and the cause which I shall produce will still be true, provided that it is verified and fully satisfied by the events which do take place in our sea; for ultimately one single true and primary cause must hold good for effects which are similar in kind. I shall, then, tell you the story of the effects which I know to exist, and assign to them the cause that is believed by me to be true; and you, gentlemen, shall produce others noticed by you in addition to these of mine, and then we shall see whether the cause I am about to adduce can account for them also.

Three periods of the tides — diurnal, monthly, and annual.

I say, then, that three periods are observed in the flow and ebb of the ocean waters. The first and principal one is the great and conspicuous daily tide, in accordance with which the waters rise and fall at intervals of some hours; these intervals in the Mediterranean are for the most part about six hours each — that is, six hours of rising and six more of falling. The second period is monthly, and seems to originate from the motion of the moon; it does not introduce other movements, but merely alters the magnitude of those already mentioned, with a striking difference according as the moon is full, new, or at quadrature with the sun. The third period is annual, and appears to depend upon the sun; it also merely alters the daily movements by rendering them of different sizes at the solstices from those occurring at the equinoxes.

Differences occurring in the diurnal period.

We shall speak first about the diurnal period, as it is the principal one, and the one upon which the actions of the moon and the sun are exercised secondarily in their monthly and annual alterations. Three varieties of these hourly changes are observed; in some places the waters rise and fall without making any forward motion; in others, without rising or falling they move now toward the east and again run back toward the west; and in still others, the height and the course both vary. This occurs here in Venice, where the waters rise in entering and fall in departing. They do this at the end of a gulf extending east and west and terminating on open shores where the water has room to spread out upon rising; if their course were interrupted by mountains or by very high dikes, they would rise and sink against these without any forward motion. Elsewhere the water runs to and fro in its central parts without changing height, as happens notably in the Straits of Messina between Scylla and Charybdis, where the currents are very swift because of the narrowness of the channel. But in the open Mediterranean and around its islands, such as the Balearics, Corsica, Sardinia, Elba, Sicily (on the African side), Malta, Crete, etc., the alterations of height are very small but the currents are quite noticeable, especially where the sea is restrained between islands, or between these and the continent.

Now it seems to me that these actual and known effects alone, even if no others were to be seen, would very probably persuade anyone of the mobility of the earth who is willing to stay within the bounds of nature; for to hold fast the basin of the Mediterranean and to make the water contained within it behave as it does surpasses my imagination, and perhaps that of anyone else who enters more than superficially into these reflections.

SIMP. These events, Salviati, did not just commence; they are very ancient, and have been observed by innumerable men, many of whom have contrived to give one reason or another to account for them. Not far from here there is a great Peripatetic who gives for them a cause recently dredged out of one of Aristotle's texts which had not been well understood by his interpreters. From this text, he deduces that the true cause of these movements stems from nothing else but the various depths of the seas. The deepest waters, being more abundant and therefore heavier, expel the waters of lesser depth; these, being raised up, then try to descend, and from this continual strife the tides are derived.

A cause for the tides adduced by a certain modern philosopher.

Then there are many who refer the tides to the moon, saying that this has a particular dominion over the waters; lately a certain prelate† has published a little tract wherein he says that the moon, wandering through the sky, attracts and draws up toward itself a heap of water which goes along following it, so that the high sea is always in that part which lies under the moon. And since when the moon is below the horizon, this rising nevertheless

Cause of the tides attributed to the moon by a certain prelate.

Girolamo Borro
and other Peri-
patetics refer
tides to the
temperate heat
of the moon.

returns, he tells us that he can say nothing to account for this effect except that the moon not only retains this faculty naturally in itself, but in this case has also the power to confer it upon the opposite sign of the zodiac. Others, as I think you know, say that the moon also has power to rarefy the water by its temperate heat, and that thus rarefied, it is lifted up. Nor are those lacking who . . .

SAGR. Please, Simplicio, spare us the rest; I do not think there is any profit in spending the time to recount them, let alone the words to refute them. If you should give assent to any of these or to similar triflings, you would be wronging your own judgment — just when, as we know, it has been much unburdened of error.

SALV. I am a little more easygoing than you, Sagredo, and I shall put in a few words for Simplicio's benefit if he thinks that some probability attaches to the things he has been telling us.

Reply to the
inanities ad-
duced as causes
of the tides.

Simplicio, I say that waters which have their external surfaces higher expel those that are lower, but not that those which are deeper do so; and the higher waters, having driven away the lower, quickly come to rest and equilibrium. Your Peripatetic must believe that all the lakes in the world (which remain placid) and all the seas where the tide is imperceptible must have perfectly level beds; I was so naïve as to persuade myself that even

Islands are an
indication of the
unevenness of
the sea bottoms.

if there were no other soundings, the islands which rise above the water would be a very obvious indication of the unevenness of the bottoms. You might tell your prelate that the moon travels over the whole Mediterranean every day, but the waters are raised only at its eastern extremity and for us here at Venice.

As for those who make the temperate heat of the moon able to swell the water, you may tell them to put a fire under a kettle of water, hold their right hands in this until the heat raises the water a single inch, and then take them out to write about the swelling of the seas. Or ask them at least to show you how the moon rarefies a certain part of the water and not the remainder, such as this here at Venice, but not that at Ancona, Naples, or Genoa.

Let us just say that there are two sorts of poetical minds — one kind apt at inventing fables, and the other disposed to believe them.

SIMP. I do not think that anyone believes fables when he knows them to be such; and as to the opinions about the cause of the tides (which are numerous), since I know that there is only one true and primary cause for one effect, I understand perfectly that at most one can be true, and all the rest must be false and fabulous. Perhaps the true one is not even among those which have been produced up to date. I rather believe this to be so, since it would be remarkable if the true cause should shed so little light as not to show through the darkness of so many false ones. But I must say, with that frankness which is permitted here among ourselves, that to introduce the motion of the earth and make it the cause of the tides seems to me thus far to be a concept no less fictitious than all the rest I have heard. If no reasons more agreeable to natural phenomena were presented to me, I should pass on unhesitatingly to the belief that the tide is a supernatural effect, and accordingly miraculous and inscrutable to the human mind — as are so many others which depend directly upon the omnipotent hand of God.

The truth has not so little light as not to be perceived through the darkness of falsehoods.

SALV. You argue very prudently, and also in agreement with Aristotle's doctrine; at the beginning of his *Mechanics,* as you know, he ascribes to miracles all things whose causes are hidden. But I believe you do not have any stronger indication that the true cause of the tides is one of those incomprehensibles than the mere fact that among all things so far adduced as *verae causae* there is not one which we can duplicate for ourselves by means of appropriate artificial devices. For neither by the light of the moon or sun, nor by temperate heat, nor by differences of depth can we ever make the water contained in a motionless vessel run to and fro, or rise and fall in but a single place. But if, by simply setting the vessel in motion, I can represent for you without any artifice at all precisely those changes which are perceived in the waters of the sea, why should you reject this cause and take refuge in miracles?

Aristotle attributes to a miracle those effects of which the causes are not known.

SIMP. I shall have recourse to miracles unless you dissuade me from it by other natural causes than the motion of the containers of the waters of the sea. For I know that the latter containers do not move, the entire terrestrial globe being naturally immovable.

SALV. But do you not believe that the terrestrial globe could be made movable supernaturally, by God's absolute power?

SIMP. Who can doubt this?

SALV. Then, Simplicio, since we must introduce a miracle to

achieve the ebbing and flowing of the oceans, let us make the earth miraculously move with that motion by which the oceans are naturally moved. This operation will indeed be as much simpler and more natural among things miraculous, as it is easier to make a globe turn around (which we see so many of them do) than to make an immense bulk of water go back and forth more rapidly in some places than in others; rise and fall, here more, there less, and in other places not at all, and to make all these variations within the same containing vessel. Besides, these are many miracles, while the other is only one. Add to this that the miracle of making the water move brings another miracle in its train, which is that of holding the earth steady against the impulses of the water. For these would be capable of making it vacillate first in one direction and then in the other, if it were not miraculously retained.

SAGR. Let us suspend judgment for a while as to the folly of the new opinion which Salviati wants to explain to us, Simplicio, and not be so quick to class it with those ridiculous older ones. As to the miracle, let us likewise have recourse to that only after we have heard arguments which are restricted within the bounds of nature. Though, indeed, to my mind all works of nature and of God appear miraculous.

SALV. That is the way I feel about it, and saying that the natural cause of the tides is the motion of the earth does not exclude this operation from being miraculous.

Now, returning to our discussion, I reply and reaffirm that it has never previously been known how the waters contained in our Mediterranean basin can make those movements which they are seen to make, so long as this basin and containing vessel rests motionless. What renders the matter puzzling is daily observed, as I am about to describe; therefore, listen carefully.

Showing the impossibility of tides occurring naturally if the earth is motionless.

We are here in Venice, where the waters are now low; the sea is quiet, the air tranquil; the water is commencing to rise, and at the end of five or six hours it will have gone up ten spans or more. This rise is not made by the original water being rarefied, but by water newly arriving here — water of the same kind as the original water, with the same salinity, the same density, the same weight. Ships float in it, Simplicio, without submerging a hairsbreadth further; a barrel of it weighs not a grain more or less than the same quantity of the other; it keeps the same coldness entirely unchanged; in short, it is water which has recently and visibly entered through the channels and mouths of the Lido.

Now you tell me how and whence it came here. Are there perchance hereabouts some abysses or openings in the bottom of the sea through which the earth draws in and expels the water, breathing like some immense and monstrous whale? If so, why does the water not rise likewise over a space of six hours at Ancona, Dubrovnik (*Ragugia*), and Corfu, where the increase is small or even imperceptible? Who will find a way to pour new water into an immovable vessel and have it rise only in one definite place and not in others?

Do you perhaps say that this new water is borrowed from the ocean, carried in through the Straits of Gibraltar? This will not remove the difficulties mentioned; it will only make them greater. In the first place, tell me what must be the course of that water which, entering by the strait, is conducted in six hours clear to the extreme coast of the Mediterranean, a distance of two or three thousand miles, and retraces the same space on its return? What would become of the ships scattered about on the sea? And what of those in the strait, on a continual watery precipice of immense bulk, entering through a channel no more than eight miles wide — a channel which must in six hours give passage to enough water to inundate a space hundreds of miles wide and thousands long? Where is the tiger or falcon that ever ran or flew with such speed? A speed, I mean, of 400 miles an hour or better.

It cannot be denied that there are currents running the length of the gulf, but they are so slow that a rowboat can outrun them, though not without losing headway. Besides, if this water comes in through the strait, there is another difficulty: How does it cause so much of a rise here, at so remote a place, without first raising the closer parts by a similar or greater amount? To sum up, I do not believe that either obstinacy or subtleness of wit could ever discover a reply to these difficulties and thereby be able to maintain the fixity of the earth against them, while remaining within natural limitations.

SAGR. So far I follow you very well, and I am anxiously waiting to hear how these marvels can take place unimpeded if we assume the motions already assigned to the earth.

SALV. As these effects must be consequences of the motions which

Natural and true
effects take place
without
hindrances.

belong naturally to the earth, it is not only necessary that they encounter no obstacle or impediment, but that they follow easily. Nor must they merely follow easily; they must follow necessarily, in such a way that it would be impossible for them to take place in any other manner. For such is the property and condition of things which are natural and true.

Having established, then, that it is impossible to explain the movements perceived in the waters and at the same time maintain the immovability of the vessel which contains them, let us pass on to considering whether the mobility of the container could produce the required effect in the way in which it is observed to take place. Two sorts of movement may be conferred upon a vessel so that the water contained in it acquires the property of running first toward one end and then toward the other, and rise and sink there. The first would occur when one end is lowered and then the other, for under those conditions the water, running toward the depressed part, rises and sinks alternately at either end. But since this rising and sinking is nothing but a retreat from and an approach toward the center of the earth, this sort of movement cannot be attributed to concavities in the earth itself as containing vessels of the waters. For such containers could not have parts able to approach toward or retreat from the center of the terrestrial globe by any motion whatever that might be assigned to the latter.

Two sorts of
movements of
the containing
vessel may make
the contained
water rise and
fall.

Concavities of
the earth cannot
cause an ap-
proach to or a
retreat from the
center of the
earth.

The other sort of motion would occur when the vessel was moved without being tilted, advancing not uniformly but with a changing velocity, being sometimes accelerated and sometimes retarded. From this variation it would follow that the water (being contained within the vessel but not firmly adhering to it as do its solid parts) would because of its fluidity be almost separate and free, and not compelled to follow all the changes of its container. Thus the vessel being retarded, the water would retain a part of the impetus already received, so that it would run toward the forward end, where it would necessarily rise. On the other hand, when the vessel was speeded up, the water would retain a part of its slowness and would fall somewhat behind while becoming accustomed to the new impetus, remaining toward the back end, where it would rise somewhat.

A progressive
and uneven
movement may
make the con-
tained water run
within a vessel.

These effects can be very clearly explained and made evident to the senses by means of the example of those barges which are continually arriving from Fusina filled with water for the use of this city. Let us imagine to ourselves such a barge coming along the lagoon with moderate speed, placidly carrying the water with which it is filled, when either by running aground or by striking some obstacle it becomes greatly retarded. Now the water will not thereby lose its previously received impetus equally with the barge; keeping its impetus, it will run forward toward the prow, where it will rise perceptibly, sinking at the stern. But if on the other hand the same barge noticeably increases its speed in the midst of its placid course, then the water which it contains (before getting used to this and while retaining its slowness) will stay back toward the stern, where it will consequently rise, sinking at the prow. This effect is indubitable and clear; it may be tested experimentally at any time, and there are three things about it which I want you to note particularly.

The first is that in order to make the water rise at one extremity of the vessel, there is no need of new water, nor need the water run there from the other end.

The second is that the water near the middle does not rise or sink noticeably unless the course of the barge happens to be very fast to begin with, and the object struck or other hindrance which checks it is very strong and unyielding. In such an event this might not only make all the water run forward, but cause most of it to jump right out of the barge; the same would also happen if a very violent impulse were suddenly given to it when it was traveling very slowly. But if to a gentle motion of its own there were added a moderate retardation or acceleration, the parts in the middle (as I said) would rise and sink imperceptibly, and the other parts would rise the less according as they were closer to the middle, and the more according as they were farther from it.

The third thing is that whereas the parts around the center make little change as to rising or sinking with respect to the water at the ends, yet they run to and fro a great deal in comparison with the water at the extremities.

Now, gentlemen, what the barge does with regard to the water it contains, and what the water does with respect to the barge containing it, is precisely the same as what the Mediterranean basin does with regard to the water contained within it, and what the water contained does with respect to the Mediterranean

The parts of the
terrestrial globe
are accelerated
and retarded in
their motion.

basin, its container. The next thing is for us to prove that it is true, and in what manner it is true, that the Mediterranean and all other sea basins (in a word, that all parts of the earth) move with a conspicuously uneven motion, even though nothing but regular and uniform motions may happen to be assigned to the globe itself.

SIMP. At first sight this looks like a great paradox to me, though I am no mathematician or astronomer. If it is true that the motion of the whole may be regular, and that of the parts which always remain attached to it may be irregular, then this is a paradox destroying the axiom which affirms *eandem esse rationem totius et partium.*

SALV. I shall prove my paradox, Simplicio, and then leave to you the burden of either defending the axiom against it or of bringing the two into accord. My demonstration will be brief and easy; it will depend upon things already dealt with at length in our past conversations, without introducing the slightest word to make it favor the ebb and flow.

We have already said that there are two motions attributed to the terrestrial globe; the first is annual, made by its center along the circumference of its orbit about the ecliptic in the order of the signs of the zodiac (that is, from west to east), and the other is made by the globe itself revolving around its own center in twenty-four hours (likewise from west to east) around an axis which is somewhat tilted, and not parallel to that of its

Demonstrating
how the parts of
the terrestrial
globe are accel-
erated and
retarded.

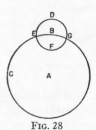

FIG. 28

annual revolution. From the composition of these two motions, each of them in itself uniform, I say that there results an uneven motion in the parts of the earth. In order for this to be understood more easily, I shall explain it by drawing a diagram.

First I shall describe around the center A the circumference of the earth's orbit BC, on which the point B is taken; and around this as center, let us describe this smaller circle DEFG, representing the terrestrial globe. We shall suppose that its center B runs along the whole circumference of the orbit from west to east; that is, from B toward C. We shall further suppose the terrestrial globe to turn around its own center B from west to east, in the order of the points D, E,

The parts of a
circle moving
regularly around
their own center
move in con-
trary motions at
different times.

F, G, during a period of twenty-four hours. Now here we must carefully note that when a circle revolves around its own center, every part of it must move at different times with contrary motions. This is obvious, considering that when the part of the circumference around the point D is moving toward the left (toward E), the opposite parts, around F, go toward the right (toward G); so that when the point D gets to F, its motion will be contrary to what it was originally when it was at D. Moreover, in the same time that the point E descends, so to speak, toward F, G ascends toward D. Since this contrariety exists in the motion of the parts of the terrestrial surface when it is turning around its own center, it must happen that in coupling the diurnal motion with the annual, there results an absolute motion of the parts of the surface which is at one time very much accelerated and at another retarded by the same amount. This is evident from considering first the parts around D, whose absolute motion will be very swift, resulting from two motions made in the same direction; that is, toward the left. The first of these is part of the annual motion, common to all parts of the globe; the other is that of this same point D, carried also to the left by the diurnal whirling, so that in this case the diurnal motion increases and accelerates the annual motion.

Mixture of the annual and diurnal motions causes the unevenness of motion in the parts of the terrestrial globe.

It is quite the opposite with the part across from D, at F. This, while the common annual motion is carrying it toward the left together with the whole globe, is carried to the right by the diurnal rotation, so that the diurnal motion detracts from the annual. In this way the absolute motion — the resultant of the composition of these two — is much retarded.

Around the points E and G, the absolute motion remains equal to the simple annual motion, since the diurnal motion acts upon it little or not at all, tending neither to left nor to right, but downward and upward. From this we conclude that just as it is true that the motion of the whole globe and of each of its parts would be equable and uniform if it were moved with a single motion, whether this happened to be the annual or the diurnal, so is it necessary that upon these two motions being mixed together there results in the parts of the globe this uneven motion, now accelerated and now retarded by the additions and subtractions of the diurnal rotation upon the annual revolution.

Now if it is true (as is indeed proved by experience) that the acceleration and retardation of motion of a vessel makes the contained water run back and forth along its length, and rise and fall at its extremities, then who will make any trouble about granting that such an effect may — or rather, must — take place in the ocean waters? For their basins are subjected to just such alterations; especially those which extend from west to east, in which direction the movement of these basins is made.

The most potent and primary cause of the tides.

Now this is the most fundamental and effective cause of the tides, without which they would not take place. But the particular events observed at different times and places are many and varied; these must depend upon diverse concomitant causes, though all must have some connection with the fundamental cause. So our next business is to bring up and examine the different phenomena which may be the causes of such diverse effects.

Different events which take place in the flowing and ebbing. First event: Water raised at one extremity returns to equilibrium by itself.

The first of these is that whenever the water, thanks to some considerable retardation or acceleration of motion of its containing vessel, has acquired a cause for running toward one end or the other, it will not remain in that state when the primary cause has ceased. For by virtue of its own weight and its natural inclination to level and balance itself, it will speedily return of its own accord; and being heavy and fluid, it will not only return to equilibrium but will pass beyond it, pushed by its own impetus, and will rise at the end where first it sank. But it will not stay there; by repeated oscillations of travel it will make known to us that it does not want the speed of motion it has received to be suddenly removed and reduced to a state of rest. It wishes this to be slowly reduced, abating little by little. In exactly this way we see that a weight suspended by a cord, once removed from the state of rest (that is, the perpendicular), returns to this and comes to rest by itself, but only after having gone to and fro many times, passing beyond this perpendicular position in its coming and going.

In the shortest vessels the oscillations are the most frequent.

The second event to be noticed is that the reciprocations of movement just mentioned are made and repeated with greater or less frequency (that is, in shorter or longer times) according to the various lengths of the vessels containing the water. In the shorter space, the reciprocations are more frequent, and they are rarer in the longer, just as in the above example of the plumb bobs the reciprocations of those which are hung on long cords are seen to be less frequent than those hanging from shorter threads.

For the third remark, you must know that it is not only a greater or lesser length of vessel which causes the water to perform its reciprocations in different times, but a greater or less depth does the same thing. It happens that for water contained in vessels of equal length but of unequal depth, the deeper water will make its vibrations in briefer times, and the oscillations will be less frequent in the shallower.

Fourth, such vibrations produce two effects in water which are worthy of being noticed and observed carefully. One is the alternate rising and falling at either extremity; the other is the horizontal moving and running to and fro, so to speak. These two different motions inhere differently in different parts of the water. The extreme ends of the water rise and fall the most; the central parts do not move up and down at all; and other parts, by degrees as they are nearer to the ends, rise and fall proportionately more than those farther from the ends. On the other hand, the central parts move a great deal in that other (progressive) movement back and forth, going and returning, while the waters in the extreme ends have none of this motion — except so far as they may in rising happen to go higher than their banks, and spill out of their original channel and container. But where the hindrance of the banks restrains them, they merely rise and fall; nor does this prevent the waters in the middle from running back and forth, as do the other parts in proportion, traveling the more or the less according as they are located farther from or closer to the middle.

The fifth particular event must be more carefully considered, because it is impossible for us to duplicate its effects by any practical experiment. It is this: In an artificial vessel like the barge mentioned previously, moving now more rapidly and again more slowly, the acceleration or retardation is always shared uniformly by the whole vessel and by each of its parts. Thus, for example, when the barge is checked in its motion, its forward parts are no more retarded than its after parts, but all share equally in the same retardation. The same happens in acceleration; that is, conferring some new cause of greater velocity upon the barge accelerates the bow in the same way as the stern. But in immense vessels, such as long sea bottoms (though these indeed are nothing more than cavities made in the solidity of the

A greater depth makes the oscillations of the water more frequent.

Water rises and sinks at the extremities of the vessel, and courses in the central parts.

The phenomena of the earth's movements cannot be represented in practice.

terrestrial globe), it nevertheless happens remarkably enough that their extremities do not increase and decrease in speed jointly, equally, and in the same instant of time. For it may happen that when one extremity of such a vessel is greatly retarded in its motion by virtue of a composition of these two motions, annual and diurnal, the other extremity may be affected by and involved in even a very swift motion.

For your easier comprehension, let us explain this by going back to the diagram previously drawn. Let us suppose a stretch of sea to be as long as one quadrant; the arc BC, for instance. Then the parts near B are, as I said before, in very swift motion

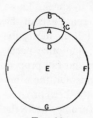

FIG. 29

because the two movements (annual and diurnal) are united in the same direction, and the parts near C are at that time in retarded motion, since they lack the forward movement depending upon the diurnal motion. If we suppose, I say, a sea bottom as long as the arc BC, we shall see at once that its extremities are moving very unequally at a given time. A stretch of sea as long as a semicircle and placed in the position of the arc BCD will have exceedingly different speeds, since the extremity B would be in very rapid motion, D in very slow motion, and the parts in the middle around C in moderate motion. In proportion as these stretches of sea were shorter, they would participate less in this strange phenomenon of having their parts diversely affected at certain times of day by speed and by slowness of motion.

Now if in the first place we see experimentally that an acceleration and a retardation shared equally by all parts of the containing vessel may indeed be the cause of the contained water running back and forth, then what must we suppose would happen in a vessel so remarkably situated that a retardation and an acceleration of motion are conferred very unevenly upon its parts? Certainly we cannot help saying that there would necessarily be perceived still greater and more marvelous causes of commotions in the water, and stranger ones. And though to many people it may seem impossible for us to test the effects of such events in artificial devices and vessels, nevertheless this is not entirely impossible; I have a mechanical model in which the effects of these marvelous compositions of movements may be observed in detail. But so far as our present purpose is concerned, what we have grasped intellectually up to this point is sufficient.

SAGR. For my part, I understand well enough that this remarkable phenomenon must necessarily exist in the ocean beds, especially in those which extend a long distance east and west; that is, along the direction of the movements of the terrestrial globe. And as the phenomenon is in a certain sense undreamed of and without parallel among the movements it is possible for us to make, it is not hard for me to believe that it may produce effects which cannot be imitated in our artificial experiments.

SALV. These things being cleared up, it is now time to examine in all their diversity the particular events which are observed experientially in the ebbing and flowing of the waters. First, it cannot be hard for us to understand why it happens that in lakes, pools, and even in small seas there is no noticeable tide. There are two impelling reasons for this. One is that because of the shortness of their basins they acquire at different hours of the day varying degrees of speed, but with little difference occurring among all their parts; they are uniformly accelerated and retarded as much in front as behind; that is, to the east as to the west. And they acquire such alterations, moreover, little by little, and not through the opposition of a sudden obstacle and hindrance, or a sudden and great acceleration in the movement of the containing vessel. The latter, with all its parts, becomes slowly and equally impressed with the same degree of velocity, and from this uniformity it follows that the contained water also receives the same impressions with little resistance or hesitation. Consequently the signs of rising and falling or of running to one extremity or the other are exhibited only obscurely. This effect is also clearly seen in small artificial vessels, in which the contained water is impressed with the same degrees of speed whenever the acceleration or retardation is made in slow and uniform increments. But in the basins of oceans which extend a great distance from east to west, the acceleration or retardation is much more noticeable and uneven when one extremity of them is in a very retarded motion and the other is moving quickly.

The second reason is the reciprocal oscillation of the water instituted by the impetus already received from the motion of its

Giving reasons for the particular events observed in the tides.

Reasons why tides occur neither in small seas nor in lakes.

container, which oscillation (as we have remarked) makes its vibrations with high frequency in small vessels. There inheres in the terrestrial movements a cause for conferring a movement upon the waters only from one twelve-hour period to another, since only once a day is the movement of the containing vessel exceedingly accelerated or retarded. Now this second cause depends upon the weight of the water, which seeks to restore it to equilibrium, and it produces oscillations of one, two, or three hours, and so on, according to the shortness of the vessel. Thus the whole movement becomes entirely insensible upon this one being combined with the first, which even by itself remains very small for small vessels. For the primary cause, which has a period of twelve hours, will not have finished impressing its disturbance when overtaken and reversed by this second one depending upon the weight of the water and having a vibration time of one, two, three, or four hours, and so on, according to the shortness and depth of the basin. Acting contrary to the first cause, this perturbs and removes that without ever allowing it to attain the height, or even the average of its motion. Any evidence of ebbing or flowing is entirely annihilated by this conflict, or is very much obscured. I say nothing of the continual changing of the wind, which by disquieting the water would not permit us to be sure of some very small rising or falling, of half an inch or less, which might actually belong to the basins and containers of bodies of water no more than one degree or so in length.

Reason why the tides are for the most part made in six-hour periods. Now, secondly, I shall resolve the question why, since there resides in the primary principle no cause of moving the waters except from one twelve-hour period to another (that is, once by the maximum speed of motion and once by its maximum slowness), the period of ebbing and flowing nevertheless commonly appears to be from one six-hour period to another. Such a determination, I say, can in no way come from the primary cause alone. The secondary causes must be introduced for it; that is, the greater or lesser length of the vessels and the greater or lesser depth of the waters contained in them. These causes, although they do not operate to move the waters (that action being from the primary cause alone, without which there would be no tides), are nevertheless the principal factors in limiting the duration of the reciprocations, and operate so powerfully that the primary cause must bow to them. Six hours, then, is not a more proper or natural period for these reciprocations than any other interval of time, though perhaps it has been the one most generally observed because it is that of our Mediterranean, which has been the only place practicable for making observations over many centuries. Even so, this period is not observed everywhere in it; in some of the narrower places, such as the Hellespont and the Aegean, the periods are much briefer, and they are also quite variable among themselves. Some say it was because of these differences and the incomprehensibility of their causes to Aristotle that he, after observing them for a long time from some cliffs of Euboea (*Negroponte*), plunged into the sea in a fit of despair and willfully destroyed himself.

In the third place we shall see very readily the reason why a sea like the Red Sea, although very long, is nevertheless quite devoid of any tide. This is so because its length does not extend from east to west, but runs from southeast to northwest. The movements of the earth being from west to east, the impulses of the water are always aimed against the meridians and not from one parallel to another. Hence in seas which extend lengthwise toward the poles and are narrow in the other direction, there is no cause of tides — unless it is that of sharing those of some other sea with which they may communicate and which is subject to large movements.

Cause for some very long seas having no tides.

We can very easily understand, in the fourth place, the reasons why the ebbing and flowing are greatest at the extremities of gulfs as to rising and falling of the waters, and least in the middle parts. Daily experience shows us this here in Venice, situated at the end of the Adriatic, where the difference commonly amounts to as much as five or six feet; but in parts of the Mediterranean distant from the extremities such changes are very small; as at the islands of Corsica and Sardinia, and on the coasts at Rome and Leghorn, where they do not exceed half a foot. We understand also why, on the other hand, where the rising and falling are small, the running to and fro is large. It is a simple thing, I say, to understand the cause of these events, because we have examples of them easily observable in all sorts of artificially manufactured vessels, in which the same effects are seen to follow naturally when we move them unevenly; that is, now accelerating and now retarding them.

Why tides are highest at the extremities of gulfs and lowest in their central parts.

Let us consider further, in the fifth place, how a given quan-

Why the course
of the waters is
faster in narrow
places than in
spacious ones.

tity of water moving slowly in a spacious channel must run very
impetuously when it has to pass through a narrow place. From
this we shall have no difficulty in understanding the cause of the
great current which is created in the narrow channel that separates Calabria from Sicily. For all the water pent up by the
extensive island and the Ionian Gulf in the eastern part of the
sea, though because of the spaciousness there it descends slowly
towards the west, yet upon being restrained in the Straits of
Messina between Scylla and Charybdis, it drops rapidly and
makes a great agitation. Something similar to this, but greater,
is said to occur between Africa and the great island of Mada-
gascar (*San Lorenzo*), when the waters of the two great Indian
and South Atlantic (*Etiopico*)† oceans, in whose midst this lies,
must be restricted in their running into the still smaller channel
between it and the coast of South Africa. The currents in the
Straits of Magellan must be extremely great, communicating
between the South Atlantic and the South Pacific oceans.

Discussion of
some less obvious
events which are
observed in the
tides.

In the sixth place, in order to give reasons for some more
recondite and curious events that are observed in this field, it
remains now for us to make another important reflection upon
the two principal causes of the tides, thereafter compounding
them and mixing them together. The first and simplest of these,
as I have often said, is the definite acceleration and retardation
of the parts of the earth from which the waters receive a determinate period, running toward the east and returning to the
west within a space of twenty-four hours. The other depends
upon the water's own weight, which, once moved by the primary
cause, tries then to restore itself to equilibrium by repeated oscillations which are not determinate as to one preëstablished time
alone, but which have differences of duration according to the
different lengths and depths of the containers and basins of the
oceans. In so far as they depend upon this second principle, some
would flow and return in one hour, some in two, in four, in six,
in eight, in ten, etc.

Now if we commence to add the first cause, which has an established period of twelve hours, to the second when it has for
example a period of five, then it will sometimes happen that the
primary and secondary causes agree in making their impulses
both in the same direction; and in such a conjunction (or, so to
speak, in such a unanimous conspiracy) the tides will be very

great. At other times it happens that the primary impulse be- comes in a certain sense contrary to that brought by the secondary; and in such encounters one impulse takes away what the other gives, so that the motion of the waters is weakened and the sea is reduced to a very peaceful and practically motionless state. At still other times, when the two principles are not in opposition nor yet entirely unified, they cause other variations in the rise and fall of the tides.

It may also happen that two very large seas which are in communication through some narrow channel are found to have, because of the mixture of the two principles of motion, a cause of flood in one at the very time the other is having the contrary movement. In this case extraordinary agitations are made in the channel through which they communicate, with opposing motions and vortexes and most dangerous churnings, of which in fact we hear continual tales and accounts. From such discordant movements, depending not only upon different situations and lengths, but even more upon the differing depths of the communicating seas, there sometimes arise various disorderly and unobservable aquatic commotions whose causes have perturbed sailors very much, and still do, when encountered in the absence either of gusts of wind or other significant atmospheric changes which might account for them.

Now these disturbances of the air must be carefully taken into consideration with the other phenomena, and regarded as a third occasional cause capable of greatly altering our observations of effects dependent upon the primary† and more essential causes. For there is no doubt that strong winds blowing continuously from the east, for instance, may sustain the waters, preventing their ebb. If then a second recurrence of the high tide, and even a third, is added at the established hours, the waters will swell up very high. In such a way, sustained for several days by the force of the wind, they may be raised much more than usual, and make extraordinary floods.

We must also take notice of another cause of movement, and this will be our seventh problem. This depends upon the great quantity of water from the rivers that empty into seas which are not vast, for which reason the water is seen to run always in the same direction in channels or straits through which such seas communicate, as happens in the Thracian Bosporus below Con-

Why in some narrow channels the water of the sea is seen to run always in the same direction.

stantinople, where the water runs always from the Black Sea
toward the Sea of Marmara (*Propontide*). For the Black Sea the
principal causes of ebb and flow are not very effective, because
of its shortness; while on the other hand very large rivers empty
into it, and this great flow of water must be passed and disgorged
through the strait, where the current is quite famous and is always toward the south. Moreover, we must take note that this
strait or channel, though it is certainly very narrow, is not subjected to any such perturbations as the strait between Scylla and
Charybdis; for the former has the Black Sea above it to the
north, with the Sea of Marmara, the Aegean Sea, and the Mediterranean adjoining it to the south — though over a long tract,
and, as we have already noted, however long a sea may be from
north to south, it is not subject to tides. But since the Sicilian
strait is situated between parts of the Mediterranean, extending
a great distance from west to east — that is, with the tidal currents — the agitations in it are very great. They would be still
greater at the Gates of Hercules, if the Straits of Gibraltar were
less open; and the currents in the Straits of Magellan are reported to be extremely strong.

This is all that occurs to me at present to tell you about the
causes of this basic diurnal period of the tides, and of their various incidental phenomena. If anything is to be brought up in
connection with these, it may be done now; then we may proceed
to the other two periods, the monthly and the annual.

SIMP. I do not think it can be denied that your argument goes
along very plausibly, the reasoning being *ex suppositione,* as we
say; that is, assuming that the earth does move in the two motions assigned to it by Copernicus. But if we exclude these movements, all the rest is vain and invalid; and the exclusion of this
hypothesis is very clearly pointed out to us by your own reasoning. Under the assumption of the two terrestrial movements, you
give reasons for the ebbing and flowing; and vice versa, arguing
circularly, you draw from the ebbing and flowing the sign and
confirmation of those same two movements. Passing to a more
specific argument, you say that on account of the water being a
fluid body and not firmly attached to the earth, it is not rigorously
constrained to obey all the earth's movements. From this you
deduce its ebbing and flowing.

In your own footsteps, I argue the contrary and say: The air

Opposing the
hypothesis of the
earth's motion
being considered
in favor of the
ocean tides.

is even more tenuous and fluid than the water, and less affixed to the earth's surface, to which the water adheres (if for no other reason) because of its own weight, which presses it down much more than the very light air. Then so much the less should the air follow the movements of the earth; hence if the earth did move in those ways, we, its inhabitants, carried along at the same velocity, would have to feel a wind from the east perpetually beating against us with intolerable force. That such would necessarily follow, daily experience informs us; for if, in riding post with no more speed than eight or ten miles an hour in still air, we feel in our faces what resembles a wind blowing against us not lightly, just think what our rapid course of eight hundred or a thousand miles per hour would have to produce against air which was free from such motion! Yet we feel nothing of any such phenomenon.

SALV. To this objection, which seems so persuasive, I reply that it is true that the air is much more tenuous and much lighter than the water, and by its lightness is much less adherent to the earth than heavy and bulky water. But the consequence which you deduce from these conditions is false; that is, that because of its lightness, tenuity, and lesser adherence to the earth it must be freer than water from following the movements of the earth, so that to us who participate completely in those movements its disobedience would be made sensible and evident. In fact, quite the opposite happens. For if you will remember carefully, the cause of the ebbing and flowing of the water assigned by us consisted in the water not following the irregularity of motion of its vessel, but retaining the impetus which it had previously received, and not diminishing it or increasing it in the exact amount by which this is increased or diminished in the vessel. Now since disobedience to a new increase or diminution of motion consists in conservation of the original received impetus, that moving body which is best suited for such conservation will also be best fitted for exhibiting the effect that follows as a consequence of this conservation. How strongly water is disposed to preserve a disturbance once received, even after the cause impressing it has ceased to act, is demonstrated to us by the experience of water highly agitated by strong winds. Though the winds may have ceased and the air become tranquil, such waves remain in motion for a long time, as the sacred poet so charmingly sings: *Qual*

Reply to the objection made against the whirling of the terrestrial globe.

Water better capable of keeping a received impetus than is air.

Light bodies
more easily
moved than
heavy ones, but
less able to retain
their motion.

l'alto Egeo, etc. The continuance of the commotion in this way depends upon the weight of the water, for as has been said on other occasions, light bodies are indeed much easier to set in motion than heavier ones, but they are also much less able to keep the motion impressed upon them, once the cause of motion stops. The air, being a thing that is in itself very tenuous and extremely light, is most easily movable by the slightest force; but it is also most inept at conserving the motion when the mover ceases acting.

It is more rea-
sonable that the
air is swept
along by the
rough surface of
the earth than by
celestial motion.

As to the air that surrounds the terrestrial globe, I shall therefore say that it is carried around by its adherence no less than the water, and especially those parts of it which are contained in vessels, these vessels being plains surrounded by mountains. And we may much more reasonably declare that such parts are carried around, swept along by the roughness of the earth, than that the higher parts are swept along by the celestial motion as the Peripatetics assert.

What I have said so far seems to me to be an adequate reply to Simplicio's objection. But I want to give him more than satisfaction by means of a new objection and another reply, founded upon a remarkable experiment, and at the same time substantiate for Sagredo the mobility of the earth.

Confirming the
earth's whirling
by a new argu-
ment, borrowed
from the air.

I have said that the air, and especially that part of it which is not above the highest mountains, is carried around by the roughness of the earth's surface. From this it seems to follow that if the earth were not uneven, but smooth and polished, there would be no reason for its taking the air along as company, or at least for its conducting it with so much uniformity. Now the surface of this globe of ours is not all mountainous and rough, but there are very large areas that are quite smooth; such are the surfaces of the great oceans. These, being also quite distant from the mountain ranges that encircle them, appear not to have any aptitude for carrying along the air above them; and whatever may follow as a consequence of not carrying it ought therefore to be felt in such places.

Simp. I also wanted to raise this same objection, which seems to me very powerful.

Salv. You may well say this, Simplicio, in the sense that from no such thing being felt in the air as would result from this globe of ours going around, you argue its immobility. But what if this thing that you think ought to be felt as a necessary consequence were, as a matter of fact, actually felt? Would you accept this as a sign and a very powerful argument of the mobility of this same globe?

SIMP. In that case it would not be a matter of dealing with me alone; for if this should happen and its cause were hidden from me, perhaps it might be known to others.

SALV. So no one can ever win against you, but must always lose; then it would be better not to play. Nevertheless, in order not to cheat our umpire, I shall go on.

We have just said, and will now repeat with some additions, that the air, as a tenuous and fluid body which is not solidly attached to the earth, seems to have no need of obeying the earth's motion, except in so far as the roughness of the terrestrial surface catches and carries along with it that part of the air which is contiguous to it, or does not exceed by any great distance the greatest altitude of the mountains. This portion of the air ought to be least resistant to the earth's rotation, being filled with vapors, fumes, and exhalations, which are materials that participate in the earthy properties and are consequently naturally adapted to these same movements. But where the cause for motion is lacking — that is, where the earth's surface has large flat spaces and where there would be less admixture of earthy vapors — the reason for the surrounding air to obey entirely the seizure of the terrestrial rotation would be partly removed. Hence, while the earth is revolving toward the east, a beating wind blowing from east to west ought to be continually felt in such places, and this blowing should be most perceptible where the earth whirls most rapidly; this would be in the places most distant from the poles and closest to the great circle of the diurnal rotation.

Now the fact is that actual experience strongly confirms this philosophical argument. For within the Torrid Zone (that is, between the tropics), in the open seas, at those parts of them remote from land, just where earthy vapors are absent, a perpetual breeze is felt moving from the east with so constant a tenor that, thanks to this, ships prosper in their voyages to the West Indies. Similarly, departing from the Mexican coast, they plow the waves of the Pacific Ocean with the same ease toward the East Indies, which are east to us but west to them. On the other hand,

Winds from the
land disturb the
seas.

voyages from the Indies eastward are difficult and uncertain, nor may they in any case be made along the same routes, but must be piloted more toward the land so as to find other occasional and variable winds caused by other principles, such as we dwellers upon terra firma continually experience. There are many and various reasons for the origin of such winds which we need not bother to bring up at present. These occasional winds blow indifferently toward all parts of the earth, disturbing seas distant from the equator and bordered by the rough surface of the earth. This amounts to saying that such seas are subjected to those disturbances of the air which interfere with the primary current of air that would be felt continually, especially on the ocean, if such accidental disturbances were lacking.

Now you see how the actions of the water and the air show themselves to be remarkably in accord with celestial observations in confirming the mobility of our terrestrial globe.

SAGR. Yet in order to cap all this, I wish also to tell you one particular which seems to me to be unknown to you, yet which confirms this same conclusion. You, Salviati, have mentioned that phenomenon which sailors encounter in the tropics; I mean that constant wind blowing from the east, of which I have heard accounts from those who have made the voyage quite often. Moreover, it is an interesting fact that sailors do not call this a "wind," but have some other name for it which slips my mind, taken perhaps from its even tenor. When they encounter it, they tie up their shrouds and the other cordage of the sails, and without ever again having any need to touch these, they can continue their voyage in security, or even asleep. Now this perpetual breeze has been known and recognized by reason of its blowing continuously without interruption; for if other winds had interrupted it, it would not have been recognized as a singular effect different from all the others. From this I may infer that the Mediterranean Sea might also participate in such a phenomenon, but that this escapes unobserved because it is frequently interrupted by other supervening winds. I say this advisedly, and upon very probable theories which occurred to me from what I had occasion to learn during the voyage I made to Syria when I went to Aleppo as consul of our nation. Keeping a special record and account of the days of departure and arrival of ships at the ports of Alexandria, Alexandretta, and here at Venice, I

Another obser-
vation borrowed
from the air in
support of the
earth's motion.

discovered in these again and again that, to my great interest, the returns here (that is, the voyages from east to west over the Mediterranean) were made in proportionately less time than those in the opposite direction, in a ratio of 25 per cent. Thus we see that on the whole the east winds are stronger than those from the west.

SALV. I am glad to know of this detail, which contributes not a little confirmation to the mobility of the earth. And though it may be said that all the water of the Mediterranean pours perpetually through the Straits of Gibraltar, having to disgorge into the ocean all the waters of so many rivers that empty into it, I do not believe that the current can be so strong that it alone could make such a remarkable difference. This is also evident from seeing that the water at Pharos runs back toward the east no less than it courses toward the west.

SAGR. I, who unlike Simplicio, have not been worrying about convincing anybody besides myself, am satisfied with what has been said regarding this first part. Therefore, Salviati, if you wish to proceed, I am ready to listen.

SALV. I am yours to command; but I should like to hear also how it looks to Simplicio, for from his judgment I can estimate how much I may expect from these arguments of mine in the Peripatetic schools, should they ever reach those ears.

SIMP. I do not want you to take my opinion as a basis for guessing at the judgments of others. As I have often said, I am among the tyros in this sort of study, and things which would occur to those who have penetrated into the profoundest depths of philosophy might never occur to me; for, as the saying goes, I have hardly greeted its doorkeeper. Yet to show some spark of fire, I shall say that as for the effects recounted by you, and this last one in particular, it seems possible to me to render quite sufficient reasons from the mobility of the heavens alone, without introducing any novelties beyond the mere converse of what you yourself have brought into the field.

It is admitted by the Peripatetic school that the element of fire and a large part of the air are carried around in the diurnal rotation from east to west by contact with the lunar sphere as their containing vessel. Now without deviating from your footprints, I should like us to establish the quantity of air participating in that motion as that part which comes down about to the

summits of the highest mountains, and would extend on down
to the earth itself if the obstacle presented by these very mountains did not hinder it. Thus, just as you declared that the air
surrounding the mountain ranges is carried around by the roughness of the moving earth, we say the converse — that all the
element of air is carried around by the motion of the heavens
except that part which is lower than the mountain peaks, this
being impeded by the roughness of the immovable earth. And
where you would say that if such roughness were removed, this
would also free the air from being caught, we may say that if this
roughness were removed, all the air would proceed in this movement. And since the surfaces of the open seas are smooth and
level, the motion of the breeze which blows perpetually from the
east continues there, and is more noticeable at places near the
equator, within the tropics, where the motion of the heavens is
most rapid.

Motion of the
water dependent
upon the celes-
tial motion.

And as this celestial movement is powerful enough to carry
the free air with it, we may say quite reasonably that it contributes this same motion to the movable water. For this is fluid,
and unattached to the earth's immobility. We may affirm this
with the more confidence in view of your own admission that
such a movement need be only very small with respect to its
effective cause, which, going around the entire terrestrial globe in
one natural day, passes over many thousands of miles per hour
(especially near the equator), while currents in the open sea
move but a very few miles per hour. In this way our voyages
toward the west would be much more convenient and rapid,
being assisted not only by the perpetual eastern breeze, but also
by the course of the waters.

Flow and ebb
may depend
upon the diurnal
movement of the
heavens.

Perhaps from that same coursing of the water, tides also may
arise; the water, striking against the variously situated shores,
might even return straight back in the opposite direction, as experience shows us in the courses of rivers. For there the water,
because of the irregularity of the banks, often meets some part
which juts out or which makes a hollow from beneath, and it
whirls around and is seen to return perceptibly. Hence it seems
to me that the same effects from which you argue the mobility
of the earth (and which mobility you offer as a cause for them)
may be sufficiently explained if we hold the earth fixed and restore the mobility to the heavens.

SALV. It cannot be denied that your argument is ingenious and carries something of probability, but I say that this is a probability in appearance only and not in reality. There are two parts to your argument; in the first, you render a reason for the continual motion of the eastern breeze, and also for the motion of the water; in the second, you wish also to obtain a cause for the tides from the same source. The first part, as I have said, has some semblance of probability, though much less than we achieve from terrestrial motion. The second part is not only entirely improbable, but is absolutely impossible and false.

The reason for the continual motion of the air and water may be more plausibly rendered by making the earth movable than by making it fixed.

As to the first, in which you say that the hollow of the lunar sphere sweeps along with it the element of fire and all the air down to the summits of the highest mountains, I say first that there is doubt whether any element of fire exists. Even assuming that it does, it is extremely doubtful whether the lunar sphere exists; or indeed, whether any of the other "spheres" do. That is to say, it is questionable whether there actually are such bodies, solid and extremely vast, or whether beyond the air there does not rather extend a continuous expanse of a substance very much more tenuous and pure than our air, and whether the planets do not wander through this, as is now commencing to be held even by most of these same philosophers.

But however that may be, there is no reason for us to believe that fire, by simple contact with a surface which you yourself consider to be remarkably smooth and even, should in its entire extent be carried around in a motion foreign to its own inclination. This has been proved throughout *Il Saggiatore,* and demonstrated by sensible experiments. Beyond this, there is the further improbability of such motions being transferred from most subtle fire to the air, which is much denser, and then from this to water.

It is improbable that the element of fire is drawn along by the lunar orb.

But that a body of very rough and mountainous surface, by revolving, should conduct along with it the contiguous air which strikes against its prominences is not merely probable, but necessary; it may be seen from experience, though I believe that even without seeing it no one would cast doubt upon it.

As for the rest, assuming that the air and even the water were conducted by the motion of the heavens, such a motion would have nothing whatever to do with the tides. For since from one uniform cause only one single uniform effect can follow, there

Flow and ebb cannot depend upon the celestial movement.

would have to be discovered in the waters a continual and uniform current from east to west, existing only in those oceans which, returning upon themselves, encircle the globe. In inland seas such as the Mediterranean, hemmed in as it is on the east, there could be no such motion. For if its waters were driven by the course of the heavens toward the west, it would have been dried up many centuries ago; besides which, our waters do not run only toward the west, but return back toward the east in regular periods. If indeed you should say, from the example of the rivers, that the course of the seas was originally from east to west only, but that the different situations of their shores might force some of the water to flow in reverse, then I shall grant you this, Simplicio; but you must take note that wherever the water is turned back for this reason, it perpetually returns again, while where it runs forward, it always keeps going in the same direction, as you may see from your example of the rivers. As to the tides, you must discover and bring forth reasons for making them run now one way and now the other at the same place — effects which, being contrary and irregular, you can never deduce from one uniform and constant cause. This, as well as overthrowing the idea of a motion being contributed to the sea by the diurnal movement of the heavens, also defeats those who would like to grant to the earth only the diurnal motion and who believe that with this alone they can give a reason for the tides. For since the effect is irregular, it is necessarily required that its causes shall be irregular and variable.

SIMP. I have nothing further to say; neither on my own account, because of my lack of inventiveness, nor on that of others, because of the novelty of the opinion. But I do indeed believe that if this were broadcast among the schools, there would be no lack of philosophers who would be able to cast doubt upon it.

SAGR. Then let us wait until that happens. In the meantime, if it is satisfactory with you, Salviati, let us proceed.

SALV. Everything that has been said up to this point pertains to the diurnal period of the tides, of which the primary and universal cause has first been proved, without which no effect whatever would take place. Next, passing on to the particular events to be observed in this diurnal period (which vary and are in a certain sense irregular), the secondary and concomitant causes upon which these depend remain to be dealt with.

Now two other periods occur, the monthly and the annual. These do not introduce new and different events beyond those already considered under the diurnal period, but they act upon the latter by making them greater or less at different parts of the lunar month and at different seasons of the solar year — almost as though the moon and sun were taking part in the production of such effects. But that concept is completely repugnant to my mind; for seeing how this movement of the oceans is a local and sensible one, made in an immense bulk of water, I cannot bring myself to give credence to such causes as lights, warm temperatures, predominances of occult qualities, and similar idle imaginings. These are so far from being actual or possible causes of the tides that the very contrary is true. The tides are the cause of them; that is, make them occur to mentalities better equipped for loquacity and ostentation than for reflections upon and investigations into the most hidden works of nature. Rather than be reduced to offering those wise, clever, and modest words, "I do not know," they hasten to wag their tongues and even their pens in the wildest absurdities.

We see that the moon and the sun do not act upon small receptacles of water by means of light, motion, and great or moderate heat; rather, we see that to make water rise by heat, one must bring it almost to boiling. In short, we cannot artificially imitate the movement of the tides in any way except by movement of the vessel. Now should not these observations assure anyone that all the other things produced as a cause of this effect are vain fantasies, entirely foreign to the truth of the matter?

Thus I say that if it is true that one effect can have only one basic cause, and if between the cause and the effect there is a fixed and constant connection, then whenever a fixed and constant alteration is seen in the effect, there must be a fixed and constant variation in the cause. Now since the alterations which take place in the tides at different times of the year and of the month have their fixed and constant periods, it must be that regular changes occur simultaneously in the primary cause of the tides. Next, the alterations in the tides at the said times consist of nothing more than changes in their sizes; that is, in the rising and lowering of the water a greater or less amount, and its running with greater or less impetus. Hence it is necessary that whatever the primary cause of the tides is, it should increase or

Alterations in effects imply alterations in the causes.

The causes of the monthly and annual periods of the tides are assigned at length.

diminish its force at the specific times mentioned. But it has already been concluded that an irregularity and unevenness in the motion of the vessel containing the water is the primary cause of the tides; therefore this unevenness must become correspondingly still more irregular from time to time (that is, must increase or diminish).

Now we must remember that the unevenness (that is, the varying velocity of the vessels which are parts of the earth's surface) depends upon these vessels moving with a composite motion, the resultant of compounding the annual and the diurnal motions which belong to the entire terrestrial globe. Of these the diurnal whirling, with its alternate addition to and subtraction from the annual movement, is the thing that produces the unevenness of the compound motion. Thus the primary cause of the uneven motion of the vessels, and hence of that of the tides, consists in the additions and subtractions which the diurnal whirling makes with respect to the annual motion. And if these additions and subtractions were always made in the same proportion with respect to the annual motion, the cause of tides would indeed continue to exist, but only a cause for their being perpetually made in the same manner. Now we must find a reason for these same tides being made greater and less at different times; hence, if we wish to preserve the identity of the cause, there is a necessity of finding changes in these additions and subtractions, making them more and less potent at producing those effects which depend upon them. But I do not see how this can be done except by making these additions and subtractions now greater and now less, so that the acceleration and retardation of the composite motion shall be made now in a greater and now in a lesser ratio.

SAGR. I feel myself being gently led by the hand; and although I find no obstacles in the road, yet like the blind I do not see where my guide is leading me, nor have I any means of guessing where such a journey must end.

SALV. There is a vast difference between my slow philosophizing and your rapid insights; yet in this particular with which we are now dealing, I do not wonder that even the perspicacity of your mind is beclouded by the thick dark mists which hide the goal toward which we are traveling. All astonishment ceases when I remember how many hours, how many days, and how many more

Monthly and annual alterations of the tides can depend only upon changes in the additions and subtractions of diurnal and annual motions.

nights I spent on these reflections; and how often, despairing of ever understanding it, I tried to console myself by being convinced, like the unhappy Orlando, that that could not be true which had been nevertheless brought before my very eyes by the testimony of so many trustworthy men. So you need not be surprised if for once, contrary to custom, you do not foresee the goal. And if you are nevertheless dismayed, then I believe that the outcome (which so far as I know is entirely unprecedented) will put an end to this puzzlement of yours.

SAGR. Well, thank God for not letting your despair lead you to the end that befell the miserable Orlando, or to that which is perhaps no less fictitiously related of Aristotle; for then everyone, myself included, would be deprived of the revelation of something as thoroughly hidden as it is sought after. Therefore I beg you to satiate my greed for it as quickly as you can.

SALV. I am at your service. We have arrived at an inquiry as to how the additions and subtractions of the terrestrial whirling and the annual motion might be made now in greater and now in lesser ratios; for it is such a diversity, and nothing else, that may be assigned as a cause for the monthly and annual changes in the size of the tides. I shall next consider three ways in which this ratio of the additions and subtractions of the earth's rotation and the annual motion may be made greater and less.

The proportion in which whirling is added to the annual motion may be altered in three ways.

First, this could be done by the velocity of the annual motion increasing and decreasing while the additions and subtractions made by the diurnal whirling remained constant in magnitude. For since the annual motion is about three times as fast† as the diurnal motion, even taking the latter at the equator, then if we were to increase it further, the addition or subtraction of the diurnal motion would make less of an alteration. On the other hand if it were made slower, this same diurnal motion would alter it proportionately more. Thus to add or subtract four degrees of speed when dealing with something which moves with twenty degrees will alter its course less than if the same four degrees were added to or subtracted from something which moved with only ten degrees of speed.

The second way would be by making the additions and subtractions greater or smaller, retaining the annual motion at the same velocity. This is very easy to see, since it is obvious that a velocity of twenty degrees (for instance) will be altered more by the addition or subtraction of ten degrees than by the addition or subtraction of four.

The third manner would be a combination of these two, the annual motion diminishing and the diurnal additions and subtractions increasing.

As you see, it was easy to get this far; yet it was indeed a laborious task for me to discover how such effects could be accomplished in nature. Yet I finally found something that served me admirably. In a way it is almost unbelievable. I mean that it is astonishing and incredible to us, but not to Nature; for she performs with the utmost ease and simplicity things which are even infinitely puzzling to our minds, and what is very difficult for us to comprehend is quite easy for her to perform.

What is very hard for us to understand is very easy for Nature to perform.

To continue, then. Having demonstrated that the proportions between the additions and subtractions of the whirling on the one hand and the annual motion on the other may be made greater and less in two manners (I say two, because the third is a composite of the others), I add now that Nature does make use of both; and I add further that if she made use of but one of them, then one of the two periodic alterations of the tide would necessarily be removed. The monthly periodic changes would cease if there were no variation due to the annual motion, and if the additions and subtractions of the diurnal rotation were kept always equal, then the annual periodic alterations would be missing.

If the annual movement did not alter, the monthly period would cease.

If the diurnal motion did not alter, the annual period would cease.

SAGR. Then do the monthly alterations of the tides depend upon changes in the annual motion of the earth? And the annual alterations in the ebb and flow are derived from the additions and subtractions of the diurnal rotation? Now I am more confused than ever, and farther from any hope of being able to comprehend how this complication comes about, more intricate to my mind than the Gordian knot. I envy Simplicio, from whose silence I deduce that he understands everything and is free from the confusion that beclouds my imagination.

SIMP. I really believe that you are confused, Sagredo, and I also think I know the cause of your confusion. In my opinion this originates from your understanding a part of what Salviati has set forth, and not understanding another part. And you are also correct about my not being confused at all, though not for the reason you suppose; that is, that I understand the whole thing.

Quite the contrary; I understand nothing whatever of it, and confusion lies in the multiplicity of things — not in nothing.

SAGR. You see, Salviati, how the checkrein that has been applied to Simplicio in the past sessions has gentled him, and changed him from a skittish colt into an ambling nag.

But please, without more delay, put an end to this suspense for both of us.

SALV. I shall do my best to overcome my obscure way of expressing myself, and the sharpness of your wits will fill up the dark places.

There are two events whose causes we must investigate; the first concerns the variation which occurs in the tides over a monthly period, and the other belongs to the annual period. We shall speak first of the monthly, and then deal with the annual; and we must first resolve the whole according to the axioms and hypotheses already established, without introducing any innovations either from astronomy or from the universe to help out the tides. We shall demonstrate that the causes for all the various events perceived in the tides reside in things previously recognized and accepted as unquestionably true. Thus I say that one true, natural, and even necessary thing is that a single movable body made to rotate by a single motive force will take a longer time to complete its circuit along a greater circle than along a lesser circle. This is a truth accepted by all, and in agreement with experiments, of which we may adduce a few.

In order to regulate the time in wheel clocks, especially large ones, the builders fit them with a certain stick which is free to swing horizontally: At its ends they hang leaden weights, and when the clock goes too slowly, they can render its vibrations more frequent merely by moving these weights somewhat toward the center of the stick. On the other hand, in order to retard the vibrations, it suffices to draw these same weights out toward the ends, since the oscillations are thus made more slowly and in consequence the hour intervals are prolonged. Here the motive force is constant — the counterpoise — and the moving bodies are the same weights; but their vibrations are more frequent when they are closer to the center; that is, when they are moving along smaller circles.

Let equal weights be suspended from unequal cords, removed from the perpendicular, and set free. We shall see the weights

A most true hypothesis is that revolutions in smaller circles take place in shorter times than those in larger circles, as clarified by two examples.

First example.

Second example.

on the shorter cords make their vibrations in shorter times, being things that move in lesser circles. Again, attach such a weight to a cord passed through a staple fastened to the ceiling, and hold the other end of the cord in your hand. Having started the hanging weight moving, pull the end of the cord which you have in your hand so that the weight rises while it is making its oscillations. You will see the frequency of its vibrations increase as it rises, since it is going continually along smaller circles.

And here I want you to notice two details which deserve attention. One is that the vibrations of such a pendulum are made so rigorously according to definite times, that it is quite impossible to make them adopt other periods except by lengthening or shortening the cord. Of this you may readily make sure by experiment, tying a rock to a string and holding the end in your hand. No matter how you try, you can never succeed in making it go back and forth except in one definite time, unless you lengthen or shorten the string; you will see that it is absolutely impossible.

The other particular is truly remarkable; it is that the same pendulum makes its oscillations with the same frequency, or very little different—almost imperceptibly—whether these are made through large arcs or very small ones along a given circumference. I mean that if we remove the pendulum from the perpendicular just one, two, or three degrees, or on the other hand seventy degrees or eighty degrees, or even up to a whole quadrant, it will make its vibrations when it is set free with the same frequency in either case; in the first, where it must move only through an arc of four or six degrees, and in the second where it must pass through an arc of one hundred sixty degrees or more. This is seen more plainly by suspending two equal weights from two threads of equal length, and then removing one just a small distance from the perpendicular and the other one a very long way. Both, when set at liberty, will go back and forth in the same times, one by small arcs and the other by very large ones.

From this follows the solution of a very beautiful problem, which is this: Given a quarter of a circle — I shall draw it here in a little diagram on the ground — which shall be AB here, vertical to the horizon so that it extends in the plane touching at the point B; take an arc made of a very smooth and polished concave hoop bending along the curvature of the circumference

Two particular
events observ-
able in pendu-
lums and their
oscillations.

Remarkable
problem of
moving bodies
descending along
a quadrant and
of things de-
scending along
any chord of an
entire circle.

ADB, so that a well-rounded and smooth ball can run freely in
it (the rim of a sieve is well suited for this experiment). Now
I say that wherever you place the ball, whether near to or far
from the ultimate limit B — placing it at the point C, or at D,
or at E — and let it go, it will arrive at
the point B in equal times† (or insen-
sibly different), whether it leaves from
C or D or E or from any other point you
like; a truly remarkable phenomenon.

FIG. 30

Now add another, no less beautiful
than the last. This is that along all chords
drawn from the point B to points C, D,
E, or any other point (taken not only in
the quadrant BA, but in the whole circumference of the entire circle), the
same movable body will descend in absolutely equal times. Thus
in the same time which it takes to descend along the whole diameter erected perpendicular to the point B, it will also descend
along the chord BC even when that subtends but a single degree,
or yet a smaller arc.

And one more marvel: The motions of bodies falling along the
arcs of the quadrant AB are made in shorter times than those
made along the chords of the same arcs, so that the fastest motion, made in the shortest time, by a movable body going from
the point A to the point B will be along the circumference ADB
and will not be that which is made along the straight line AB,
although that is the shortest of all the lines which can be drawn
between the points A and B. Also, take any point in that same
arc (let it be, for instance, the point D), and draw two chords
AD and DB; then the moving body leaving from the point A will
get to B in less time going along the two chords AD and DB
than going along the single chord AB. The shortest time of all
will be that of its fall along the arc ADB, and similar properties
are to be understood as holding for all lesser arcs taken upward
from the lowest limit B.

SAGR. Enough; no more; you are confusing me so with marvels,
and are distracting my mind in so many directions, that I fear
only a small part of it will remain free and clear for me to apply
to the main subject we are dealing with — which, I regret to say,
is too obscure and difficult as it is. I beg you, as a favor to me,

that when we have finished with the theory of the tides there shall be other days when you will again honor this house of mine and of yours, to discuss the many other problems that have been left dangling. Perhaps they will be no less interesting and elegant than these which we have been treating in the days just past, and which ought to be finished today.

SALV. I shall be at your disposal, though we shall have to have more than one or two sessions if, in addition to the questions reserved to be separately dealt with, we wish to add the many that pertain both to local motion and to the motions natural to projectiles — subjects dealt with at length by our Lincean Academician.

Getting back to our original purpose, we were explaining that for things moved circularly by some motive force which is kept continually the same, the times of circulation are preëstablished and determined, and impossible to lengthen or shorten. Having given examples of this and brought forth sensible experiments which we can perform, we may affirm the same to be true of our experience of the planetary movements in the heavens, for which the same rule is seen to hold: Those which move in the larger circles consume the longer times in passing through them. We have the most ready observations of this from the satellites of Jupiter, which make their revolutions in short times. So there is no question that if, for example, the moon, continuing to be moved by the same motive force, were drawn little by little into smaller circles, it would acquire a tendency to shorten the times of its periods, in agreement with that pendulum which in the course of its vibrations had its cord shortened by us, reducing the radius of the circumference traversed. Now this example which I gave you concerning the moon actually takes place and is verified in fact. Let us remember that we had already concluded with Copernicus that it is not possible to separate the moon from the earth, about which it unquestionably moves in a month. Let us likewise remember that the terrestrial globe, always accompanied by the moon, goes along the circumference of its orbit about the sun in one year, in which time the moon revolves around the earth almost thirteen times. From this revolution it follows that the moon is sometimes close to the sun (that is, when it is between the sun and the earth), and sometimes more distant (when the earth lies between the moon and the

The earth's annual motion along the ecliptic irregular because of the motion of the moon.

sun). It is close, in a word, at the time of conjunction and new moon, it is distant at full moon and opposition, and its greatest distance differs from its closest approach by as much as the diameter of the lunar orbit.

Now if it is true that the force which moves the earth and the moon around the sun always retains the same strength, and if it is true that the same moving body moved by the same force but in unequal circles passes over similar arcs of smaller circles in shorter times, then it must necessarily be said that the moon when at its least distance from the sun (that is, at conjunction) passes through greater arcs of the earth's orbit than when it is at its greatest distance (that is, at opposition and full moon). And it is necessary also that the earth should share in this irregularity of the moon. For if we imagine a straight line from the center of the sun to the center of the terrestrial globe, including also the moon's orbit,† this will be the radius of the orbit in which the earth would move uniformly if it were alone. But if we locate there also another body carried by the earth, putting this at one time between the earth and the sun and at another time beyond the earth at its greatest distance from the sun, then in this second case the common motion of both along the circumference of the earth's orbit would, because of the greater distance of the moon, have to be somewhat slower than in the other case when the moon is between the earth and the sun, at its lesser distance. So that what happens in this matter is just what happened to the rate of the clock, the moon representing to us that weight which is attached now farther from the center, in order to make the vibrations of the stick less frequent, and now closer, in order to speed them up.

From this it may be clear that the annual movement of the earth in its orbit along the ecliptic is not uniform, and that its irregularity derives from the moon and has its periods and restorations monthly. Now it has already been decided that the monthly and annual periodic alterations of the tides could derive from no other cause than from varying ratios between the annual motion and the additions to it and subtractions from it of the diurnal rotation; and that such alterations might be made in two ways; that is, by altering the annual motion and keeping fixed the magnitudes of the additions, or by changing the size of these and keeping the annual motion uniform. We have now detected the first of these two ways, based upon the unevenness of the annual motion; it depends upon the moon, and has its period monthly. Thus it is necessary that for this reason the tides should have a monthly period within which they become greater and smaller.

Now you see how the cause of the monthly period resides in the annual motion, and at the same time you see what the moon has to do with this affair, and how it plays a role without having anything to do with oceans or with waters.

SAGR. If a very high tower were shown to someone who had no knowledge of any kind of staircase, and he were asked whether he dared to scale such a supreme height, I believe he would surely say no, failing to understand that it could be done in any way except by flying. But being shown a stone no more than half a yard high and asked whether he thought he could climb up on it, he would answer yes, I am sure; nor would he deny that he could easily climb up not once, but ten, twenty, or a hundred times. Hence if he were shown the stairs by which one might just as easily arrive at the place he had adjudged impossible to reach, I believe he would laugh at himself and confess his lack of imagination.

You, Salviati, have guided me step by step so gently that I am astonished to find I have arrived with so little effort at a height which I believed impossible to attain. It is certainly true that the staircase was so dark that I was not aware of my approach to or arrival at the summit, until I had come out into the bright open air and discovered a great sea and a broad plain. And just as climbing step by step is no trouble, so one by one your propositions appeared so clear to me, little or nothing new being added, that I thought little or nothing was being gained. So much the more is my wonder at the unexpected outcome of this argument, which has led me to a comprehension of things I believed inexplicable.

Just one difficulty remains from which I desire to be freed. If the movement of the earth around the zodiac in company with the moon is irregular, such an irregularity ought to have been observed and noticed by astronomers, but I do not know that this has occurred. Since you are better informed on these matters than I am, resolve this question for me and tell me what the facts are.

SALV. Your doubt is very reasonable, and in response to the objection I say that although astronomy has made great progress over the course of the centuries in investigating the arrangement and movements of the heavenly bodies, it has not thereby arrived at such a state that there are not many things still remaining undecided, and perhaps still more which remain unknown. It is likely that the first observers of the sky recognized nothing but a general motion of all the stars — the diurnal motion — but I think it was not long before they discovered that the moon is inconstant about keeping company with the other stars. Years would have passed before they had distinguished all the planets, however. In particular, I believe that Saturn, on account of its slowness, and Mercury, because of being rarely seen, were the last objects to be recognized as vagrant and wandering. Many more years probably passed before the stoppings and retrograde motions of the three outer planets were observed, and their approaches and retreats from the earth, which occasioned the need to introduce eccentrics and epicycles — things unknown even to Aristotle, who makes no mention of them. How long did Mercury and Venus, with their remarkable phenomena, keep astronomers in suspended judgment about their true locations, to mention nothing else? Thus even the ordering of the world bodies and the integral structure of that part of the universe recognized by us was in doubt up to the time of Copernicus, who finally supplied the true arrangement and the true system according to which these parts are ordered, so that we are certain that Mercury, Venus, and the other planets revolve about the sun and that the moon revolves around the earth. But we cannot yet determine surely the law of revolution and the structure of the orbit of each planet (the study ordinarily called planetary theory); witness to this fact is Mars, which has caused modern astronomers so much distress. Numerous theories have also been applied to the moon itelf since the time when Copernicus first greatly altered Ptolemy's theory.

Now to get down to our particular point; that is, to the apparent motions of the sun and moon. In the former there has been observed a certain great irregularity, as a result of which it passes the two semicircles of the ecliptic (divided by the equinoctial points) in very different times, consuming about nine days more† in passing over one half than the other; a dif-

There may be many things still undiscovered by astronomy.

Saturn, because it is slow, and Mercury, because it is rarely seen, were among the last to be discovered.

Detailed structures of the planetary orbits still not resolved.

The sun gets through one half of the zodiac nine days faster than the other.

ference which is, as you see, very conspicuous. It has not yet been observed whether the sun preserves a regular motion in passing through very small arcs, as for example those of each sign of the zodiac, or whether it goes at a pace now somewhat faster and now slower, as would necessarily follow if the annual motion belongs only apparently to the sun and really to the earth in company of the moon. Perhaps this has not even been looked into.

As to the moon, its cycles have been investigated principally in the interest of eclipses, for which it suffices to have an exact knowledge of its motion around the earth. The progress of the moon through particular arcs of the zodiac has accordingly not been investigated in thoroughgoing detail. Therefore the fact that there is no obvious irregularity is insufficient to cast doubt upon the possibility that the earth and the moon are somewhat accelerated at new moon and retarded at full moon in traveling through the zodiac; that is, in going along the circumference of the earth's orbit. This comes about for two reasons; first, that the effect has not been looked for, and second, that it cannot be very large.

Tides are very
small with
respect to the
vastness of the
oceans and the
speed of the ter-
restrial globe.

Nor is there any need for the irregularity to be very large in order to produce the effect that is seen in the alterations of the size of the tides. For not only the changes, but the tides themselves, are small with respect to the magnitude of the bodies in which they occur, though with respect to us and to our smallness they seem to be great things. Adding or deducting one degree of speed where there are naturally seven hundred or a thousand cannot be called a large change, either in what confers it or in what receives it; and the water of our sea, carried by the diurnal whirling, travels about seven hundred miles per hour. This is the motion common to it and to the earth, and therefore imperceptible to us. The motion which is made sensible to us in currents is not even one mile per hour (I am speaking of the open sea, and not of straits), and it is this that alters the great, natural primary motion.

Still, such a change is considerable with respect to us and to our ships. A vessel that can make, say, three miles per hour in quiet water under the power of its oars, will have its travel doubled by such a current favoring it instead of opposing it. This is a very notable difference in the motion of the boat, though it is quite small in the movement of the sea, which is changed by only one seven-hundredth. I say the same of its rising and falling one, two, or three feet — scarcely four or five feet even at the extremity of a basin two thousand or more miles long, where its depth is hundreds of feet. Such a change is much less than if, in one of the barges bringing sweet water to us, this water should rise in the prow by the thickness of a leaf at an arrest of the barge. From this I conclude that very small alterations with respect to the immense size and extreme speed of the oceans would be sufficient to make great changes in them in relation to the minuteness of ourselves and our phenomena.

SAGR. I am fully satisfied as to this part. It remains for you to explain to us how these additions and subtractions deriving from the diurnal whirling are increased or diminished, upon which alterations you hinted would depend the annual period of growth and diminution in the tides.

SALV. I shall use all my resources to make myself understood, but the difficulty of the phenomena themselves and the great abstractness of mind needed to understand them intimidate me.

The irregularity of the additions and subtractions which the diurnal rotation makes upon the annual motion depends upon the tilting of its axis to the plane of the earth's orbit, or ecliptic. By this tilting, the equator crosses the ecliptic and is inclined and oblique to it with the same slope as that of the axis. The magnitude of the additions amounts to as much as the entire diameter of the equator when the center of the earth is at the solstitial points, but outside of those it amounts to less and less according as the center approaches the equinoctial points, where such additions are least of all.† This is the whole story, but it is wrapped in the obscurity which you perceive.

SAGR. Rather in that which I do not perceive, since so far I do not understand a thing.

SALV. That is just what I expected; nevertheless, we shall see whether the drawing of a little diagram will not shed some light on it. It would be better to represent this effect by means of solid bodies than by a mere picture; however, we may get some assistance from perspective and foreshortening. So let us show, as before, the circumference of the earth's orbit, the point A being supposed to be at one of the solstices and the diameter AP being the common section of the solstitial colure and the plane of the earth's orbit, or ecliptic. Suppose the center of the terrestrial

Cause of the unevenness of the subtractions and additions of the diurnal rotation upon the annual motion.

globe to be located at this point A; its axis, CAB, tilted to the
plane of the earth's orbit, falls in the plane of the said colure,
which passes through the axes of both equator and ecliptic. To
avoid confusion, we shall show only the equatorial circle, indi-
cating this with the letters DGEF, whose common section with
the plane of the earth's orbit will be the line DE, so that one
half of the equator, marked DFE, will be below the plane of the
earth's orbit, and the other half, DGE, will be above it.

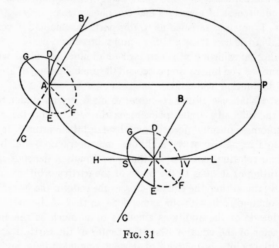

FIG. 31

It is now supposed that the revolution of the equator is in the
order of the points D, G, E, F, and that the motion of the center
is toward E. The center of the earth being at A, its axis CB
(which is perpendicular to the equatorial diameter DE) falls as
we said in the solstitial colure, the common section of this with the
earth's orbit being the diameter PA; hence this line PA will be
perpendicular to DE, because the colure is perpendicular to the
earth's orbit. Therefore DE will be tangent to the earth's orbit
at the point A, so that in this position the motion of the center
along the arc AE, which amounts to one degree per day, would
vary but little; it would even be as if it were along the tangent
DAE. And since the diurnal rotation, carrying the point D
through G to E, is increased over the motion of the center (which
moves practically along this same line DE) by as much as the

whole diameter DE, while on the other hand the other semicircle EFD is diminished by the same amount in its motion, the additions and subtractions at this point (that is, at the time of the solstice) will be measured by the entire diameter DE.

Next we shall see whether they are of the same magnitude at the times of the equinoxes. Transporting the center of the earth to the point I, one quadrant away from the point A, let us take the same equator GEFD, its common section DE with the ecliptic, and its axis CB at the same tilt. Now the tangent to the ecliptic at the point I will no longer be DE, but a different one, cutting this at right angles. This will be marked HIL, in the direction of which will be the motion of the center I, proceeding along the circumference of the earth's orbit. Now in this situation the additions and subtractions are not measured any more by the diameter DE, as they were at first, for since this diameter does not extend along the line of the annual motion HL, but rather cuts it at right angles, D and E add and subtract nothing.

The additions and subtractions must now be taken along that diameter which falls in the plane perpendicular to that of the earth's orbit and cutting it in the line HL; let this be the diameter GF. The additive motion will then be made by the point G along the semicircle GEF, and the subtractive motion will be the balance, along the other semicircle FDG. Now this diameter being not in the same line as the annual motion, HL, but cutting it as is seen in the point I (with the point G being elevated above and F depressed below the plane of the earth's orbit), the additions and subtractions are not determined by its entire length. Rather, they must be that fraction of it taken between the parts of the line HL which are cut off between the perpendiculars drawn upon it from the points G and F, which would be two lines GS and FV. Hence the measure of the additions is the line SV, and this is less than GF or DE, which was the measure of the additions at the solstice A.

According, then, to the placement of the center of the earth at any other point of the quadrant AI, we draw the tangent at such a point and drop perpendiculars upon it from the ends of the equatorial diameter determined by the plane through this tangent vertical to the plane of the ecliptic; and such a part of this tangent, which will be always less toward the equinoxes and greater toward the solstices, will give us the magnitudes of the additions and subtractions. Then as to how much the least additions differ from the greatest, this is easy to determine; between these there is the same variation as between the whole axis (or diameter) of the globe and that part of it which lies between the polar circles. This is less than the whole diameter by one-twelfth, approximately, assuming that the additions and subtractions are made at the equator; in other latitudes they are less in proportion as their diameters are diminished.

That is all I can tell you about the matter, and perhaps it is as much as can be comprehended within our knowledge—which, as is well known, can be only of such conclusions as are fixed and constant. Such are the three general periods of the tides, since these depend upon invariable causes which are unified and eternal. But with these primary and universal causes there are mixed others which, though secondary and particular, are capable of making great alterations; and these secondary causes are partly variable and not subject to observations (the changes due to winds, for example), and partly, though determinate and fixed, are not observed because of their complication. Such are the lengths of the sea basins, their various orientations in one direction or another, and the many and various depths of the waters. Who could possibly formulate a complete account of these except perhaps after very lengthy observations and reliable reports? Without this, what could serve as a sound basis for hypotheses and assumptions on the part of anyone who, from such a combination, wished to furnish adequate reasons for all the phenomena? And, I might add, for the anomalies and particular irregularities that can be perceived in the movements of the waters?

I am content to have noticed that incidental causes do exist in nature, and that they are capable of producing many alterations; I shall leave their minute observation to those who frequent the various oceans. I merely call to your attention, in bringing this conversation of ours to a close, that the precise durations of the ebbing and flowing are changed not only by the lengths and depths of the basins, but I believe that noteworthy variations are also introduced by the juncture of various stretches of ocean which differ in size and in situation or, let us say, in orientation. Such a contrast occurs right here in the Adriatic Gulf, which is much smaller than the rest of the Mediterranean and is placed at such a different orientation that whereas the latter has its closed end in the eastern part at the shores of Syria, the former is closed at its western part. And since it is at the extremities that by far the greatest tides occur — indeed, nowhere else are there very great risings and fallings — it may very well be that the times of flood at Venice occur during the ebbings of the other sea. The Mediterranean, being much larger and extending more directly from west to east, in a certain sense dominates the Adriatic. Hence it would not be surprising if the effects that depend upon the primary causes were not verified in the Adriatic at the appointed times and corresponding to the proper periods, as well at least as they would be in the rest of the Mediterranean. But this matter would require long observations which I have not made in the past, nor shall I be able to make them in the future.

SAGR. It seems to me that you have done a great deal by opening the first portal to such lofty speculations. In your first general proposition, which seems to me to admit of no refutation, you have explained very persuasively why it would be impossible for the observed movements to take place in the ordinary course of nature if the basins containing the waters of the seas were standing still, and that on the other hand such alterations of the seas would necessarily follow if one assumed the movements attributed by Copernicus to the terrestrial globe for quite other reasons. If you had given us no more, this alone seems to me to excel by such a large margin the trivialities which others have put forth that just to think of those once more makes me ill. And I am much astonished that among men of sublime intellect, of whom there have been plenty, none have been struck by the incompatibility between the reciprocating motion of the contained waters and the immobility of the containing vessels, a contradiction which now seems so obvious to me.

SALV. What is more to be wondered at, once it had occurred to the minds of some to refer the cause of the tides to the motion of the earth (which showed unusual perspicacity on the part of these men), is that in seizing at this matter they should have caught on to nothing. But this was because they did not notice that a simple and uniform motion, such as the simple diurnal motion of the terrestrial globe for instance, does not suffice, and that an uneven motion is required, now accelerated and now

A simple motion of the terrestrial globe is not sufficient to produce the tides.

retarded. For if the motion of the vessels were uniform, the contained waters would become habituated to it and would never
make any mutations.

Likewise it is completely idle to say (as is attributed to one
of the ancient mathematicians) that the tides are caused by the
conflict arising between the motion of the earth and the motion
of the lunar sphere, not only because it is neither obvious nor
has it been explained how this must follow, but because its
glaring falsity is revealed by the rotation of the earth being not
contrary to the motion of the moon, but in the same direction.
Thus everything that has been previously conjectured by others
seems to me completely invalid. But among all the great men who
have philosophized about this remarkable effect, I am more

astonished at Kepler† than at any other. Despite his open and
acute mind, and though he has at his fingertips the motions attributed to the earth, he has nevertheless lent his ear and his
assent to the moon's dominion over the waters, to occult properties, and to such puerilities.

SAGR. It is my guess that what has happened to these more reflective men is what is happening at present to me; namely,
inability to understand the interrelation of the three periods,
annual, monthly, and diurnal, and how their causes may seem
to depend upon the sun and the moon without either of these
having anything to do with the water itself. This matter, for a
full understanding of which I need a longer and more concentrated application of my mind, is still obscure to me because of its
novelty and its difficulty. But I do not despair of mastering it
by going back over it by myself, in solitude and silence, and
ruminating on what remains undigested in my mind.

In the conversations of these four days we have, then, strong
evidences in favor of the Copernican system, among which three
have been shown to be very convincing — those taken from the
stoppings and retrograde motions of the planets, and their approaches toward and recessions from the earth; second, from
the revolution of the sun upon itself, and from what is to be observed in the sunspots; and third, from the ebbing and flowing
of the ocean tides.

SALV. To these there may perhaps be added a fourth, and maybe
even a fifth. The fourth, I mean, may come from the fixed stars,
since by extremely accurate observations of these there may be

discovered those minimal changes that Copernicus took to be imperceptible. And at present there is transpiring a fifth novelty from which the mobility of the earth might be argued. This is being revealed most perspicuously by the illustrious Caesar Marsili, of a most noble family at Bologna, and a Lincean Academician. He explains in a very learned manuscript that he has observed a continual change, though a very slow one, in the meridian line. I have recently seen this treatise, and it has much astonished me. I hope that he will make it available to all students of the marvels of nature.

Caesar Marsili observes the meridian to be in motion.

SAGR. This is not the first time that I have heard mention of the subtle learning of this gentleman, who has shown himself to be the zealous protector of all men of science and letters. If this or any other of his works is made public, we may be sure in advance that it will become famous.

SALV. Now, since it is time to put an end to our discourses, it remains for me to beg you that if later, in going over the things that I have brought out, you should meet with any difficulty or any question not completely resolved, you will excuse my deficiency because of the novelty of the concept and the limitations of my abilities; then because of the magnitude of the subject; and finally because I do not claim and have not claimed from others that assent which I myself do not give to this invention, which may very easily turn out to be a most foolish hallucination and a majestic paradox.

To you, Sagredo, though during my arguments you have shown yourself satisfied with some of my ideas and have approved them highly, I say that I take this to have arisen partly from their novelty rather than from their certainty, and even more from your courteous wish to afford me by your assent that pleasure which one naturally feels at the approbation and praise of what is one's own. And as you have obligated me to you by your urbanity, so Simplicio has pleased me by his ingenuity. Indeed, I have become very fond of him for his constancy in sustaining so forcibly and so undauntedly the doctrines of his master. And I thank you, Sagredo, for your most courteous motivation, just as I ask pardon of Simplicio if I have offended him sometimes with my too heated and opinionated speech. Be sure that in this I have not been moved by any ulterior purpose, but only by that of giving you every opportunity to introduce lofty thoughts, that I might be the better informed.

SIMP. You need not make any excuses; they are superfluous, and especially so to me, who, being accustomed to public debates, have heard disputants countless times not merely grow angry and get excited at each other, but even break out into insulting speech and sometimes come very close to blows.

As to the discourses we have held, and especially this last one concerning the reasons for the ebbing and flowing of the ocean, I am really not entirely convinced; but from such feeble ideas of the matter as I have formed, I admit that your thoughts seem to me more ingenious than many others I have heard. I do not therefore consider them true and conclusive; indeed, keeping always before my mind's eye a most solid doctrine† that I once heard from a most eminent and learned person, and before which one must fall silent, I know that if asked whether God in His infinite power and wisdom could have conferred upon the watery element its observed reciprocating motion using some other means than moving its containing vessels, both of you would reply that He could have, and that He would have known how to do this in many ways which are unthinkable to our minds. From this I forthwith conclude that, this being so, it would be excessive boldness for anyone to limit and restrict the Divine power and wisdom to some particular fancy of his own.

SALV. An admirable and angelic doctrine, and well in accord with another one, also Divine, which, while it grants to us the right to argue about the constitution of the universe (perhaps in order that the working of the human mind shall not be curtailed or made lazy) adds that we cannot discover the work of His hands. Let us, then, exercise these activities permitted to us and ordained by God, that we may recognize and thereby so much the more admire His greatness, however much less fit we may find ourselves to penetrate the profound depths of His infinite wisdom.

SAGR. And let this be the final conclusion of our four days' arguments, after which if Salviati should desire to take some interval of rest, our continuing curiosity must grant that much to him. But this is on condition that when it is more convenient for him, he will return and satisfy our desires — mine in particular — regarding the problems set aside and noted down by me to submit to him at one or two further sessions, in accordance with our agreement. Above all, I shall be waiting impatiently to hear the

elements of our Academician's new science of natural and constrained local motions.

Meanwhile, according to our custom, let us go and enjoy an
hour of refreshment in the gondola that awaits us.

End of the Fourth and Final Day

NOTES

p. 5 **Pythagorean.** Pythagoras, a semilegendary figure of the sixth century B.C., was credited by Copernicus with the suggestion of a heliocentric astronomy. Such a system was said to have been developed by Philolaus, a Pythagorean philosopher contemporary with Socrates. Modern scholars have shown that although the Pythagoreans supposed the earth to move, they did not attribute to it a motion around the sun and hence are not entitled to be considered forerunners of Copernicus.

p. 6 **Peripatetics.** The term applied to followers of Aristotle because of that philosopher's custom of strolling about the Lyceum while discoursing with his disciples. The ensuing play on words here sets the tone for the entire *Dialogue* in dealing with the philosophers of Galileo's day.

p. 7 **Sagredo,** born at Venice in 1571, was a pupil of Galileo's at Padua and perhaps his closest friend. A confirmed bachelor, devoted to the enjoyment of life, he never tired of enjoining Galileo to take better care of his health and to stay out of trouble by keeping his discoveries to himself. His valued practical counsel stood Galileo in good stead, and he was frequently able to assist the scientist through his connections in high places. He was himself a competent scientific amateur who enjoyed constructing and manipulating experimental apparatus, was well schooled in philosophy, and was a brilliant conversationalist. Sagredo served as intermediary between Galileo and Welser in the correspondence on sunspots (see pp. 53 ff. and 345 ff., and the related notes.) From 1608 to 1611 he served at Aleppo as consul of the Republic of Venice, and it is believed that Galileo might have remained at Padua if his friend had not received that appointment. Sagredo died in 1620. In the *Dialogue* he represents the educated layman for whose favorable opinion the two experts are striving.

p. 7 **Salviati** was born at Florence in 1582, of an ancient and noble family of that city. Little is known of his life. He is believed to have studied under Galileo at Padua, and from Galileo he received nomination to membership in the Lincean Academy (first note to p. 20). It was at his Villa delle Selve, near Signa, that Galileo wrote the text of his *Letters on the Solar Spots,* which book was dedicated to Salviati; Galileo highly valued his hospitality at this quiet retreat and accomplished much of his work there. Salviati died in 1614 during a sojourn in Spain, having gone there to recover his peace of mind after a humiliation at the hands of one of the Medici over a matter of precedence. In the *Dialogue,* Salviati represents Galileo himself as the expert in science.

p. 7 **philosopher.** The name given to this interlocutor is that of a famous sixth-century commentator on the works of Aristotle. Doubtless the character here portrayed represents a composite of the professional and amateur philosophers and literary men whom Galileo had encountered. The traditional story that Galileo intended Simplicio to represent Maffeo Barberini (Pope Urban VIII) cannot be supported. Such an act would have been a preposterous piece of insolence serving no purpose except malice, whereas good relations existed between the two when the *Dialogue* was being written. Simplicio, of course, represents in this work the expert in philosophy and the adversary of Salviati.

p. 9 **Aristotle,** founder of the philosophy which dominated Western thought throughout the Middle Ages, was born at Stagira in 384 B.C. He was a pupil of Plato's and the tutor of Alexander the Great. He died in 322 B.C., leaving works on logic, metaphysics, and science which show him to have been one of the most astute and versatile geniuses of all time.

p. 9 **Claudius Ptolemy,** who flourished at Alexandria about A.D. 150, compiled in its definitive form the geocentric astronomical system of antiquity and contributed to it many concepts without which the relatively refined observations of his period could not have been reconciled with the assumption of a fixed earth. His doctrine required that all celestial appearances be accounted for by uniform circular motions, as did Aristotle's.

Ptolemy's system is expounded in the *Almagest,* which, together with the *De Revolutionibus* of Copernicus (see next note) has been translated into English in *Great Books of the Western World,* vol. 16 (Chicago, 1952). In that translation Ptolemy's preface has been counted as a chapter; citations of the *Almagest* in these notes are to Halma's French text.

p. 9 **Nicholas Copernicus** was born in 1473 at Torun, in Poland. His great classic of astronomy, *De Revolutionibus Orbium Coelestium,* was not published until 1543, the first printed copy being placed in his hands on his deathbed. His system had been essentially completed thirty years previously, and knowledge of the existence of his manuscript had spread among the learned; though he was repeatedly urged to publish it, his discretion long prevailed over all persuasion. A canon of the Church, he dedicated the work to Pope Paul III, and for more than seventy years no ban was placed upon it. Hostility to the Copernican system was at first more prevalent among Protestants than among Catholics, Luther in particular having condemned Copernicus as a madman. The Copernicus system is briefly outlined on pp. 322-326. Like Ptolemy, Copernicus insisted upon perfectly circular motions for the planets, and was thereby forced to preserve some of the artificial devices which encumbered the geocentric systems; cf. second note to p. 53 and first note to p. 65.

p. 9 **invariant.** Literally "impassible"; the concept is that of "incapable of playing the role of patient in any action." The translation "invariant" has been chosen in order to suggest something which remains constant under all attempts to influence or modify it. Where the antonym is required, the translation "variable" has been used in preference to "variant," since the latter term has a connotation of state rather than of potentiality.

p. 10 **demonstrations.** Cf. Aristotle, *De Caelo* I, 1, 268a, 7-20. The word "texts" which Simplicio uses here and elsewhere in citing references to Aristotle reflects the custom of commentators of the period. The standard modern notation is adopted in these notes. Except as otherwise indicated, all direct quotations from Aristotle have been taken from *The Student's Oxford Aristotle,* edited by W. D. (Sir David) Ross (Oxford University Press, 1942).

p. 10 **Perfect.** In the Aristotelian philosophy the word "perfect" has the sense of "complete" rather than that of "supremely excellent," but Galileo's word *perfetto* was also in general use with the latter sense. Galileo was undoubtedly playing on this ambiguity in order to weaken Simplicio's position in the minds of his readers. Hence any translation of the discussion on pp. 9-15 is to some extent arbitrary and is likely to make the arguments appear capricious.

p. 10 **ad pleniorem scientiam:** "For a more complete knowledge." This phrase was customarily used to introduce additional material after a sufficient proof of the point in question had been given.

p. 10 **text.** *De Caelo* I, 1, 268b, 3-9.

p. 11 **Senate.** The anecdote occurs in Macrobius, *Saturnalia,* I, 6. Papirius's mother was told by him that the secret debate being held in the Senate concerned the question whether it would be better to allow one man two wives, or one woman two husbands. The natural result was a large and eloquent delegation of townswomen to argue for the latter alternative before an astonished Senate.

p. 12 **realizing it.** A reference to the Socratic doctrine that unconscious knowledge exists in the memory and may be drawn out by questioning; see also note to p. 191.

p. 14 **etc.** There is no abridgment of the text here; Galileo very frequently employs this method of abbreviation, preserved in this translation as characteristic of the style of the original.

p. 15 **elsewhere.** This passage is rather confusing; it is precisely this definition which Salviati has already used and which is given by Aristotle in the very sentence containing his statement that all natural bodies are movable. ("All natural bodies and magnitudes we hold to be, as such, capable of locomotion; for nature, we say, is their principle of movement." *De Caelo* I, 2, 268b, 15-17.) The other definition occurs in *Physica* II, 1, 192b, 22-23: "Nature is a source or cause of [a thing] being moved and of being at rest in that to which it belongs primarily." The idea behind Salviati's argument is clear enough, but its expression remains puzzling; explanations offered by Favaro and by Strauss would remove only a part of the difficulty. The puzzle might be disposed of by transferring the word "elsewhere" to a place following the word "definition" in the next sentence. No authority exists for such a correction, but compare the end of Salviati's speech on p. 32.

p. 19 **implies.** Cf. *De Caelo* III, 2, 301a, 5ff.

p. 20 **disordering.** This argument is substantially that given by Copernicus, *De Revolutionibus,* bk. i, ch. 8.

p. 20 **Plato.** The passage which Galileo probably had in mind occurs in the *Timaeus,* 38–39, and commences: "Now, when all the stars . . ." Here, however, Galileo has taken greater liberties with the interpretation of Plato's text.

p. 20 **Academician.** That is, Galileo; whenever he is referred to in the *Dialogue,* it is by this or a similar phrase. The *Accademia dei Lincei* ("academy of the lynxlike"—i.e., "sharp-eyed") was a distinguished body of scientists and mathematicians founded at Rome in 1603 by Prince Cesi. Galileo became a member in 1611, an honor in which he took great pride.

p. 20 **speed.** Galileo's word *velocità* lacks the technical character which the term "velocity" has in later physics; hence in this translation it has often been rendered by the word "speed." Galileo's parallel word *tardità* (here translated "slowness") shows us how much his conception of motion was still hampered by certain ancient ideas about qualities and contraries. These are exemplified today in our speaking of heat and cold as if they were two different sorts of things rather than arbitrary categories applied to measures of one physical entity.

p. 21 **[Let any . . . the latter.]** This passage (and all subsequent matter in the text enclosed in square brackets) did not appear in the original edition of the *Dialogue,* but was added by Galileo in his own copy of the first edition which is now in the library of the Seminary of Padua. The text followed here is that of the definitive National Edition compiled by Professor Favaro (Florence, 1897), where such additions by the author are adjoined as footnotes.

p. 22 **yards.** The *braccio* of Galileo's time, though here translated "yard," was somewhat less than two feet. In order to avoid altering the numbers in the text or introducing unfamiliar units of measurement, arbitrary translations have been made of the names of units employed by Galileo. The resulting distortions of distance are of little importance, since most of these terms occur only in illustrative examples. Fortunately the words "inch" and "mile" correspond rather well to the two Italian measurements used in the *Dialogue* which often refer to actual distances. The following table, based upon information given by Strauss and Pagnini, will enable the reader to restore the original quantities when he so desires.

ITALIAN	APPROXIMATE VALUE	TRANSLATION
dito	thumb's breadth	inch
palmo	four inches	span
pied	eight inches	foot
braccio	21 to 22 inches	yard
canna	39 inches	ell
miglio	5,375 feet	mile

p. 22 **impetus.** For Galileo this was not a mathematically defined concept, but an intuitive idea of some quality possessed by a moving body and capable of being conserved or communicated to other bodies. Sometimes he speaks of "impetus" as though it were synonymous with "velocity," but in those instances (e.g., p. 24) he is dealing with identical bodies or with bodies of equal mass.

p. 24 **equal times.** *Physica* VII, 4, 249a, 20.

p. 29 **wonderful.** (A lacuna occurs in the original edition in this passage; the reading given here is based upon a grammatical alteration without the addition of conjectural words.) There is no sound basis for Galileo's statement that a "truly wonderful agreement" had been found between actual observations and the calculations described by him. The correct relation between the orbits and periods of the planets had been given by Kepler; cf. Foreword, p. xv, and note to p. 269. Galileo's mistaken belief that he had discovered such a relation as he here describes may have originated in calculations made many years previously, before his realization that in uniformly accelerated motion the increments of velocity are proportional to the times. Newton, in the third of his four published *Letters to Bentley,* remarked that if the gravitational power of the sun was halved during the straight fall, and restored at the instant of orbital rotation, the effect described by Galileo would be realized.

p. 30 **pulse beats.** See second note to p. 223.

p. 31 **uniform.** Cf. *Physica* VIII, 8, 265a, 34 ff. The ensuing passage is one of several in the opening section of the *Dialogue* which are often adduced to show the supposed inability of Galileo to throw off the spell of the "perfection of the circle"; cf. Einstein's remark on p. xi. Such passages may perhaps better be regarded as part of his strategy in neutralizing the hostility of Aristotelian opponents by utilizing their own arguments for cosmic circular motions. Elsewhere in his works he derided the notion that any geometrical figure was endowed with special physical qualities. It should be noted that the word *forze* used here by Galileo is not intended to introduce the concept of forces, but rather that of the strengths of the "natural inclinations." Galileo uses the words *violenza* and *virtù* for external force or constraint, a concept of which he had a clear (if intuitive) apprehension much too strong to be overcome by his adherence to certain mistaken ideas belonging to his predecessors; see p. 215 and note thereto.

p. 32 **orbit.** Literally "orb"; a reference, not to the path of the moon as we think of it, but to the crystalline sphere in which the moon was supposed to be embedded (Foreword, p. ix). To use the translation "orbit" seems preferable to requiring the reader to keep constantly in mind the old Ptolemaic concepts. "Arc of the moon's orbit" (literally, "hollow of the moon's orb") is a frequently recurring phrase which refers to the supposed containing vessel of all elemental material. The element of fire was supposed to go naturally straight up, but to be confined by the sphere in which the moon was fixed.

p. 33 **eadem est ratio totius et partium:** "The reasoning which applies to the whole applies also to the part." *De Caelo* I, 3, 270a, 11. This axiom of Aristotle's occurs also with various grammatical modifications adapting it to the contexts in which it appears.

p. 34 **contra negantes principia non est disputandum:** "One must not argue against him who denies axioms"; cf. *Physica* I, 2, 185a, 3. Salviati's arguments which provoke this utterance are essentially those of Copernicus as given in bk. i, chs. 8 and 9, *De Revolutionibus.*

p. 34 **per accidens:** "Merely by coincidence." Cf. *De Caelo* II, 14, 296b, 15-16.

p. 38 **Aristotle writes.** Cf. *De Caelo* I, 3, 270a, 14-17. Here the translation is from Galileo's Italian paraphrase.

p. 42 **sorites.** The word *cornuto* has been translated "forked" rather than "horned," because of the association in English between the latter term and the dilemma. As Strauss remarks, "the Cretan paradox is no sorites; he calls it a *Scheinbeweis* (pseudoproof), in admirable anticipation of modern logicians. The name "sorites" properly belongs to the chain argument, in the classic example of which one disproves the existence of a heap of wheat by removing one grain at a time as inconsequential to the heap.

p. 43 **heaviness.** In order to avoid suggesting that Galileo had anticipated the Newtonian implications of the word "gravity," the word *gravità* has been translated "heaviness" where the idea of cause is involved (with one or two necessary exceptions). Where the word is used as the name of a quality only (for instance, when used in opposition to "levity"), the word "gravity" is employed as the English translation.

p. 43 **rare.** A lacuna occurs in the Italian text here, which has been removed by altering the grammar of the sentence. Concerning Cremonino, who is mentioned in the margin, see second note to p. 69 and note to p. 112.

p. 48 **Abila and Calpe** were the ancient names of the Pillars of Hercules; Abila is a hill in North Africa near Ceuta, and Calpe is the Rock of Gibraltar. The legend mentioned exists in two forms. This one is Pliny's, whereas Strabo has it that the Mediterranean already existed as an inland sea when the Atlantic Ocean broke through.

p. 51 **new stars.** Novas of great brilliance (supernovas) appeared in 1572 and 1604. The former appeared in Cassiopeia and is known as "Tycho's star"; it was so bright that it remained visible in broad daylight for several weeks, and by night for eighteen months. The nova of 1604 was in Serpentarius, though Galileo in the *Dialogue* consistently refers to it as having been in Sagittarius.

p. 52 **Anti-Tycho.** A book by Scipio Chiaramonti (1565-1652) which was published in 1621. Galileo, who was not an admirer of Tycho, praised the book in his *Saggiatore* of 1623. Kepler predicted that Galileo would come to regret any endorsement of Chiaramonti, as indeed he did; see pp. 279–318.

p. 52 **Tycho Brahe** (1546-1601), a Dane, is often called the first truly modern astronomer because of the extensiveness and accuracy of his observations as well as the painstaking skill with which he designed, constructed, and manipulated his large and costly instruments. His anti-Ptolemaic theory was still geocentric; he had the planets revolve about the sun, which in turn went around the earth.

Apropos of the seeming oddity of referring to Tycho by his given name it may be remarked that the same is true of Galileo, whose family name was Galilei. The Italians generally refer to their greatest men in this manner, for example Dante (Alighieri), Leonardo (da Vinci), and Michelangelo (Buonarroti).

p. 52 **telescope.** The story of Galileo's construction of the first telescope ever applied to astronomical observations seems worth repeating here in his own words.

"About ten months ago a report reached me that a Dutchman had constructed a telescope, by the aid of which visible objects, although at a great distance from the eye of the observer, were seen distinctly as if near; and some proofs of its most wonderful performances were reported, which some gave credence to, but others contradicted. A few days after, I received confirmation of the report in a letter written from Paris by a noble Frenchman, Jacques Badovere, which finally determined me to give myself up first to inquire into the principle of the telescope, and then to consider the means by which I might compass the invention of a similar instrument, which after a little while I succeeded in doing through deep study of the theory of refraction; and I prepared a tube, at first of lead, in the ends of which I fitted two glass lenses, both plane on one side, but on the other side, one spherically convex and the other concave."

This account was given in his *Sidereus Nuncius* (The Sidereal Messenger), first published in March, 1610; the above translation is by E. S. Carlos (London, 1880). In *Il Saggiatore,* published thirteen years later, Galileo asserted that a single day had sufficed him to reason out the conditions necessary for magnification to be produced by such an instrument. Cf. *Discoveries and Opinions of Galileo* (New York, 1957), pp. 28–29, 244.

p. 53 **spots.** Galileo may have been the first to observe sunspots, though this is dubious. He was certainly anticipated by Johann Fabricius and by the Jesuit father Christopher Scheiner in publishing on the subject. The former's announcement (*De Maculis in Sole* . . ., Wittenberg) appeared in June, 1611, though neither Scheiner nor Galileo seems to have seen it. Scheiner's first publication took the form of three letters addressed to Mark Welser, prefect of Augsburg, early in 1612, under the pseudonym *Apelles latens post tabulam.* Welser sent these to Galileo for comment; see pp. 345 ff. Galileo's three answering letters were published by the Lincean Academy under the title *Istoria e Dimostrazioni alle Macchie Solari* . . . (Rome, 1613), mentioned on the next page. An abridged English translation appears in *Discoveries,* pp. 89–144.

p. 53 **separate.** Although this passage appears as a direct quotation in the original, it appears rather to be a rough summary by Galileo of the views published by Scheiner in his letters to Welser, mentioned in the preceding note.

p. 53 **eccentric.** The Ptolemaic theory made use of two devices for providing the supposedly circular orbits of the sun, moon, and planets with appropriate irregularities so that they might correspond with actual observations. One of these was the epicycle (first note to p. 65), and the other was the eccentric. The latter provided for circular orbits with the earth displaced from their precise centers. This arrangement was particularly well adapted to account for the apparent motion of the sun. The combination of epicycles with eccentrics resulted in three types of "center" in the Ptolemaic system: the center of the universe, at which the earth was located; centers of deferents (eccentric circles carrying epicyclic centers); and centers of "equants," from which the motion was uniform and circular. By the accumulation of such makeshifts it was possible to perform accurate calculations and predictions, though only very laboriously. Meanwhile these complications stood in the way of any rational celestial mechanics. Copernicus felt the equant to be an improper device because it produced circular motion nonuniform with respect to its true center; he therefore restricted himself at first to combinations of epicycles but later introduced eccentrics as well.

p. 55 **argument.** See also p. 32; cf. *De generatione animalium* III, 10, 760b, 31.

p. 60 **natura nihil frustra facit:** "Nature does nothing uselessly." Galen, *De usu partium,* x, 14.

p. 65 **epicycle.** (See also third note to p. 53.) To account for the fact that the planets appeared from time to time to stop and turn back in their motions (as shown on pp. 342 ff.), Ptolemy provided the planetary orbs with subsidiary orbs which rotated as they were carried around. In such a framework the entire surface of the planet would of course pass before the eyes of an external observer. Ptolemy called these subsidiary orbs "epicycles"; Copernicus attributed to the moon not merely one, but two epicycles, the center of one being located on the circumference of the other.

p. 65 **Contraterrenes.** The Pythagoreans held that the earth had an exact counterpart which revolved about the "central fire" in such a way as never to be seen by us; cf. *De Caelo* II, 13, 293a, 20 ff.

p. 66 **meridian.** Literally, "in this or that belly of its dragon." The path of the moon is tilted with respect to the ecliptic by some five degrees; hence its course among the stars appears to undulate. This suggested to the ancients the idea of a dragon; they placed the head and tail of the dragon at what are now called the moon's nodes (the points where its orbit cuts the ecliptic). Its most northerly position was called the upper belly of the dragon, and its most southerly position the lower belly.

p. 67 **light.** The secondary light of the moon had been observed and partly explained by Leonardo da Vinci, though not in any work that had been published at that time. It is of interest to compare Leonardo's remarks with those of Galileo; the former made the common error of supposing the principal reflection to come from our oceans. See *The Notebooks of Leonardo da Vinci*, ed. Edward MacCurdy (New York, 1939), pp. 295-296.

p. 69 **hardness.** Strauss remarked that he had never been able to find this view set forth in any published work of Aristotle, though it was generally held by the later Peripatetics to be a necessary consequence of his system. In this connection *De Caelo* II, 7, 289a, 12-16 and II, 9, 291a, 18-22 are possibly of interest. Salviati has voiced his opinion on p. 52; see also pp. 120, 443. Sagredo has presented the Peripatetic arguments on p. 43.

p. 69 **Padua.** Cesare Cremonino, a famed exponent of Aristotle and the leading authority of his time on Alexander of Aphrodisias (Sicily), the latter having been a noted third-century Peripatetic philosopher. Cremonino was several times under suspicion of heresy, and in a memorandum of the Inquisition his name was mysteriously linked with Galileo's when the former was being investigated in 1611. The records of Cremonino's interrogation by this body appear not to be available. Galileo was for the most part on good terms with Cremonino personally, though they were diametrically opposed intellectually.

p. 71 **Il Saggiatore** (The Assayer). Galileo's celebrated reply to his opponents, and in particular to Father Grassi. It was published in 1623, and was a devastating conclusion to the long controversy about the nature of comets. At the same time it set forth the philosophical basis of experimental science, explained many of Galileo's discoveries and views, and won him widespread acclaim and support. *Lettere Solari* refers to Galileo's book on the sunspots published in 1613; cf. second note to p. 53. An English translation of *The Assayer* appears in abridged form in *Discoveries*, pp. 231–280, and in complete form in *The Controversy on the Comets of 1618* (Philadelphia, 1960), pp. 151–336.

p. 74 **activity.** Here, as in several other places, no source has been identified for the passage cited by Galileo as a direct quotation. Because of Galileo's inclination to paraphrase freely when quoting, it is often difficult to ascertain his sources.

p. 75 **invisible.** It is of interest to note that the phenomenon here described by Galileo has been advanced to account for a new anomaly observed in the skies. The planet Pluto presents such a small visual diameter that its known mass would require it to have an impossibly high density. Sir James Jeans suggested that this might be an instance of specular reflection; that is, the visibility of only a small portion of the entire surface because of extreme smoothness.

p. 77 **more vividly:** This phrase was substituted by Galileo in his own copy for the expression "greater light" in the original edition, and the end of the preceding speech was similarly altered.

p. 81 **tenebrae sunt privatio luminis:** "Darkness is the absence of light." Cf. *De anima* II, 7, 418b, 18 ff.

p. 90 **forty times.** This is an odd remark for Galileo to make, since he understood perfectly well that the ratio of surfaces rather than of volumes should be used for this purpose; cf. p. 67. The correct figure is about fourteen times.

p. 91 **theses.** *Disquisitiones mathematicae de controversiis ac novitatibus astronomicis* (Ingolstadt, 1614). A book written at the instigation of Scheiner (first note to p. 53) by his pupil Locher. This book looms large in the discussions of the Second Day. It was written at a time when Galileo was still on good terms with Scheiner and before he had antagonized the Jesuits in the literary feud which commenced in 1619 and was waged over the comets of the previous year.

p. 92 **Cleomedes** was editor of a compendium of Greek works under the title *Cyclica consideratio meteorum* (1539).
Vitellio (Witelo) was the author of a classic treatise on perspective; of Polish origin, he lived in Italy toward the end of the thirteenth century.
Macrobius was a fourth-century Roman philosopher and the author of a commentary on Cicero's *Somnium Scipionis* containing the idea here mentioned.
a modern. Franciscus Aquilonius, who had published a treatise on optics in 1613 in which Kepler's fundamental work published in 1604 was completely ignored.

p. 96 **classic spots.** Literally, "antique spots." Describing his first telescopic discoveries, Galileo had written: "Now those spots which are fairly dark and rather large are plain to everyone and have been seen throughout the ages; these I shall call the 'large' or 'ancient' spots, distinguishing them from others that . . . had never been seen by anyone before me." (*Discoveries*, p. 31.)

p. 98 **such thing?** This question was omitted from the first edition of the *Dialogue* by mistake and was subsequently supplied on an erratum slip. The slip is lacking in Galileo's copy, and in restoring the question to the text he added the preceding observation.

p. 102 **dove of Archytas.** A celebrated automaton of antiquity, constructed by a Pythagorean famous as a statesman, mathematician, astronomer, and skilled artisan.

p. 103 **propositions.** Despite their innocuous character, these passages were seized upon as one of the textual points offensive to the Church. The commission appointed by the Pope to examine the *Dialogue* noted eight such points which may be stated briefly as follows:
1. That the imprimatur of Rome was put on the title page without proper authority.
2. That the preface was printed in different type and thus vitiated, that the closing argument was put in the mouth of a simpleton, and that it was not fully discussed.
3. That Galileo often treated the motion of the earth as real and not hypothetical.
4. That he treated this subject as undecided.
5. That he contemned opponents of the Copernican opinion.
6. That he asserted some equality between the Divine and the human mind in geometrical matters.
7. That he represented it to be an argument for the truth that Ptolemaics become Copernicans, but not vice versa.
8. That he ascribed the tides to motion of the earth which was nonexistent.

p. 106 **quintessence.** A fifth substance as distinguished from the four elements (earth, water, air, and fire). Celestial bodies were supposed to be composed of this unearthly substance, called *aither* by Aristotle; cf. *De Caelo* I, 3, 270b, 21-25.

p. 108 **method.** Literally "the disturbed method," referring to the so-called disturbed proportions of Euclid (*Elements,* bk. v., especially Def. 18 and Prop. 22). Here Simplicio ostentatiously shows off his mathematical vocabulary with ludicrous irrelevance.

p. 109 **place.** *De generatione animalium* V, 1, 780b, 21.

p. 109 **Joachim.** A Cistercian bishop of the twelfth century whose works were generally assigned a prophetic significance; cf. Dante, *Paradiso* xii, 139-141. In this and the ensuing passages Galileo exhibits a freedom from superstition which was rare indeed for his time. Kepler himself subscribed to certain doctrines of astrological prediction, and even Newton occupied much of his life in alchemical investigations.

p. 112 **Alexander.** A noted Aristotelian philosopher and commentator who flourished about A.D. 200. Strauss considered it likely that Pendasio (d. 1603) was the philosopher who wrote the letter referred to, and cited also F. Fiorentino's conjecture that this was Zabarella (d. 1589). It is not impossible, however, that the philosopher may have been Cremonino (see notes to pp. 69, 320), whose dates, principal studies, dialectical skill, and known intellectual cynicism all accord well with such a possibility.

p. 114 **writer.** Cf. *De Caelo* II, 13, 293b, 31 ff.

p. 118 **are.** The sense of this sentence is made clearer by inserting at this point the words "opposite but."

p. 118 **Jupiter.** Galileo discovered four of Jupiter's satellites and named them the "Medicean stars" in honor of the Grand Ducal family of Tuscany. This discovery was of great importance in undermining the Aristotelian doctrines and in lending plausibility to the Copernican theory in the minds of Galileo's contemporaries.

p. 119 **years.** Copernicus calculated the precession of the equinoxes to have a period of 25,816 years; the ancient estimate had been 36,000 years. Cf. *De Revolutionibus,* bk. iii, ch. 6.

p. 122 **primum mobile.** The highest sphere in the ancient cosmology, lying beyond that of the fixed stars. It was supposed to revolve in twenty-four hours, sweeping along with it the fixed stars and (against their supposed natural tendency) the planets and the moon; cf. p. 117.

p. 123 **frustra fit per plura quod potest fieri per pauciora:** "It is pointless to use many to accomplish what may be done with fewer"; cf. p. 117.

p. 124 **aeque bene:** equally well. This passage and the ensuing reply appear to have been inserted to answer Christopher Clavius (see note to p. 360), who inserted the phrase in question before giving his critique of Copernicus; cf. Clavius, *In Sphaeram Ioannes de Sacrobosco* (Rome, 1581) pp. 434 ff.

p. 124 **refutations.** Cf. *De Caelo* II, 14, 296a, 27 – 296b, 12. The translation in the text is from Galileo's Italian paraphrase.

p. 126 **point-blank.** That is, without elevating or depressing the gun with respect to the horizon.

p. 128 **Wursteisen.** Born in 1544 at Basle, Christian Wursteisen died in 1588, making it most unlikely that Galileo was indebted to him for his first acquaintance with the Copernican doctrine, as some writers have inferred. He was the author of a commentary on Peurbach's *Theory of the Planets* containing passages which probably led Galileo to believe him a Copernican. Since the story is here placed in Sagredo's mouth rather than Salviati's, and since not even Sagredo claims to have heard the lectures, there is no reason to suppose this account autobiographical.

p. 134 **natural.** The argument as given here lacks a premise to the effect that all natural motions are (at least potentially) eternal. Salviati seems to be leaning on the last clause of his preceding speech for this premise, which no astute Aristotelian would have granted; he proceeds to equivocate over the use of the word "eternal," and when Simplicio (rather poorly) states its Aristotelian use, accuses the latter of his own equivocation.

p. 136 **paragraph 97.** (See first note to p. 10.) *De Caelo* II, 14, 296a, 34 ff.

p. 140 **petitio principii:** "begging the question." The name of a formal fallacy in logic, consisting of assuming the very thing that is to be proved.

p. 140 **ignotum per aeque ignotum:** "The unknown by means of something equally unknown."

p. 140 **middle term.** In Aristotelian logic, that term which is used in each of the two premises but which is absent from the conclusion.

p. 146 **vires acquirunt eundo:** "Gain strength as they go." The phrase alludes to Virgil's famous passage about gossip (*Aeneid,* iv, 175).

p. 147 **happen.** Apparently referring to the discussion on p. 23. This speech of Simplicio's seems unnecessarily stupid, but perhaps the real intent was to emphasize the insistencé of philosophers upon reasoning out even the most commonplace phenomena of observation.

p. 147 **boundless.** This completes the statement of Galileo's inertial law, partly anticipating Newton's First Law of Motion; see further, note to p. 165.

p. 148 **spontaneously.** Since the concept of "natural" as against "forced" (forcible, constrained) motion frequently recurs in the *Dialogue,* a clarifying passage may be quoted from Aristotle: "But since 'nature' means a source of movement within the thing itself, while a force is a source of movement in something other than it or in itself *quâ* other, and since movement is always due either to nature or to constraint, movement which is natural, as downward motion is to a stone, will be merely accelerated by an external force, while an unnatural movement will be due to the force alone." *De Caelo* III, 2, 301b, 17-23.

p. 150 **non entium nullae sunt operationes:** "The nonexistent performs no actions."

p. 157 **hoops** (*ruzzole*). Not literally hoops, but wooden discs about six inches in diameter and one inch thick, rolled on the ground either by hand or by strings wound around them by the players.

p. 157 **Socrates's.** Socrates spoke of the source of his inspiration as his "demon." Sagredo amusingly develops his riposte by offering to be a source of inspiration to Simplicio through use of the Socratic method of questioning.

p. 158 **square ones.** *Mechanica*, ch. 8, 851b, 15 ff. This work is not genuinely Aristotle's, though attributed to him traditionally.

p. 160 **marbles** (*chiose*). Salusbury states that *chiosa* was the name of a game played by rolling bullets down a slanted rock. Strauss follows Favaro in describing *chiose* as rounded lead objects molded by children for use as play money and the like.

p. 161 **tennis players.** The Italian tennis prevalent in Galileo's day, and said to have been still popular at the close of the nineteenth century, was played with a much larger ball than ours, between two teams of indefinite but equal number, on a large court with a center stripe but without a net.

p. 161 **bowlers.** Here the reference is to the national game of Italy, called *boccie* (or *bocce*) ball. It is very similar to lawn bowls except that the playing ground may be quite irregular and rough, or may be indoors. The "given mark" is the *pallino,* a small ball which is first to be bowled in each round.

p. 164 **motion.** *De motu naturaliter accelerato,* which appeared in the *Discorsi e Dimostrazioni Matematiche intorno a due Nuove Scienze* (Leyden, 1638), but was probably in essentially final form by 1609. An English translation occupies pp. 160 ff., *Dialogues Concerning Two New Sciences* (New York, 1914; English translation by Crew and de Salvio).

p. 164 **Archimedes,** the greatest of ancient mathematicians and the founder of mechanics, died 212 B.C. See further, note to p. 204. The work here referred to is *De lineis spiralibus.*

p. 165 **straight one.** The ensuing demonstration, despite its errors, is of particular interest in the light of later discussion of the same problem, treated at length by Alexandre Koyré in *A Documentary History of the Problem of Fall* (Philadelphia, 1955). Koyré was unwilling to accept Galileo's disclaimers in the text and in his subsequent correspondence which indicate that Galileo did not take the speculation very seriously. The implication that the body would come to rest at the center certainly contradicts the passages on pp. 22–23 and 227 relating to bodies falling through a tunneled earth; and the implication of a circular inertia which would justify the use of equal arcs CF, FG, etc. as measures of the stone's travel during equal times contradicts the discussion of the tangential character of inertial motions a few pages later, where even Simplicio accepts that view (p. 192). On the other hand, this speculation is remarkable as an attempt by Galileo to generalize his principle of the relativity of motion (pp. 114, 186, 248) and to treat the behavior of a body at rest on a rotating tower as equivalent to that of an unsupported projectile given equal motion at the outset. It is likely that in this speculation, Galileo first reached his three conclusions (equality of speed, shape of path, and uniformity of motion) and then attempted to construct a demonstration to fit them. Since it did not quite come off, he presented it as a diversionary "play within a play" and spoke of the results as "curiosities." When the French mathematician Pierre Fermat criticized the passage and remarked that the line should be a spiral, Galileo replied (through Fermat's pupil and friend, Pierre Carcavy) that his argument was not intended seriously, and that in any case the path of a falling body near the earth would be parabolic, as he later demonstrated (*Two New Sciences*, pp. 244 ff.).

p. 166 **prettier.** This remark, though based upon an erroneous demonstration, is particularly noteworthy for the light it throws on the deepest scientific predilections of the author. Galileo's attempt thus to discover an equivalence among "natural" motions is a philosophical anticipation of the elimination of the old concept of gravitational "force" in modern physics.

p. 174 **ingenious.** Clement Clementi, *Enciclopaedia amplissimo* . . . (Rome, 1624). As the marginal note indicates, the remark is sarcastic; the "handbook" was a large quarto, and its contents were not conclusions but philosophical disputations compiled by a verbose Jesuit.

p. 175 **piece.** An enormous advance in physical thought is represented by this separation of motions and by the discovery that analysis could treat each motion as independent of and inoperative upon the other. When Salviati later objects, he is merely voicing the prevailing thought among philosophers and physicists. Sagredo's rejoinder embodies an outstanding point in the Galilean revolution in physics.

p. 178 **problem.** Salviati's errors here appear to have been intentionally put in by Galileo, since he has Sagredo correct many of them in his next speech.

p. 180 **carry true.** This passage is usually regarded as further evidence that Galileo considered the inertial path of the ball to be circular, but it is more likely that he had in mind the analogies of the musket and the quadrant aboard ship (pp. 249–250, 375) and the composition of motions in a moving gun (p. 176). Salviati, in the preceding sentence and the ensuing attempt to prove that the error could not be detected by measurement, assumes a tangential path for the shot.

p. 181 **500 yards.** Obviously only the roughest approximation is made here, and the rest of the demonstration is quantitatively worthless, though it serves to indicate that the deviation would be small. By his initial assumption that the experiment takes place at the equator, Galileo pretends to give his opponents an advantage in the maximum linear speed of the earth. His own subsequent reasoning vitiates this "advantage," and shows that he merely wished to make an approximate calculation which would not be valid where any appreciable curvature of the earth was involved.

p. 181 **chords.** By "chord" is meant double the sine of half the angle. The tables referred to occur at the end of *De Revolutionibus*, bk. i, ch. 12.

p. 182 **low.** In the original edition the words "high" and "low" are reversed, and the error remains uncorrected in Galileo's copy. The sense of Sagredo's remark is that if existing gunners were to shoot on a stationary earth, the habits they had formed under actual conditions would betray them. Clearly the author had misgivings about his previous "proof" (p. 181) that there would be no difference in the two cases.

p. 188 **Ptolemy.** This argument is cited by Copernicus (*De Revolutionibus,* bk. i, ch. 7, last paragraph) as being Ptolemy's. But Copernicus was actually paraphrasing Ptolemy so as to bring into this argument from centrifugal action certain consequences which Ptolemy (*Almagest,* bk. i, ch. 6, second paragraph; tr. Halma, Paris, 1813) attributed only to a freely falling earth. Many subsequent authors, perhaps relying on Copernicus, ascribe the argument to Ptolemy at least by implication; e.g., Alisandro Picco- lomini, *De la sfera del mondo* (Venice, 1559, f. 16 *recto*) and Francesco Giuntini, *La sfera del mondo* (Lyons, 1582, p. 115); Strauss cites in this connection Maestlin, *Epitome astronomica* (Heidelberg, 1582).

p. 191 **nostrum scire sit quoddam reminisci:** "Our knowledge is a kind of recollection." This Socratic doctrine is a recurrent theme in Galileo's dialogues as well as in Plato's.

p. 194 **leave it.** Pagnini considers this passage to reveal a step of considerable importance toward the application of the laws of heavy bodies to the principles of celestial mechanics. Even though Galileo was prevented from going farther by his error in supposing that no tangential velocity could overcome any centripetal attraction (see note to p. 201), he still might have deduced the idea of an orbit from a composition of the "natural tendency to move toward the center" and the "impetus to move along the straight tangent" mentioned in the preceding speech.

p. 195 **ad destruendum sufficit unum:** "A single instance is sufficient for dis- proof."

p. 196 **tangent.** It is very curious that just when his supposed theory of circular inertia (cf. note to p. 165) would have been a useful thing for him to in- voke, Galileo commenced instead to speak of a tangential impulse. Hap- pily, the difficulties thus created stimulated his ingenuity to produce an analysis which proves that his mathematical insights were of the highest order. The whole spirit of the ensuing discussion is precisely that of the calculus, despite its shortcomings from the standpoint of physics.

p. 198 **diminution.** Weight has nothing to do with the matter, as Salviati later points out (p. 202). But since the prevailing view was Aristotle's (second note to p. 202), it was necessary to raise this question and treat it seriously.

p. 199 **times.** Strauss remarks that this probably is the first scientific attempt to use the abscissae and ordinates of a single diagram explicitly for magni- tudes of two different sorts (time and velocity). This fact alone would place the *Dialogue* in the highest rank of scientific importance. The sub- sequent use of this same diagram to represent spaces of fall (and presum- ably of horizontal motion) on p. 201, however, results in serious confu- sion.

p. 201 **whatever.** In this discussion Galileo once more shows his mathematical insights to be of the highest order though he errs in the application of them. His analysis is ingenious and approximately right so far as it concerns the relative diminutions with respect to time of the space traversed and the velocity achieved by a falling (accelerated) body. But the tangential velocity — the centrifugal component — is not similarly accelerated; it has an instantaneous and constant value. In making Simplicio fix a value for the relation of the tangential velocity to the velocity of fall, Salviati stumbles into the error of treating the former as a function of the latter, whereas it is in fact quite independent. To put the matter another way, what Salviati does is to seek an elapsed time so small that the ratio which Simplicio has supplied will be inadequate for its intended purpose. This leads him directly into the confusion mentioned in the preceding note. Had Simplicio imposed a ratio specified as existing between the two ve- locities at a given instant, or had he simply supplied a large and arbitrary velocity for the motion along the tangent without reference to that of falling, Salviati would have been forced to face the actual physical prob- lem involved. [In the related diagram (p. 199) it should be remarked that point M does not lie on the arc but is merely utilized to identify it because of its accidental proximity. It should also be remarked that the "parallels" KL, HI, etc. should really be directed toward the center of the circle, but are here shown as in the original; although Salviati uses the word "wheel" twice in the discussion, he is obviously thinking of fall to the earth, and considers the centripetal lines as parallel.]

p. 202 **tangent.** The original edition has the word "secant" in this place. This obvious slip, noted by Favaro, remained uncorrected in Galileo's copy.

p. 202 **weights.** *Physica* IV, 8, 216a, 12–16. Galileo is said to have refuted this by the classical experiment of dropping very unequal weights from the Leaning Tower of Pisa. No less interesting is his logical "proof" that Aristotle was in error. This will be found in *Two New Sciences,* pp. 62 ff.; it was partly anticipated by J. B. Benedetti (1530–1590).

p. 203 **sphaera tangit planum in puncto:** "The sphere touches a plane in one point." The ensuing discussion is an episode in the age-old controversy between philosophers, mathematicians, and scientists, touching the true role of mathematical reasoning.

p. 204 **Peripatetic.** The fallacious proof here discussed had been set forth by Francesco Buonamico of Pisa, one of Galileo's early teachers. Cf. Giovanni Barenghi, *Considerazioni sopra il Dialogo* (Pisa, 1638), p. 11.

p. 204 **prove.** Galileo's admiration for Archimedes was boundless. When he himself once succeeded in reconstructing what was probably a method used by the ancient mathematician, he adopted a very different attitude from that of Simplicio's Peripatetic. This was in one of his earliest papers, *La Bilancetta,* inventing the hydrostatic balance. There Galileo begins by remarking that no one who has read "the writings of that divine man (which moreover are extremely easy to understand, so that all other geniuses are inferior to that of Archimedes)" would ever believe that the crude device traditionally ascribed was actually his method of detecting the imposture of the goldsmith who alloyed the gold in Hiero's crown. In place of the clumsy traditional method, Galileo offered an elegant and exact solution worthy of Archimedes, based upon principles and theorems announced by the latter.

p. 208 **sphaera aenea,** etc. A bronze sphere, etc.; cf. note to p. 203.

p. 213 **upward.** This remark is at least a partial adumbration of Newton's Third Law of Motion.

p. 214 **effect.** This observation, which casts some light upon Galileo's methods of reasoning in physics, is also of interest as an anticipation of one of J. S. Mill's canons of inductive logic (the so-called Method of Difference); in the last paragraph on p. 445 Galileo sets forth another (the Method of Concomitant Variations).

p. 215 **single unit.** This paragraph shows that Galileo recognized the product *mass x velocity* as a measure of impulse, his only concept of mass being that of weight. A hint of the concept of *force* as the product of mass and acceleration may also be seen here, if we are liberal in interpreting the ideas of "resisting restraint" and of "conferring velocity." Credit for the explicit definition of these concepts and for their mathematical expression must, however, be reserved for Newton.

p. 216 **less?** Here Galileo supplies an excellent approximation to Newton's Second Law of Motion, within the limitations mentioned in the preceding note.

p. 217 **diminished.** Pagnini remarks that although this conclusion is true the proof is defective, and that Galileo speaks as if the centripetal acceleration were determined by the segments of the secants FG and DE rather than by the angles CAE and CAG. And although the inverse relationship of centrifugal force to radius was correctly described, Galileo wrongly believed the force to vary with the linear velocity rather than its square.

p. 218 **theses.** This book has been previously mentioned (note to p. 91); for the other work cited, see third note to p. 247.

p. 218 **Sagittarius.** See note to p. 51.

p. 219 **equator.** Literally, "under the equinoctial." Galileo refers all such phenomena to the ancient celestial coördinates; in order to facilitate reading, these have been translated into ordinary terrestrial terms whenever possible.

p. 219 **miles.** A German mile is the 5,400th part of the circumference of the equator; Strauss observes that in Galileo's time this was a reckoning device rather than a practical measurement.

p. 221 **uniform.** The balance of this discussion is particularly interesting in connection with Galileo's cosmogonical speculations on p. 29.

p. 221 **writings.** Cf. first note to p. 164. The passage here indicated may be found in *Two New Sciences*, pp. 173-175.

p. 223 **weights.** Cf. second note to p. 202.

p. 223 **five seconds.** This statement was taken as a literal experimental result by many readers of the *Dialogue*, despite the phrases by which it is introduced in the text as a mere arbitrary basis for calculation. Fortunately we know precisely how Galileo regarded it, for his friend G. B. Baliani wrote to him to question the figure, and in reply, Galileo told him that for the refutation of the statement under discussion here, the exact time was of no consequence. He then went on to explain how one might determine the acceleration due to gravitation experimentally, if Baliani cared to trouble with it. Galileo does not assert that he had ever done so, and since he was then blind, it is improbable that he ever did. Galileo was interested only in the general relation of spaces traversed to times elapsed; since no national standards of length then existed in Italy, it was natural enough not to seek an expression in conventional units for the acceleration in free fall. See Galileo's letter to Baliani, 1 August 1639 (*Opere*, XVIII, 77). The ensuing calculation in the text is also erroneous for another reason; Galileo assumed that the acceleration would be constant throughout the fall, instead of varying inversely with the square of the distance, as ultimately discovered by Newton. Concerning the assumed distance to the moon, see second note to p. 298, below.

p. 224 CALCULATIONS. The left-hand vertical column contains the series of trial divisors for extraction of the square root of the large number; the last of these appeared incorrectly as 24240 in the original text, and is given here as corrected by Strauss. Immediately below the square are the partial remainders, in a manner analogous to those entered in division problems as explained below (first note to p. 297). The square root, taken from the final digits of the figures in the left-hand column, is entered at the lower right. Its successive quotients by 60 are written beneath it, the remainders being carried into the text as minutes and seconds.

p. 227 FIGURES IN MARGIN. In order to agree with the discussion in the text this column should commence with zero, contain only one figure ten, and end with zero. The number of intervals would then be twenty, as desired, and the sum of the figures would be one hundred. Galileo appears to have confused the number of intervals with the number of figures representing speeds at the ends of intervals, resulting in this inconsistent representation.

p. 228 **continuously.** The admirable discussion which ensues is, according to Strauss, the first instance of an integration based upon pure mathematical reasoning and applied to mechanics (though geometrical integrations as such had been performed centuries earlier).

p. 230 **equal times.** This statement is only approximately true, as Galileo suspected, and he repeats his qualifying remark on p. 450. The discussion which follows shows once again Galileo's keen power of observation and his ingenuity in explaining physical phenomena.

p. 231 LATIN QUOTATIONS. The quotations commencing here are not always faithful to their imputed source. To some extent they were altered by Galileo because he had to forgo the use of certain diagrams occurring in the book from which he was quoting. Other errors apparently crept in too, either in copying from the source or in printing the *Dialogue*. For the most part the Latin as given by Strauss is followed here, while the English translations are substantially those made by Salusbury.

p. 232 **buovoli** (modern Italian, *bòvoli;* a kind of edible snail). The amusing recital here of the many irrelevancies in the engraved plates which were common in scientific books of the period heightens the contrast between the latter and the works of Galileo. Here we may see the beginnings of the modern severe and unadorned treatment of scientific matters, which before Galileo was practically limited to mathematical treatises.

p. 232 **than one.** In the original edition the figures 72 and 200 are given in place of 12 and 36; in his own copy of the book, Galileo has entered 36 and 100 on the correction slip; the correct figures were sent by him to his favorite pupil Benedetto Castelli in a letter dated May 17, 1632, to be inserted in a special copy of the book sent to the Jesuit College at Rome. In this letter Galileo speaks of the figures in the first edition as misprints, but in view of his other set of corrections and an otherwise unaccountable slip in this very letter, it appears more likely that they were erroneous calculations of his own, he having compared his previous determinations with those of his opponent instead of confining himself to his adversary's own internal inconsistency, which was all that was justified by the context of the passage. Conjecturally, the steps may be reconstructed as follows. First, Galileo reasoned that his opponent would have the ball take more than twelve days to travel a distance equal to the diameter of the moon's orbit; referring to p. 226 he found it calculated that this time would be less than four hours (see further below), and accordingly he took Locher's implied diameter to be 72 times what it should be. Treating the figure 72 as a measure of diameter instead of a ratio, he deduced an error of more than 200 to 1 in the size of his opponent's supposed orbit of the moon. Later he suspected this figure; looking back at the calculation on p. 226 he realized that he had mistakenly based the comparison upon his figure for uniform travel over the diameter at the speed already attained at the center. Accordingly he noted the figures 36 and 100 on the correction slip at the end of the book, thus reducing the comparison to the two conflicting computations for the radius of the orbit (four hours vs. six days). Despite the fact that a computation of accelerated travel over the diameter would have given him a still further advantage, he was satisfied with what he had and could not add the new calculations and explanations to the printed book. Subsequently he recognized the true nature of his error and sent the proper correction to Castelli, though he neglected to go back and enter it in his own copy. The final reasoning is, of course, that the author made the ball take more than 12 days to fall through the diameter of a circle which it went around in one day; it should have done this in less than one-third of one day, so the magnitude of the error exceeds 36 to 1.

p. 233 **quandoque bonus, etc.** *Quandoque bonus dormitat Homerus:* "If Homer, usually good, nods for a moment" (Horace, *Ars Poetica,* 359; transl. Wickham). Simplicio's remarks starting on p. 203 are now sarcastically recalled by Sagredo.

p. 234 **downward.** It is possible to take this passage as meaning that in his own mind Galileo identified the cause of falling with the cause of planetary circulation. It may mean only that the two are equally mysterious, but considering his speculations on p. 29, and the general context in this place, it would not be absurd to credit him with suspecting that a true comprehension of gravity would yield also an understanding of planetary motion.

p. 234 **gravity.** One must remember that "gravity" here is not to be taken in the sense which it has had since the time of Newton, and that Salviati was quite justified in his ensuing remarks; see also next note.

p. 235 **"assisting."** The assisting spirits were angels who guided the planets in their courses; Kepler himself was not above invoking such forces. Abiding spirits (informing intelligences) were the internal moving principles of animate beings. Occurrence of the word "gravity" in the company of such jargon illustrates the emptiness of this word at that time.

p. 243 **over.** Cf. first note to p. 219.

p. 243 **view.** "Not improbable" means here "not implausible, though incorrect." For Aristotle's treatment of an analogous argument, see *De Caelo* II, 13, 295b, 16 ff.

p. 247 **revolutions.** This remark seems preposterous unless Salviati merely refers to a circle which slides as it rolls. Such a possibility would be obvious even to Simplicio if flatly stated. But Galileo believed he had a subtle proof that such an effect could occur without involving sliding; see *Two New Sciences*, pp. 21–25, especially p. 24.

p. 247 **De tribus novis stellis** *quae annis 1572, 1600, 1609 comparuere* (Cesena, 1628). The author was Chiaramonti (see first note to p. 52).

p. 247 FOOTNOTE. The rather puzzling interchange to which this note refers is probably to be explained as follows:

    1) The unnamed follower of Copernicus advances the analogy of a rolling cartwheel for the earth's two motions.

    2) The author of the booklet ridicules him for not seeing that this would require the earth to be much larger than it is, or its orbit much smaller than Copernicus thought.

    3) Salviati supposes the author's reasoning to be this: The Copernican must adopt Copernicus's measurements for the earth and its orbit, and in so doing he makes the former too small and the latter too large for his cartwheel analogy.

    4) Seeing the passage in the book merely confirms Salviati in his misapprehension; for, instead of reading the context, he merely looks at the words "smaller" and "larger," which he sees to be applied just as Simplicio has said. Having previously found gross errors committed by the same author, he takes this as just one more blunder.

    5) Not until the *Dialogue* is published does Galileo catch the author's real sense, at which time he annotated his copy as shown in the footnote.

p. 253 **an hour.** The actual speed is more than twenty times as great. The distances of all heavenly bodies except the moon were grossly underestimated at this time. Ptolemy had put the mean distance from the earth to the sun at 1,210 terrestrial radii, a figure which was accepted without substantial change up to Galileo's time; the correct figure is 23,439. Aristarchus, by his method of dichotomy (note to p. 274), had put the distance of the sun at 18 to 20 times that of the moon; Ptolemy in turn gave a mean value of 59 terrestrial radii for the latter (see further, note to p. 298). The solar distance thus implied agreed rather well (perhaps not entirely by accident) with a separate determination made by Ptolemy and based upon certain erroneous suppositions about the vertices of the cones of shadow cast by the moon and by the earth. This coincidence of values seemed to place the sun's location beyond doubt for Ptolemy's successors; Copernicus, when he attempted to rectify it, placed the sun even closer. See further, note to p. 359.

p. 254 **orbis magnus.** A term coined by Copernicus to denote the earth's orbit; see appendix, p. 635, Florian Cajori's edition of Newton's *Principia* (Univ. of Calif. Press, Berkeley, 1934).

p. 262 **third motion.** Copernicus assigned to the earth a special motion designed to keep its axis parallel to itself throughout the yearly movement. Galileo recognized that no special motion was required to account for this phenomenon; see further, pp. 398 ff.

p. 263 **globes.** The rings of Saturn were never recognized as such by Galileo. The changing shape of Saturn much puzzled him, and he attempted to explain it by assigning to that planet two satellites very close to its body. The correct description and explanation was not given until 1655, by Huygens.

p. 266 **six planets.** The moon was considered the nearest planet, and no planets beyond Saturn were then known.

p. 269 **Kepler.** Johannes Kepler (1571-1630) was a warm admirer of Galileo. Kepler discovered that the planets move in elliptical rather than circular orbits and worked out the laws of their motion, thus paving the way for Newton's law of universal gravitation. As Tycho's pupil and friend he resented Chiaramonti's *Anti-Tycho* (first note to p. 52).

The first dozen words quoted here occur in Kepler's *De stella nova in pede Serpentarii* (Prague, 1606), p. 86. The quotation as given in the text is an inversion and a condensation of Kepler's much more pungent remark: "Philosophers thus busy themselves removing from Copernicus's eye this mote of immense stellar distance while concealing in their own eye the much greater beam . . of an incredible stellar velocity, surpassing Copernicus in absurdity to the extent that it is harder to stretch the property beyond the model of the thing than to augment the thing without the property."

p. 274 **Aristarchus of Samos** (ca. 300-230 B.C.) is considered to have been first to formulate a coherent heliocentric theory. His chief contribution to astronomy was the "method of dichotomy" for determining the relative distances of the sun and moon from the earth; this consists in determining their exact positions when the moon is exactly one-half illuminated. Although nothing is wrong in theory with the method, the practical difficulty of making the determinations accurately with primitive instruments rendered the ancient findings very misleading (cf. note to p. 253).

p. 278 **in puncto regressus mediat quies:** "At the point of returning, rest intervenes." *Physica* VIII, 8, 262a, 12-14; 263a, 1-2.

p. 279 **Lorenzini.** Antonio Lorenzini da Montepulciano, author of a discourse on the 1604 nova printed at Padua in 1605. Although Galileo is not mentioned in it by name, he was its target of attack. The reference to foreign opinion is based upon a passage in Kepler's *De stella nova* taking Galileo and a number of other Italian mathematicians to task for not refuting Lorenzini's *De numero, ordine et motu coelorum* (Paris, 1606).

p. 280 **number.** Thirteen are named, but two of them (Peucer and Schuler) used the same data. Most of the figures given in the text came originally from Tycho's *Astronomiae instauratae progymnasmata* (Uraniborg, 1602) The original edition of the *Dialogue* contains many mistakes in giving the data and the calculations. In the present text Galileo's own figures as cited by Favaro from a fragment of the manuscript are used as corrections without special mention. Later corrections by Strauss and others are supplied in italics, usually without further comment; when corrections affect the text of the speeches or would greatly alter the calculations, they are attended by notes. The observers mentioned, omitting Tycho (see note to p. 52), are as follows.

Paul Hainzel, an amateur astronomer of Augsburg and a close friend of Tycho's. A famous quadrant 17½ feet in radius, which is said to have required forty men for its emplacement at Goeppingen, was employed by Hainzel for his observations.

Caspar Peucer, of Wittenberg, the son of a famous physician bearing the same name who corresponded with Hainzel and the Landgrave about the new star.

The Landgrave of Hesse, William IV, a famous patron of science and amateur astronomer.

Wolfgang Schuler, a friend of the younger Peucer, was a professor at the University of Wittenberg.

Thaddeus Hagek, physician to the king at Prague, wrote a book about this famous nova which was published at Frankfurt in 1574. It was Hagek who first acquainted Tycho with the manuscript in which Copernicus's system was circulated among the learned before publication.

Elias Camerarius, a professor at Frankfurt.

Adam Ursinus of Nürnburg, author of a number of astrological works, wrote of it in his *Prognosticatio anni 1574;* he believed this new star to be sublunar.

Jerome Muñoz, professor of mathematics and Hebrew at the University of Valencia.

Cornelius Gemma of Louvain, son of the eminent astronomer Gemma Frisius. Gemma wrote briefly on the nova during its first appearance in 1572, and afterward at length in his *De divinis mundi characterismis* (Antwerp, 1575).

Georg Busch, a painter and amateur astronomer of Erfurt, who argued that the nova was sublunar.

Erasmus Reinhold, son of the compiler of the famous Prutenic Tables (of planetary movements), was a physician at Saalfeld. Tycho exposed his appropriation without acknowledgment of the Landgrave's observations.

Francis Maurolycus, Bishop of Messina, one of the first to observe the new star.

p. 297 CALCULATIONS. As customary at the time, Galileo does not show the successive products in the process of division; multiplication and subtraction are carried out simultaneously. An explanation of this example will serve for those to come. The last five digits in each product, which were ignored throughout the calculations, have here been printed in strike-out type. The divisor being multiplied by the first partial quotient (that is, 58 by 5), the product (290) was subtracted from the first three digits of the dividend (347), and the remainder (57) was set down. Next, a trial quotient was sought for 573, the digit 3 being not "brought down" as in our practice, but simply read as belonging after the digits 57. The quotient 9 being selected, its product into 58 was then taken, giving 522; this in turn was subtracted from 573, leaving 51. The digit 1 was entered in the highest available space having the proper decimal position, which placed it after the digits 57, and the digit 5 was forced by the same rule into a new line, beneath the 7. If the division were to be carried out with respect to the entire dividend (347313294), the next partial remainder would thus be 511, and the above process would be continued by dividing 58 into that.

p. 297 **154° 45′.** Obviously this should be 154° 35′; rectification of the sine from 42657 to 42920 would, however, alter the entire calculation and would affect the text which follows, so the original errors have been preserved.

p. 298 **product.** The original reads "quotient." (Corrected by Favaro.)

p. 298 **52 radii.** This is approximately the distance which Copernicus had determined for perigee (the closest approach of the moon to the earth). In a sense the comparison made here with Ptolemy's figure is not a fair one, since in his most essential calculations Ptolemy uses a mean of 59 radii and not 33 as here implied. The latter figure, widely used by astronomers of the time, belonged more appropriately to philosophical discussions of the boundary of the "elemental sphere" excluding all heavenly bodies than it did to serious scientific considerations; it represented Ptolemy's finding for the lunar perigee at quadrature. Ptolemy gives 54 radii as the perigee at opposition and conjunction; his value of 59 radii for the mean distance of the moon at syzygies was quite good, and Copernicus's figure of about 60⅓ radii was almost precisely right. The extremely wide variation of lunar distances according to Ptolemy's theory was unsubstantiated by any observed changes in apparent diameter of the moon; it arose from his employment of both an epicycle and an eccentric (rejected by Copernicus in favor of two epicycles; cf. first note to p. 65). It is true that Chiaramonti having deliberately chosen the lowest of Ptolemy's determinations for arguing about the position of the nova, Galileo was amply justified in adopting his opponent's ground for the refutation. In fact he was practically obliged to do so, for under either Ptolemy's or Copernicus's calculations of the farthest departure of the moon from the earth, this pair of observations and the ensuing one would have made the new star sub-

lunar; Ptolemy gave 64 terrestrial radii as this maximum, and Copernicus 68⅓ radii. See *Almagest,* bk. v, ch. 13, and *De Revolutionibus,* bk.
iv, chs. 17 and 22.

p. 300 **97845.** Strauss and Favaro note that this should have been 97827. The
tables used by Galileo were those mentioned in the second note to p. 181.

p. 302 **his side.** This is far from true, as Galileo was probably aware, and the
sophistry he employs to make his assumption appear favorable is most
amusing. Calculation shows that seven radii would have been more favorable, in the sense of reducing the total of the corrections required to make
all the "investigations" agree. That total would then be 658 minutes instead
of the 756 minutes (see p. 307) required on the assumption of 32 radii.
Galileo introduced a further, though minor, error by computing as if all
the errors had been made by the more southerly observer in every instance
instead of dividing the errors equally between each pair.

p. 302 **second calculation.** Strauss points out that the values used in this calculation are not those given in the table on p. 294, which would have yielded
a negative parallax for the star. Following Chiaramonti, Galileo used a
calculated value for Schuler's upper altitude of the star.

p. 304 **4034.** This chord should be 4304 (Strauss, Favaro).

p. 305 **36643.** Should be 36623 (Strauss, Favaro).

p. 306 **30 58,672/100,000.** The denominator should obviously be 300,000, and
the end of the sentence should read "a little less than 30 1/5 radii."

p. 310 **clocks.** The clocks of Galileo's time were quite unsatisfactory for astronomical purposes such as this. The prevailing type was the wheel clock,
the escapement of which is described on p. 449. It had not yet occurred to
Galileo to utilize the pendulum for the escapement of a clock, though he
had used it for the measurement of time as described below. A few months
before his death he hit upon the idea of its application to clocks, but
because of his blindness he could not carry it into execution. He dictated
a design to his son Vincenzo, who made correct drawings but did not complete an actual model. Credit for the successful construction of the pendulum clock therefore belongs to Huygens, who published his invention some
sixteen years later. Galileo's first application of the pendulum, while he
was still a medical student, was called the *pulsilogia,* and consisted of a
board bearing a peg to which was attached a bob swung on a cord. On the
board at appropriate places were written various diagnostic descriptions
of a patient's pulse. The physician had only to stop the cord with his
thumb so as to bring the swinging bob into sychronism with the pulse, and
read off the diagnosis directly ("sluggish," "feverish," etc.). Galileo's
method of measuring small intervals of time was to fill a large vessel with
water which could escape through a very small orifice into an empty
vessel which had previously been dried and weighed. Removing his thumb
from the orifice at the start of an experiment — for example, the dropping
of a ball along an inclined plane — and replacing it when the ball had
reached any desired point, he could, by weighing the water which had
escaped, determine the elapsed time by comparing this with the weight of
water escaping in a known time. (See *Two New Sciences,* p. 179.)

p. 314 **FIGURE 19A.** Commencing with the calculation on Peucer's observations,
the original edition introduces the letters IAC and IEC in place of IOT
and IFT without supplying any correspondingly lettered diagram. The
National Edition does the same; Strauss alters the letters to correspond
with the previous diagram, but without comment. Figure 19A, which is
here introduced to correspond with the text, is copied from a manuscript
of Galileo's reproduced in the National Edition, vol. vii, p. 528; this manuscript page contains a number of these calculations and a somewhat similar
description. Presumably when Galileo prepared his copy for the printer he
copied a part of the calculations from one manuscript and a part from
another, without noticing the differences in the diagrams. The diagram

used in these later calculations corresponds to the figure employed by Chiaramonti himself in *De Tribus Novis Stellis* (Cesena, 1628), p. 143.

p. 319 **unbounded.** Cf. *De Caelo* I, 6 and 7. This was very dangerous ground for Galileo to tread; Giordano Bruno's conviction and execution had depended largely upon his having espoused the view that the universe was infinite.

p. 320 **at them.** Galileo had had direct experience of this; Cremonino at Padua and Libri at Pisa are known to have refused even to look through the telescope, and tradition has it that several professors absented themselves from Galileo's supposed public demonstration that falling bodies move independently of their weights.

p. 328 **has illuminated me.** This passage seems to imply that Salviati, who speaks for Galileo, has yielded to reason as "a clearer light than usual." The context tends to support that idea. Nearly all scholars, however, agree that Galileo's real intention in the final clause was to refer obliquely to the Church's edict as that "clearer light than usual" which had ultimately shown him that Copernicus (and reason) were in error. It is therefore probable that the phrase "than I have been" should be read "—as indeed I have been," even though the printed text can not be literally so translated.

p. 328 **nuper me** etc.: "upon the shore I lately viewed myself
When the sea stood still, unruffled by the winds."
—Virgil, *Bucolics,* ii, 25 f.

p. 334 **to us.** Copernicus, *De Revolutionibus,* bk. 1, ch. 10, mentions these hypothetical explanations as having been offered by others, but does not commit himself.

p. 342 **center.** The device by which Copernicus had eliminated the equant was greatly admired by those who held it to be axiomatic that all heavenly motions were uniformly circular.

p. 343 FIGURE 21. This figure has been slightly modified in redrawing for this edition in order to make the lines easier for the eye to follow. The small arcs *XY* and *ST* here show the course of the planet continuing instead of reversing its direction, and the text has been modified to correspond.

p. 344 **Apollonius of Perga** flourished about 200 B.C. He was among the greatest of the ancient geometers, his chief contribution being the theory of conic sections.

p. 345 **spots.** See first note to p. 53.

p. 346 **Apelle.** See first note to p. 53.

p. 350 **meridian.** AOC has been identified on the previous page as the axis of the ecliptic, but from this point it is spoken of as the projection of our meridian. This was a curious slip on the part of Galileo. Apparently all he meant to say was that the maximum curvature of the path of the sunspots occurs when the axis of the sun's rotation points toward us or away from us; but having drawn his diagram so that in such a position the entrance and exit of the spots fell on a horizontal line, he forgot that this line represented the plane of the ecliptic and spoke of it as if it were our equatorial plane.

p. 354 **appearances.** This passage has led some modern critics to charge Galileo with a blunder it would have been impossible for him to make. See, for example, Strauss, p. 556, note 48, and Taylor, *Galileo and the Freedom of Thought* (London, 1938), p. 135; these authors remark that if the sun went around the earth with its axis always parallel to itself, the sunspots would appear to travel exactly as described. This is true only if the daily motion of the sun around the earth is merely apparent; that is, only if we grant the earth a diurnal rotation. But Galileo is speaking here of an absolute fixed-earth theory, and from that standpoint one cannot maintain that the sun's axis preserves a constant direction and at the same time admit the observed variations in the paths of the sunspots over the course of a year without absurdities from the standpoint of dynamics. These were clear to Galileo, who knew that angular momentum is conserved. It is true that Simplicio's argument is correct in that the appearances are capable of occuring under either theory, nor does Salviati deny this, but he does show that to assign both motions to the sun results in great complications, while the appearances are easily explained by assigning the motions to the earth.

p. 355 **third movement.** See note to p. 262; this phrase refers to the Copernican terminology, and not to the numerical order of the motions listed. The matter is discussed further on pp. 398 ff.

p. 356 **Terram igitur . . .** This passage from Locher's *Disquisitiones* (note to p. 91) is here so abbreviated as to make the speech which follows it incomprehensible. The essential part of the paragraph cited is as follows: "The earth and the moon travel in one year from east to west between Mars and Venus, the center [of the earth] tracing out the *orbis magnus* or *orbis annuus.*"

p. 359 **radii.** (See also note to p. 253.) Copernicus gave 1,179 terrestrial radii as the maximum distance (apogee) of the sun (*De Revolutionibus,* bk. iv, ch. 19); his figure for the mean distance was 1,142 radii (*ibid.,* ch. 21). Ptolemy's determination was 1,210 radii (*Almagest,* bk. v, ch. 15). The source of Galileo's 1,208 radii was Locher's *Disquisitiones,* p. 25.

It will be noted that Galileo's ideas of the stellar distances fell far short of the truth. Yet they were a great advance over the misconceptions prevailing among other astronomers of his time, and Galileo rendered a notable service to astronomy by setting forth his opinions and the reasons for them.

p. 360 **ten-millionth.** The original reads "hundred-millionth" because of a mistake in setting down the cube of 220 in Salviati's next speech which was corrected by Galileo in his own copy.

p. 360 **al-Fergani** (Alfragan) flourished about A.D. 800.

al-Battani (Albategnius) died A.D. 928; he was the most famous of the Arab astronomers.

Thabit ben Korah (Qurra) (836-901) was the leading Arabic editor of Ptolemy.

Christopher Clavius (1537–1612) was the leading Jesuit mathematician at Rome and author of a commentary on Sacrobosco; cf. first note to p. 124 and note to p. 414.

p. 362 **star.** Sagredo's objection is about the only one which Salviati could have countered. Actually the method described, though ingeniously conceived, is rather impractical because of atmospheric disturbances, the apparent motion of stars, and other interfering factors. It is remarkable that Galileo succeeded in obtaining any useful results by the procedure outlined here. Struve, writing two centuries later, remarked on the impossibility of ever observing the true angular diameter of a star, and showed that even those determined by the most perfect telescopes were spurious.

p. 362 **Astronomical Letters.** *Epistolae astronomicae* (Uraniborg, 1596). The original edition of the *Dialogue* has "chapter" in place of "page."

p. 363 **It is Sagredo.** Sagredo wrote to Galileo in 1612 regarding refraction within the eye; cf. *Opere* XI, p. 350.

p. 364 **impinge.** Literally, "from which the visual rays emerge"; in this passage, as elsewhere, the translation is accommodated to modern concepts of vision.

p. 366 **observed this.** Cf. note to p. 119.

p. 367 **man.** As is apparent from the reply, this epithet is not intended for Simplicio personally; in the Italian text the familiar pronoun appears instead of the polite form invariably employed among the interlocutors in addressing each other. The "foolish man" is Scheiner (or his pupil Locher).

p. 372 **ad hominem: Against the man.** The fallacy called *argumentum ad hominem* consists in directing the argument against the person uttering a proposition, or against other propositions known to be held by him, rather than against the proposition in dispute.

p. 372 **variation.** Bessel, in 1837, first detected the parallax of a fixed star due to the earth's annual motion. Using the star 61 Cygni he found a parallax of the order of three-tenths of one second, amply justifying the conjectures of Galileo here and in the following pages.

p. 373 **anti-Copernicans.** The reference is to Francesco Ingoli (1578-1649), author of a tract disputing the Copernican system, and later secretary of the Propaganda Fide. In 1616 he had addressed a communication to Galileo on this subject, which Galileo answered with an extensive letter (not published during his lifetime) comprising the present and other arguments for the Copernican opinion.

p. 388 **Vega.** It is creditable to Galileo that he selected Vega (α Lyrae) as one of the most promising stars for the detection of parallax; two centuries later it was selected by Struve and extensively observed for this purpose.

p. 392 **perpendicular.** The words "from the perpendicular" (*dal perpendicolo*) do not appear in the original edition; they were added by Favaro, and, having thus become part of the text of the National Edition, they are included here. The addition appears superfluous, since the angle of tilt is doubly described in this very passage. The identical phrase occurring in the original edition is used again by Galileo three pages later when he refers back to this passage, and Favaro makes no addition at that point.

p. 400 **William Gilbert,** physician to Queen Elizabeth, is considered the first great English experimental scientist. His book, *De magnete, magneticisque corporibus* (London, 1600) is a classic of systematic observation.

p. 405 **other.** Galileo viewed the tilting of the needle as an exertion of greater force against the end pulled down rather than as a directional effect.

p. 405 **armature.** A concave hemisphere of thin iron applied to the lodestone, or a conical iron jacket fitted to it. (Gilbert, *De magnete*, bk. ii, ch. 17.)

p. 405 **parted with it.** The reference is to an actual event which is quoted below at length from Mary Allan-Olney's *The Private Life of Galileo* (London, 1870). It may be remarked that Galileo's study of these matters commenced as early as 1602, and that "Cosmo" should be "Cosimo."

"In the year 1607 Galileo made various observations on the loadstone. . . . These observations he imparted to his friend Secretary Picchena, who in his turn imparted them to Prince Cosmo. The young prince sent to say he would like to possess such a loadstone as the one Galileo had, weighing about half a pound Tuscan. The hint was plain enough. Galileo wrote back to say that the loadstone and all else belonging to him was at the prince's disposal, but that a friend of his possessed a loadstone infinitely more worthy of the Serene notice, which might probably be parted with for a consideration. From the correspondence which ensued we learn that the Grand Duke was no more above bargaining than any pedlar in Tuscany. It is with pain that we see Galileo, the man to whom the secrets of the heavens were so shortly to be revealed, actually lending himself to small subterfuges for the sake of saving his Serene pupil's father a few crowns. At the same time it is fair to state that this is the sole instance of the tortuous, higgling spirit, which we feel to be more fitting to a dealer at the rag-fair in Piazza San Giovanni than to the father of experimental philosophy. The friend to whom this unique loadstone belonged was Sagredo. Galileo concealed his name, for what reason we are unable to guess, merely affirming that he (Sagredo) had been offered 200 gold crowns by a German jeweller, who had wished to buy the loadstone for the Emperor, but that he had declared he would only part with it for as much gold as it would carry fastened to the end of an iron wire, viz., more than 800 crowns; or, in plain Tuscan, its price was 400 crowns. Galileo had invented a story about a Polish gentleman to account for his curiosity respecting Sagredo's loadstone. To account for the delay in Picchena's answer, he found it necessary to state that this Polish gentleman, his pupil, was staying in Florence for a time. It is probable that Sagredo did not wish to part with the loadstone, and therefore put a fancy price upon it. Galileo found to his mortification that the negotiation would have been expedited by his telling the truth at once, as Sagredo would have felt himself honoured by

Prince Cosmo's acceptance of the loadstone as a free gift. The bargain was concluded after four months' haggling over the price. Galileo, fearing that his friend Sagredo would feel that his interests had quite been lost sight of when he came to know who the Polish gentleman was, begged Picchena to ask his Serene Highness to give 100 doubloons instead of 100 gold crowns, which was the price agreed upon."

p. 406 **verae causae:** true causes; that is, actual physical entities or actions as distinguished from hypothetical constructions serving as agents to implement a scientific theory.

p. 406 **hammers.** A reference to the legendary tradition that Pythagoras discovered the arithmetical relations underlying the theory of harmony by noticing the difference in pitch of the tones of four hammers of differing weights striking upon an anvil.

p. 411 **Gilbert.** Sagredo mistakenly attributes the discovery of the vertical dip of the compass needle to Gilbert. Though Gilbert describes the effect, he attributes its first discovery to the "skilled navigator and ingenious artificer Robert Norman," who announced it in England in 1576.

p. 413 **rise.** Galileo originally intended the *Dialogue* to center on his theory of the tides; cf. note to p. 416.

p. 414 **Joannes de Sacrobosco** (John Holywood) was the first and most important medieval writer on spherical astronomy. Of English birth, he died in 1256 at Paris where he was professor of astronomy. His book *Sphaera Mundi* went through countless editions and was a standard work until the seventeenth century. The passage here criticized by Sagredo occurs just before Sacrobosco's proof that the earth is the center of the universe.

p. 416 **mobility.** This section of the *Dialogue* is essentially a reworking and expansion of Galileo's *Discorso sopra il flusso e reflusso del mare,* which he transmitted to Cardinal Orsini in 1616 as a part of his unsuccessful attempt to moderate the Church's opposition to the Copernican theory. He was so excessively fond of this explanation of the tides that he once intended to bestow upon this entire book a title similar to the above. Galileo's explanation of the tides, as will be seen, depends upon the varying velocity of a point on the earth's surface due to the composition of its rotation and its revolution about the sun. This could not account for the actually observed periodicity of the tides, and it is likely that if the same argument had been brought forward by someone else, Galileo would have rejected the refuge to which he himself here has recourse; namely, the influence of the length and depth of each sea basin. The daily periods are intimately related to the motion of the moon and require nearly an hour more than a day; this fact had long been observed, and was the main reason for so many authors having resorted to the moon's authority over the waters as an explanation. Galileo was quite justified in rejecting the basis upon which this explanation was generally offered; one can hardly doubt that in his own mind he was thereby doing precisely the thing he so much admired in Copernicus — that is, refusing to abandon a rational explanation simply because the evidence of his senses appeared to contradict it (cf. pp. 328 and 335). So far as Galileo's "primary cause" of the tides is concerned, it was quite rational under the assumption of an absolute reference system (in this case the fixed stars, which he regarded as motionless); for a discussion of this matter, see Ernst Mach, *Science of Mechanics* (Open Court Publishing Co., La Salle & London, 1942), pp. 262–264. Mach's view is criticized, and the origin and fate of Galileo's tidal theory is examined, in *Physis* v. III, pp. 185–194.

p. 419 **prelate.** Marcantonio de Dominis, in *Euripus sive sententia de fluxu et refluxu maris* (Rome, 1624).

p. 420 **Girolamo Borro,** professor of medicine and philosophy at Pisa, had proposed warmth from the moon as an explanation of the tides in *Del flusso e reflusso del mare e dell'inondatione del Nilo* (Florence, 1583).

p. 421 **Mechanics.** Not Aristotle's; cf. note to p. 158.

p. 434 **Etiopico.** It was customary in Galileo's time to refer to everything in Africa south of Egypt as "Ethiopia." Maps of the seventeenth century sometimes show the ocean on both sides of South Africa as the "Ethiopian Ocean."

p. 435 **primary.** Erroneously "secondary" in the original edition; the correction is by Favaro.

p. 438 **Qual l'alto Egeo etc.** "As the deep Aegean, when the north wind ceases that swept it, rests not, but retains in its waves the sound and the motion."—Torquato Tasso, *Jerusalem Liberated,* xii, 63. Galileo's reference to Tasso here as "the sacred poet" is in sharp contrast to his unfavorable opinion of Tasso when he compared that poet with Ariosto in earlier years.

p. 447 **fast.** This understatement is a result of Galileo's mistaken assumption about the distance of the sun (note to p. 253) and hence about the size of the earth's orbit.

p. 451 **equal times.** As usual, Galileo shows his thoroughgoing experimental caution by noting an almost imperceptible discrepancy. Actually the curve which has the property of tautochronism is not the circle but the cycloid, a curve first studied and described by Galileo. The discovery and proof of this fact constituted one of the famous challenge problems of the seventeenth century; it was proposed in 1687 and solved first by Jakob Bernoulli. The cycloid is also the curve of quickest descent (not the circle as stated in the ensuing discussion); this even more famous challenge problem, proposed in 1696 by Johann Bernoulli, was solved by Leibniz on the day he received it. Newton, when appealed to by the British mathematicians, solved the problem at once; though his solution was published anonymously, Bernoulli instantly divined its source.

p. 453 **orbit.** It certainly does great credit to Galileo's acumen that, despite his rejection of the moon's influence on the tides, he was able to find a rational explanation for the appearances which had led others to attribute such an influence to the moon — they being even more wrong in accepting it than Galileo was in rejecting it. Galileo's description of the common orbit of the earth and moon may sound contradictory in its phrasing. Its logical basis is the Ptolemaeo-Copernican concept of orb and epicycle, the orb remaining the "true" path of a planet despite the epicyclic excursions of the latter. The passage here has been very freely translated; the actual words are: *Imperocchè, se noi intenderemo una linea retta prodotta dal centro del Sole per il centro del globo terrestre, e prolungata sino all'orbe lunare, questa sarà il semidiametro dell'orbe magno, nel quale la Terra, quando fusse sola, si moverebbe uniformemente; ma se nel medesimo semidiametro collocheremo un altro corpo da esser portato, ponendolo una volta tra la Terra e il Sole,* etc.

p. 455 **nine days more.** Here Galileo misinterprets as an irregularity in speed what is largely a consequence of the shape of the earth's orbit. He is often criticized for holding fast to the idea of a circular orbit in the face of such evidence, especially when he had Kepler's research at his fingertips. But it should be remembered that his authorization to write the *Dialogue* was limited to the discussion of arguments for Ptolemy and Copernicus who both assumed circular orbits. Even had he accepted Keplerian ellipses, it would have been a strategic error to introduce them here. Not only would it have reduced the plausibility of the earth's motion for his contemporaries, both professional and lay readers, but it would have antagonized further the Catholic authorities, who had banned the Protestant Kepler's *Epitome of the Copernican System.*

p. 457 **least of all.** Unfortunately for Galileo's theory, it is the reverse which holds true for solstitial and equinoctial tides; the latter are most extreme because of receiving the maximum effects of the sun's gravitational pull.

Cf. Newton, *Principia,* bk. iii, prop. xxiv (Univ. of Calif. Press edition, p. 437). The difference being minor, and being often offset locally by the effects of seasonal storms, it is no discredit to Galileo that he uncritically embraced this further deduction from his ingenious but mistaken theory of the tides.

p. 462 **Seleucus** was a Babylonian who flourished about 150 B.C. He was one of the few followers of the ancient heliocentric view of Aristarchus, and Plutarch attributes to him the opinion that the tides were caused by the motion of the earth.

p. 462 **Kepler.** Galileo's criticism of Kepler here has been misunderstood and to some extent misinterpreted. Fundamentally, Galileo's objection to Kepler's tidal theory was twofold; first, it ignored the purely mechanical hypothesis suggested to Galileo by the double motion of the earth, and second, it endowed the moon with a particular attraction for the earth's waters. Kepler reasoned that if the earth should cease to attract its waters, they would flow to the moon, and he deduced that as the moon passed round the earth, it drew the waters toward the equator. Modern readers are prone to overlook the fact that the moon's dominion over the waters was an ancient superstition rather than a scientific anticipation of Newton's general gravitational law. Thus it had for Galileo all the defects of the "occult qualities" invoked by philosophers as causes. Both Kepler's theory and Galileo's had the defect of implying a single daily tide, and both men appealed to local and accidental phenomena to explain the double period of ebb and flow. To Galileo as a physicist, the purely mechanical basis of his explanation was decisive in its favor. On the other hand, Kepler was much closer to a correct view of gravitation than others who ascribed the tides to attraction of the waters by the moon, and Galileo was remiss in failing to pursue that idea. In the preface to Kepler's *Astronomia Nova* of 1609, he spoke of gravity as a mutual bodily attraction such that the earth attracts a stone much more than the stone attracts the earth, and he conjectured that if the earth and moon were not restrained somehow, they would fly together, the moon moving 53 parts to the earth's one. He conjectured further that if two stones were placed anywhere in the universe outside the sphere of force of other bodies, they would move toward one another. Galileo generally refrained from speculating about such matters, as lying beyond the reach of experiment, and his physics remained a terrestrial physics except to the extent that he advocated the interpretation of celestial observations in terms of terrestrial analogies. His admiration for Kepler was always tempered by a distrust of the German astronomer's geometrical and harmonic mysticism.

p. 463 **Caesare Marsili** sent to Galileo, shortly before the publication of the *Dialogue,* a treatise in which he declared that he had detected a shift in the meridian of the Church of St. Peter at Bologna, where the direction of the meridian had been engraved in the floor. Marsili's observations were not conclusive of any motion of the earth; Strauss remarks that the alteration which has occurred even today would be imperceptible by means of instruments available in the seventeenth century.

p. 464 **doctrine.** This is the famous passage setting forth the favorite argument of Urban VIII against the conclusiveness of this "proof" of the motion of the earth. To place it in the mouth of Simplicio was the only possible course for Galileo to pursue when he was expressly commanded to include it. Yet the fact that he did so was a point in his indictment; see note to p. 103, point 2.

INDEX

References to marginal notes are indicated by the letter "m," and notes in the appendix are indicated by italics.